DESIGN OF STEEL BINS FOR STORAGE OF BULK SOLIDS

PRENTICE-HALL INTERNATIONAL SERIES IN
CIVIL ENGINEERING AND
ENGINEERING MECHANICS

William J. Hall, editor

DESIGN OF STEEL BINS FOR STORAGE OF BULK SOLIDS

EDWIN H. GAYLORD, JR.

Professor Emeritus of Civil Engineering
University of Illinois at Urbana-Champaign

CHARLES N. GAYLORD

Professor Emeritus of Civil Engineering
University of Virginia

PRENTICE-HALL, INC.
Englewood Cliffs, New Jersey 07632

Library of Congress Cataloging in Publication Data

Gaylord, Edwin Henry.
Design of steel bins for storage of bulk solids.

Includes bibliographical references and index.
1. Bins—Design and construction. 2. Bulk solids—Storage. 3. Building, Iron and steel. I. Gaylord, Charles N. II. Title.
TH4498.G38 1984 690′.53 83-17734
ISBN 0-13-201368-1

Editorial/production supervision and interior design: Nancy Milnamow
Manufacturing buyer: Anthony Caruso

Printed in the United States of America

10 9 8 7 6 5 4 3 2 1

ISBN 0-13-201368-1

Prentice-Hall International, Inc., *London*
Prentice-Hall of Australia Pty. Limited, *Sydney*
Editora Prentice-Hall do Brasil, Ltda., *Rio de Janeiro*
Prentice-Hall Canada Inc., *Toronto*
Prentice-Hall of India Private Limited, *New Delhi*
Prentice-Hall of Japan, Inc., *Tokyo*
Prentice-Hall of Southeast Asia Pte. Ltd., *Singapore*
Whitehall Books Limited, *Wellington, New Zealand*

To the

memory of

MARGARET GAYLORD

and to

DOROTHY GAYLORD

CONTENTS

PREFACE xiii

NOTATION xv

1 BINS AND ACCESSORIES 1

1-1. Containers for Bulk Solids *1*
1-2. Types of Flow in Bins *2*
1-3. Liners *4*
1-4. Feeders and Conveyors *4*
1-5. Screw Conveyors *5*
1-6. Apron Conveyors *7*
1-7. Vibratory Conveyors *7*
1-8. Belt Conveyors *8*
1-9. Hopper Gates *11*
1-10. Apron Feeders *12*
1-11. Belt Feeders *12*
1-12. Reciprocating Plate Feeders *13*
1-13. Vibrating Feeders *13*
1-14. Bar-Flight Feeders *14*
1-15. Screw Feeders *14*
1-16. Rotary Table Feeders *16*
1-17. Rotary Plow Feeders *16*
1-18. Rotary Drum and Vane Feeders *16*

1-19. Selection of Feeder *17*
1-20. Mechanical Dislodgers *19*
1-21. Pneumatic Inducers *19*
1-22. Vibrators *20*
1-23. Vibrating Hoppers *22*
1-24. Continuous Level Indicator *23*
1-25. Paddle Unit *23*
1-26. Pressure-Sensitive Diaphragm Sensor *24*
1-27. Level Control by Conductivity *24*
1-28. Screw Elevator *24*
1-29. Bucket Elevators *24*

2 MATERIALS OF CONSTRUCTION 27

2-1. Mechanical Properties of Metals *27*
2-1. Steels for Bins, Tanks, and Hoppers *28*
2-3. Clad Steels *29*
2-4. Non-ASTM Steels *29*
2-5. Structural Bolts *29*
2-6. Welding Electrodes *31*

3 PROPERTIES OF BULK SOLIDS 37

3-1. Pressure-Density Relations *37*
3-2. Particle Size *38*
3-3. Angle of Repose *38*
3-4. External Angle of Friction *39*
3-5. Internal Angles of Friction *40*
3-6. Flow Function *42*
3-7. Flow Properties *44*
3-8. Conditions Affecting Flow Properties *45*
3-9. Tables of Properties *51*

4 FUNCTIONAL DESIGN OF BINS 55

4-1. Funnel-Flow Bins *55*
4-2. Mass-Flow Bins *56*
4-3. Expanded-Flow Bins *58*
4-4. Type of Material *59*
4-5. Segregation *60*
4-6. Blending *60*
4-7. Hoppers *61*
4-8. Outlets *62*

4-9. Size of Outlet *64*
4-10. Outlets to Prevent Piping *71*
4-11. Impact Pressure Due to Filling *73*
4-12. Hoppers for Mass-Flow Bins *75*
4-13. Hoppers for Funnel-Flow Bins *79*
4-14. Live Storage *80*
4-15. Rate of Discharge *82*

5 LOADS **85**

5-1. Introduction *85*
5-2. Stresses in a Solid *85*
5-3. States of Stress in Bins *87*
5-4. Pressures in Bins *90*
5-5. Janssen's Formula *92*
5-6. Noncircular Cross Sections *96*
5-7. Reimbert Formulas *97*
5-8. Pressures from Dust-Like Materials *97*
5-9. Shallow Bins *99*
5-10. Vertical Compression in Cylinder Wall *100*
5-11. Funnel-Flow Bins *106*
5-12. Pressures on Hopper Wall *106*
5-13. Walker Formulas for Mass Flow *109*
5-14. Jenike-Johanson Theory for Mass Flow *114*
5-15. Mass-Flow Bin Tests *122*
5-16. Wind Loads *129*
5-17. Snow Loads *135*
5-18. Earthquake Loads *136*

6 DESIGN OF STRUCTURAL COMPONENTS **142**

6-1. Introduction *142*
6-2. Beams *142*
6-3. Compression Members *148*
6-4. Combined Bending and Axial Compression *153*
6-5. Curved Members *154*
6-6. Buckling of Flat Plates *157*
6-7. Postbuckling Strength of Compressed Plates *159*
6-8. Design of Beams Based on Postbuckling Strength of Elements *161*
6-9. Design of Compression Members Based on Postbuckling Strength of Elements *163*

6-10. Stiffened Flat Plates in Compression *166*
6-11. Axially Compressed Plates with Longitudinal and Transverse Stiffeners *171*
6-12. Effective Widths of Wide Flanges in Tension *172*
6-13. Local Buckling of Axially Compressed Cylindrical Shells *172*
6-14. Buckling of Cylindrical Shells in Bending *179*
6-15. Buckling of Axially Compressed Longitudinally Stiffened Cylindrical Shells *179*
6-16. Buckling of Circular Cylinders Under Axial Compression and Internal Pressure *183*
6-17. Laterally Loaded Plates *185*
6-18. Plates Under Nonuniform Lateral Load *189*
6-19. Laterally Loaded Plates—Large-Deflection Theory *193*
6-20. Rings *200*

7 **STRUCTURAL DESIGN OF BIN ROOFS** **207**

7-1. Introduction *207*
7-2. Types of Roofs *207*
7-3. Self-Supporting Spherical Domes *210*
7-4. Self-Supporting Conical Roofs *214*
7-5. Supported Roofs *219*
7-6. Supported Cone and Umbrella Roofs *221*
7-7. Cone Roofs with Simply Supported Rafters *222*
7-8. Cone Roofs with Rafters Fixed at Ring *240*
7-9. Supported Dome Roofs *245*
7-10. Unsymmetrically Loaded Dome Roofs *250*
7-11. Bracing of Supported Roofs *257*

8 **STRUCTURAL DESIGN OF BINS** **259**

8-1. Theory of Shells *259*
8-2. Stresses in Circular Cylinder Wall *259*
8-3. Buckling of Bins with Circumferential Lap Joints *264*
8-4. Wall Stresses Due to Unsymmetrical Load *266*
8-5. Wind Buckling of Circular Cylinders *270*
8-6. Membrane Stresses in Hopper *271*
8-7. Compression Ring *273*
8-8. Stresses at Cylinder-Cone Junction *275*
8-9. Tension Ring *277*
8-10. Column-Supported Bins *279*

8-11. Membrane Forces in Rectangular Bin Hoppers *285*
8-12. Rectangular-Bin Walls *289*
8-13. Temperature Stresses in Cylindrical-Bin Walls *292*
8-14. Flat-Bottom Bins *294*
8-15. Design for Earthquake Resistance *294*
8-16. Openings in Bin Walls *294*
8-17. Concentrated Loads on Bin Shell *295*
8-18. Corrugated-Sheet Bins *295*
8-19. Recommendations for Design *296*
8-20. Fabrication and Erection *298*

9 FOUNDATIONS 337

9-1. Allowable Bearing Pressures *337*
9-2. Flat-Bottom Bins *339*
9-3. Footings on Sand *341*
9-4. Footings on Clay *341*
9-5. Raft Foundations *342*
9-6. Pier Foundations *342*
9-7. Pile Foundations *343*
9-8. Continuous Footings *344*

SUBJECT INDEX 349

AUTHOR INDEX 357

PREFACE

This book presents a detailed treatment of the design of steel bins, bunkers, silos, and tanks for the storage of bulk solids (also called granular materials). Properties of bulk solids and conditions which affect their flowability are discussed. Functional design, including the advantages and disadvantages of funnel-flow bins, mass-flow bins, and expanded-flow bins and their suitability for various solids, sizes of outlets to prevent arching and piping, and various other topics are covered in detail. Much of this material is from unpublished reports by A. W. Jenike and J. R. Johanson of research sponsored by the American Iron and Steel Institute. Theories of funnel-flow and mass-flow pressures are explained and compared with results of tests, and recommendations are made regarding their applicability. Tables to facilitate the use of the Jenike-Johanson mass-flow theories have been prepared. Effects of hopper geometry are explained in detail, and feeders and other accessories as they affect the design of bins are also discussed.

New formulas for the design of self-supporting dome and conical roofs are presented, and the design of rafter-supported roofs is discussed in detail. Formulas for buckling of axially compressed cylinders and flat plates, both stiffened and unstiffened, are compared with results of tests, and recommendations are made regarding their applicability. Formulas for transversely loaded flat plates as they are used in rectangular bins and bunkers are also discussed. In addition to numerous examples to illustrate the use of various theories and formulas as they are presented, there are three detailed design examples, one of a column-supported funnel-flow bin, one of a skirt-supported mass-flow bin, and one of a rectangular bunker.

An American Iron and Steel Institute task group reviewed the manuscript as it developed. The first chairman was D. S. Wolford; he was succeeded by J. N. Macadam. Among members of the task group who were especially helpful in offering detailed constructive criticisms of first drafts of various chapters are A. J. Oudheusden, Reece Stuart III, J. H. Trammel, and M. J. Weigel. A. J. Colijn and S. L. Chu, who were not task-group members, also helped. The authors also acknowledge the assistance of E. W. Gradt, former staff representative of the institute, and A. C. Kuentz, who succeeded him. However, neither the foregoing individuals nor the American Iron and Steel Institute or any of its members warrant the accuracy of the material in this book or any recommendation contained in it.

Edwin H. Gaylord, Jr.
Charles N. Gaylord

NOTATION

A	area of outlet, area of impact of falling solid, cross-sectional area of bin
A_e	effective area
A_f	area of compression flange
A_g	gross area
A_r	cross-sectional area of ring
B	width of rectangular bin, width of footing
C	perimeter of bin
C_d	overpressure coefficient
C_p	wind-pressure coefficient (shape factor)
C_w	warping constant
D	diameter of bin
E	modulus of elasticity
F	allowable stress
F_a	allowable axial compressive stress
F_b	allowable bending stress
F_c	vertical compression in bin wall
F_v	allowable shearing stress
F_y	yield stress, yield point
FF	flow function
H	height of bin above hopper, height to top of stored solid, horizontal wind or seismic force
I	moment of inertia, structure importance factor
I_p	polar moment of inertia
J	torsion constant

K effective length coefficient
K_a active-pressure coefficient
K_p passive-pressure coefficient
L length, length of long side of rectangular bin, unbraced length of beam, length of footing
M bending moment
M_0 moment on end of rafter
M_t torsional moment
N_h hoop-tension stress resultant
N_m meridional stress resultant in hopper
N_y vertical stress resultant in cylindrical shell
N_1 meridional principal-stress resultant
N_2 latitudinal principal-stress resultant
P axial force
P_{cr} critical (buckling) force
Q concentrated load on ring or rafter, weight-flow rate
R radius of ring, hydraulic radius, radius of column circle in column-supported bin
S elastic section modulus, site-structure resonance coefficient
T tensile force, temperature change, fundamental period of vibration
V shear force, velocity
Z plastic section modulus, seismic-zone coefficient
a radius of spherical dome, length of rectangular plate, horizontally projected length of rafter
b width of rectangular plate, flange width, width of rectangular outlet, width of ringwall
b_e effective width of compression element
c cohesion
d depth of beam, diameter of circular outlet, height of ringwall
d_f diameter of core in funnel flow
e eccentricity of outlet
f stress
f_b bending stress
f_c compressive stress, unconfined yield strength
f_v shearing stress
ff flow factor
g gravitational acceleration
h height of hopper from apex to transition, rise of dome roof
h_c height of solid above transition, height of cone of surcharge
k plate-buckling coefficient, coefficient in formula for laterally loaded plate
l length of rectangular outlet, length of stiffener
p unit pressure normal to surface, load per unit area, wind pressure
p_{cr} critical (buckling) pressure
p_h horizontal pressure in solid
p_v vertical pressure in solid

p_{vt} vertical pressure in solid at transition
p_1, p_2 principal pressures
p' pressure normal to hopper wall
q unit frictional force, load per unit of horizontally projected length of rafter, uniform load on circumference of ring, stagnation pressure
q' unit friction on hopper wall
r radius of cylindrical shell, radius of gyration
r_{eq} equivalent radius of gyration
s_1 stress at abutment of arch in solid in bin
t thickness of roof shell, thickness of angle, beam web, column web, wall of tubular member
w flat width, load per unit of horizontally projected area of roof
α angle of repose, coefficient of expansion
γ bulk density, angle of twist
δ effective angle of friction, deflection
η factor of safety
θ hopper half angle at vertex, base angle of conical or spherical dome, central angle between columns in column circle
θ_v angle of hopper valley with vertical
θ' slope of flow-funnel boundary
μ' coefficient of friction
ν Poisson's ratio
ρ mass density of air
ϕ internal angle of friction, angle of diagonal brace with horizontal
ϕ' external angle of friction

DESIGN OF STEEL BINS FOR STORAGE OF BULK SOLIDS

1

BINS AND ACCESSORIES

1-1. Containers for Bulk Solids

Containers for the storage of bulk solids are called bins, bunkers, silos, and tanks. While there is no generally accepted distinction according to these terms, shallow containers for coal, coke, ore, crushed stone, gravel, etc., are usually called bunkers, and tall containers for grain, cement, and such materials are usually called silos. *Bin* is a more inclusive term. However, tall cylindrical containers are also used to store power-plant coal and in this use are called silos.

Bins are usually circular, square, or rectangular in cross section and may be arranged singly or in groups (Fig. 1-1). For the same height, a square bin provides 27% more storage than a circular bin whose diameter equals the length of the side of the square bin, so that square or rectangular bins may be necessary to obtain the desired volume of storage where lateral space is limited. Flat-bottom bins require less height for a given volume of stored material. Their initial cost is low compared to other types, and they are in common use. They may discharge by gravity flow or by mechanical means.

Bins may be supported on columns (Figs. 1-1a and b) or on load-bearing skirts (Fig. 1-1c). Skirts may be stiffened or unstiffened. Bins may also be hung from or supported by beams in the floor framing for conveyors or other equipment. Flat-bottom bins are usually grade-supported.

The plates used for bins may be welded or bolted. A number of manufacturers supply standard bins which are erected by bolting. Bins of circular cross section have a structural advantage compared with rectangular bins be-

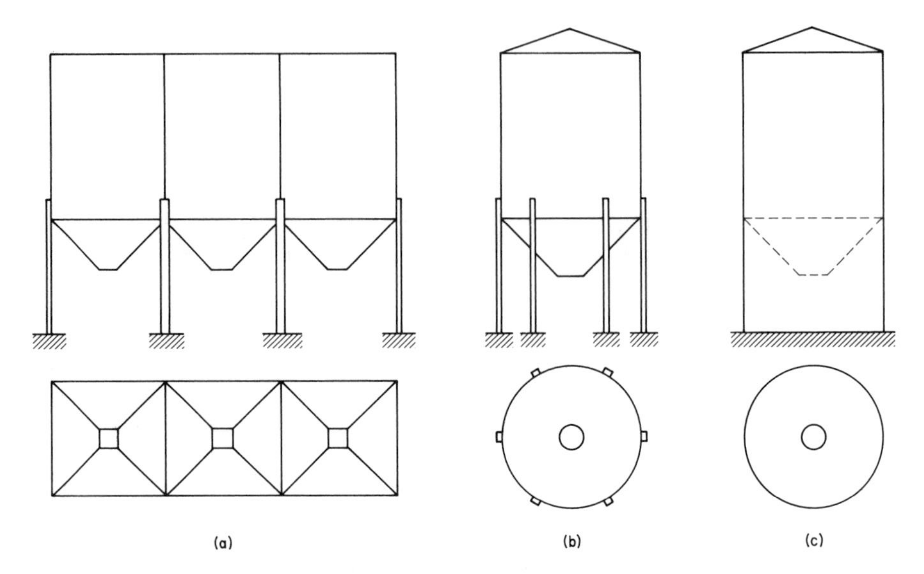

Figure 1-1 Typical bins.

cause hoop tension is the most efficient resistance to the lateral pressure from the stored material. However, the resulting relatively thin walls must not be so thin as to buckle under the vertical compression that develops from friction between the wall and the stored material or from snow or other load on the roof. Furthermore, thin-walled circular bins are sensitive to asymmetric load and should therefore generally have centered outlets, and if off-center charging is necessary, it should be designed to distribute the material as concentric with the bin axis as possible. Rectangular bins are generally less sensitive to asymmetric load and the structural effects are easier to calculate.

The plates which form the vertical and inclined walls of rectangular bunkers are usually supported by stiffeners. Vertical stiffeners are usually used except in small square, or nearly square, bunkers for which horizontal stiffeners may be more economical. Longitudinal beams may be needed in deep bins to reduce the span of the vertical stiffeners. Ties through the bunker may be needed for intermediate support of the longitudinal beams in tall bunkers. They should usually be located at the crotch lines or at intersections of the tops of the hoppers (Fig. 1-2).

1-2. Types of Flow in Bins

Depending on the shape of a bin, the roughness of its interior surfaces, and the properties of the stored material, several patterns of flow during emptying are possible.

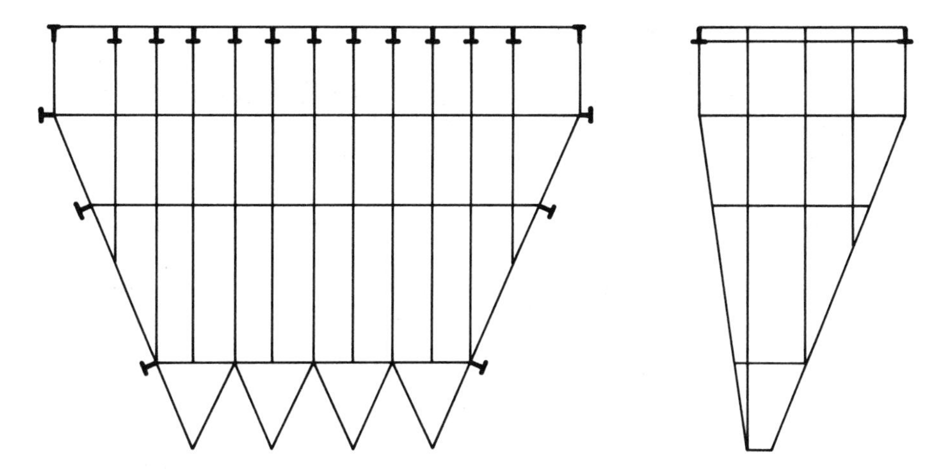

Figure 1-2 Typical coal bunker.

Mass flow is said to occur when the entire volume of stored material flows during emptying (Fig. 1-3a). A mass-flow bin usually consists of a vertical cylinder of circular, rectangular, hexagonal, or other cross section and a steep-walled hopper. The advantages and disadvantages of mass-flow bins are discussed in Sec. 4-2.

Funnel flow occurs in flat-bottom bins and bins with hoppers not steep enough to promote mass flow. The material flows toward the outlet in a channel that forms within the stored material. The channel is typically circu-

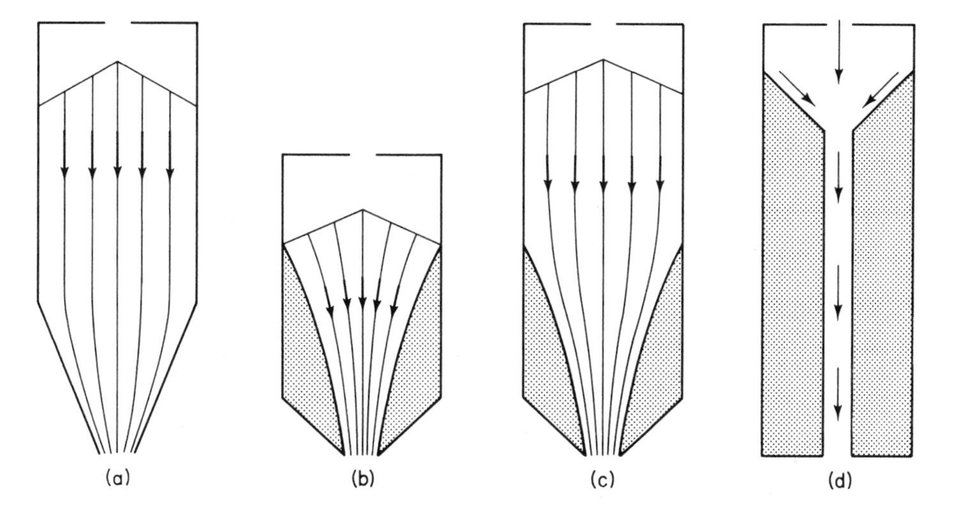

Figure 1-3 Types of flow: (a) mass flow, (b)-(d) funnel flow.

lar in cross section and usually assumes a conical shape widening upwards from the outlet. Several funnel-flow patterns are shown in Figs. 1–3b, c, and d. The intersection of the flow channel with the bin wall is called an *effective transition* because of the analogy with the transition at the hopper-cylinder junction. A funnel-flow bin usually consists of a vertical cylinder of circular, rectangular, hexagonal, or other cross section and a hopper or flat-bottom floor. The advantages and disadvantages of funnel-flow bins are discussed in Sec. 4–1.

An expanded-flow bin is a funnel-flow bin with one or more mass-flow hoppers. This type of bin is discussed in Sec. 4–3.

1-3. Liners

Liners may be required to protect the walls of bins for the storage of abrasive materials. Thin steel plates which can be replaced are satisfactory provided the time the bin is down for replacement does not seriously interrupt the operation. Stainless steel may be used as a liner or as the bin and/or hopper wall.

Coal bunkers are usually lined with gunite reinforced with wire mesh. The gunite is sprayed on to a thickness of about 2 in. and is held in place by small connectors attached to the walls. A gunite liner will usually last for the life of the bunker.

Epoxies improve the flow characteristics of a hopper and are as good as stainless steel in this respect. However, they cannot be used with acid-forming products because the acids attack the epoxy. Epoxies may be oxidized in powder or liquid form.

1-4. Feeders and Conveyors

Feeders are used to regulate the flow of material from bins and hoppers. Because a feeder only throttles the flow, the bin or hopper opening must be sized to allow the stored material to flow uniformly at the required rate of discharge. Therefore, the feeder must be designed as an integral part of the bin-feeder system. The principal types of feeders are discussed later in this chapter.

Conveyors are used to transport material delivered by a feeder to a point of discharge. There are about 80 types of conveyor systems, but only the screw, apron, vibrating, and belt types are discussed here.

The selection of a system should be based on the following factors: the properties and handling characteristics of the bulk material, the quantity and flow rate of the material, the process requirements, and the need for automatic control.

CONVEYORS

1-5. Screw Conveyors[1,2]

The most common device for handling bulk solids of fine and moderate sizes is the screw conveyor. It is economical, compact, and versatile and is assembled from standardized parts. It consists primarily of a conveyor screw, supported on bearing hangers, which rotates in a stationary trough. It may also be cantilevered. It can be mounted in a horizontal, inclined, or vertical position. In addition to its main function of moving materials, it can be modified to mix, blend, or agitate them and to meter them. The conveyor can be effectively sealed against the escape of dust or fumes and the entrance of dirt or moisture. It can be jacketed to serve as a dryer or cooler. Manufacturers' catalogues usually give the following information on materials which can be handled: maximum particle size, weight, abrasiveness, corrosiveness, flowability, and recommendations for conveyor loading.

Two types of screw conveyors are available. Helicoid screws are cold-rolled from strip steel to required diameter, pitch, and thickness (Fig. 1–4a). Some manufacturers roll a helix with the thickness of the outer edge one-half that of the inside edge. The continuous, one-piece helix, called a flight, is mounted on a pipe or solid shaft with heavy-duty end lugs and regularly spaced intermediate welds. For extreme heavy-duty applications the flighting is welded continuously to the pipe. Sectional flights are formed by cold-pressing a circular flat plate into a helix (Fig. 1–4b). These flights are either lap- or butt-welded to form a continuous helix which is secured to the conveyor pipe in the same manner as the helicoid screw.

The size of the pipe or shaft is governed by the magnitude of the transmitted torque and deflection limitations between supports. The standard clearance between the trough and flight is $\frac{1}{2}$ in.; therefore the deflection due to the weight of the pipe and the flight should not exceed $\frac{1}{4}$ in. A pitch equal to the screw diameter is considered standard. A pitch of one-half or two-thirds of the standard is recommended for inclined applications or for handling materials which may fluidize. Many variations of sectional flight conveyor screws are available to meet specific needs.

Ends of adjacent flights which have been installed in bolted sections of a trough are connected with coupling bolts and supported by intermediate

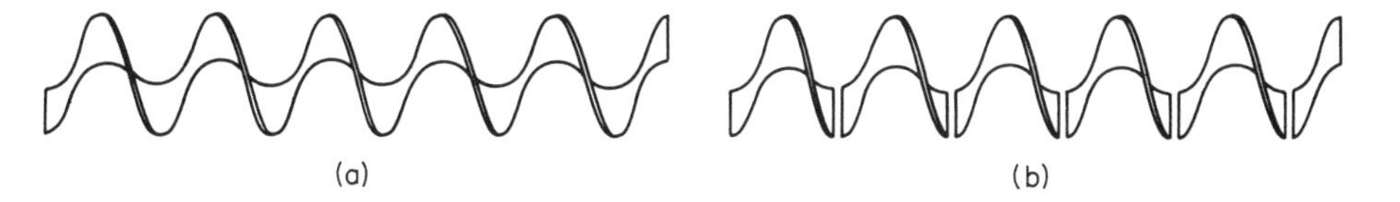

Figure 1–4 Conveyor screws: (a) helicoid, (b) sectional flight.

hanger bearings. Because these bearings are difficult to service and are exposed to a dusty and sometimes corrosive atmosphere, their spacing should be as large as is practicable. Graphite inserts or Stellite sleeves can be used to reduce maintenance.

Limiting the depth of solids in a screw conveyor to a maximum of 45% of the trough cross section protects the bearings and prevents jamming. Even with a loading of only 40–50% some backflow may occur, which reduces efficiency and thus requires greater power. Backflow increases with inclination of the conveyor, and for some solids screw efficiency decreases significantly if the slope exceeds 15°.

Special types of conveyor screws are available for moving certain kinds of materials. Ribbon-flight conveyor screws, which consist of continuous helical flighting formed from steel bar and connected to the pipe with supporting lugs, are used for sticky, gummy, or viscous materials (Fig. 1–5a). Double-ribbon flights (Fig. 1–5b) are also used for these materials; they have an advantage in that they promote more uniform flow. Tapering-flight conveyor screws may be used as feeder screws for handling friable lumpy materials or to maintain mass flow (Fig. 1–5c). Stepped-diameter conveyor screws, which consist of flights of different diameters, each with a uniform pitch and mounted in tandem on one shaft, are frequently used as feeder screws (Fig. 1–5d). The smaller diameter is located under the bin or hopper to regulate the flow of material. Stepped-pitch conveyor screws have succeeding single flights, or groups of sectional flights, increasing in pitch (Fig. 1–5e). These

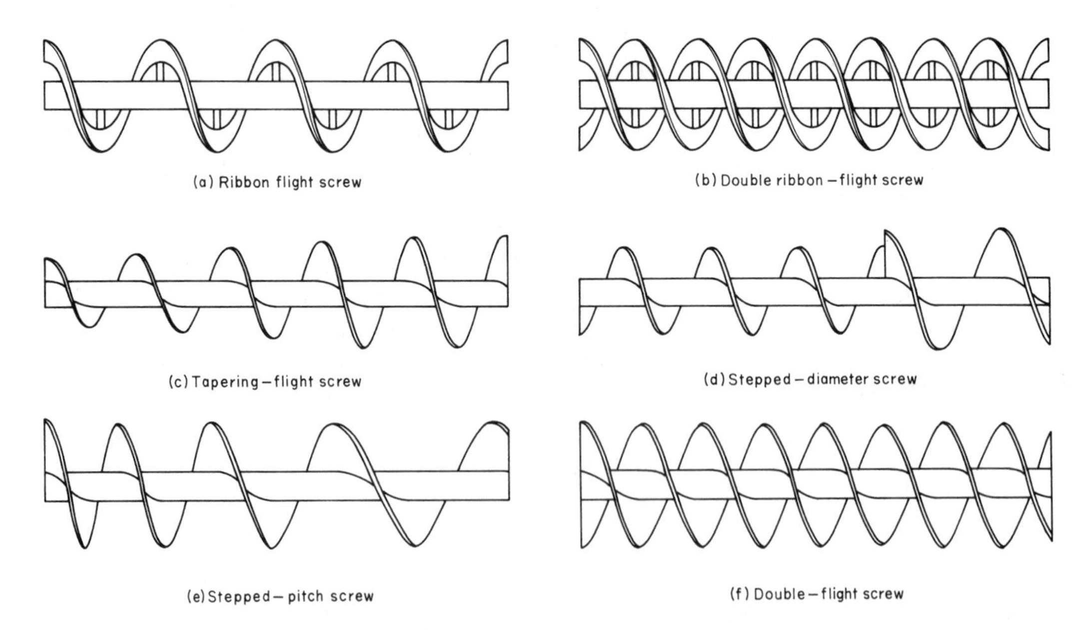

Figure 1–5 Conveyor screws.

screws draw fine, free-flowing materials uniformly from the entire length of the feed opening. Double-flight screws of uniform pitch produce a gentle flow and discharge for certain materials (Fig. 1-5f).

1-6. Apron Conveyors[1,3]

Apron conveyors will transport practically any bulk material. They can handle hot materials and absorb the impact of heavy lumps. Stone, sand, coal, ore, and similar materials can be moved with a minimum of degradation over horizontal and inclined paths or a combination of the two. Inclines of 25° or less are common. These conveyors usually operate at a speed of 5–100 fpm. In this range the life of the conveyor is increased, a minimum of maintenance is required, and material breakdown is prevented.

The conveyor consists of a series of overlapping or interlocking apron pans usually mounted on two strands of chain operating on a track. The larger-radius bead or corrugation of each pan overlaps the smaller radius of the adjacent pan to provide a tight joint. This continuous surface prevents material from wedging between adjoining pans when the apron pivots to discharge over the head terminal. Leakage of fines in handling very friable materials is a concern at the joints.

1-7. Vibratory Conveyors[4,5]

Vibratory conveyors (also called oscillatory conveyors) are used to transport bulk solids a distance of about 20–100 ft. They can deliver several hundred tons of material per hour. The most popular type (Fig. 1-6) consists of a trough T and a spring system S which supports the trough on the foundation and tunes the system to a near resonant condition. Its advantage over other types is that, because the stroke and frequency of oscillation are constant, material is delivered at a uniform rate. The structure-supporting spring must be designed for a reactive force equal to the product of the displacement and the spring constant. This type of conveyor can be used wherever the structure can absorb the spring forces. If the conveyor is located well above ground or in a structure which cannot resist the spring forces, a balanced conveyor should be used (Fig. 1-7). Here, a mass M, which is free to move, absorbs the reaction forces. This conveyor transmits much less force to the supporting structure but has several disadvantages such as higher cost, considerable variation in output, and difficulty in increasing or shortening its length.

Vibratory conveyors have the following advantages: They operate with

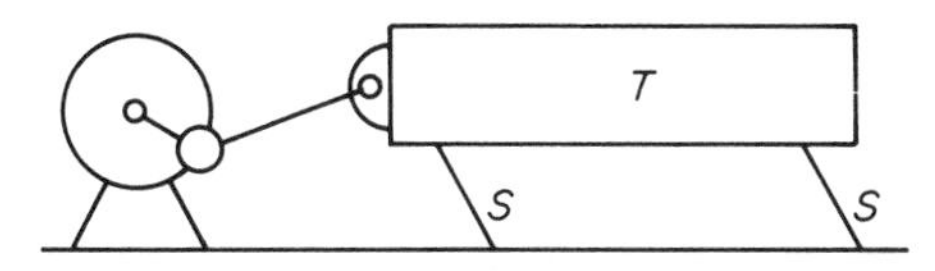

Figure 1-6 Vibratory conveyor (schematic).

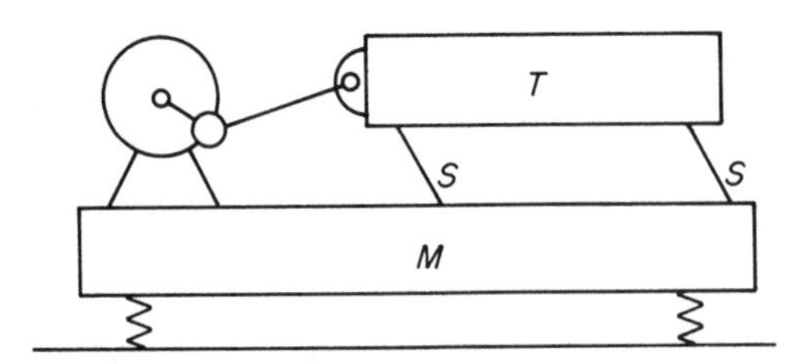

Figure 1-7 Balanced vibratory conveyor (schematic).

minimum degradation of material; material may be scalped, screened, or picked; hot or abrasive materials are easily handled; material may be dewatered and cooled and dried during conveying; a single unit may have a divided flow stream and multiple discharge points; units are self cleaning; and covered units can be made dust-tight. However, because there should be no head load from a hopper, the conveyor must be fed by a vibratory feeder which controls the feed rate from the storage bin or hopper (Sec. 1-13). The feeder can operate under high head loads and can usually be adapted to a rate-control system.

Vibratory conveyors operate at frequencies high enough to project the material from the conveying surface. The projected particle travels forward until it comes in contact with the surface again at which time the process is repeated. The material travels at speeds up to 100 fpm, depending on the combination of drive angle, frequency, and amplitude and on the properties of the material.

Vibratory conveyors operate between 300 and 600 c/min (cycles per minute) at strokes of 4–0.25 in. (varying inversely with the speed), while vibratory feeders usually operate at 1200–3600 c/min. The acceleration is usually 1–4*g* for conveyors and 3–13*g* for feeders. A high-amplitude, low-frequency vibration may produce an acceleration of 1*g* in a conveyor and a low-amplitude, high-frequency vibration an acceleration of 1*g* in a feeder.

1-8. Belt Conveyors[1,3]

Belt conveyors can handle bulk materials at high flow rates dependably and at low cost (Fig. 1-8). In the low- to medium-capacity range a flat belt in conjunction with plows will distribute material satisfactorily into a storage bin. A flat belt is usually 24–36 in. wide and operates at a speed of 300–400 fpm. It may be horizontal or inclined but must not have complex horizontal or vertical curves. Troughed belts are required for large-capacity or long-distance conveying. They operate at much higher speeds with less likelihood of spillage.

The belt carcass consists of plies of rubber-impregnated fabric separated by thin coats of rubber. The fabric may be cotton, rayon, nylon, combina-

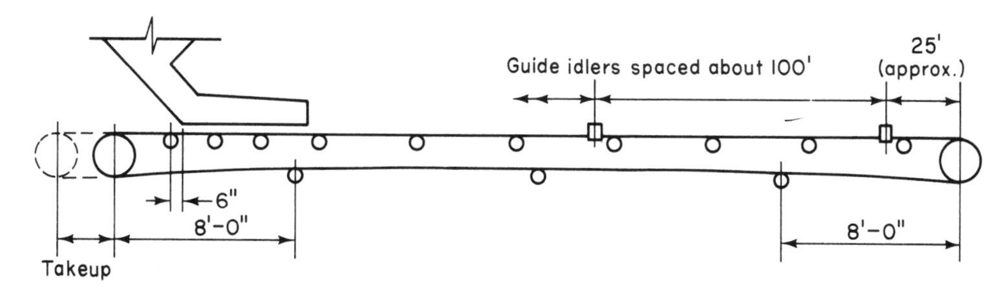

Figure 1-8 Belt conveyor.

tions such as cotton-nylon or rayon-nylon, or one of several synthetics. Belting is produced in widths of 72 in. and cut to the required width.

Steel cable belts are formed by wrapping in a fabric a layer of steel cables which have been suspended in rubber. Because of their high strength and minimum deformation, they are used for high lifts and long hauls.

Factors influencing the capacity of a conveyor belt are the following: character of the material to be conveyed, slope and speed of the belt, nature of the loading, and spacing and type of idlers. For most applications the 35° troughing idler is the most economical if the belt is designed to withstand a large amount of flexing; otherwise a 20° troughing idler may be the better choice. For light and fluffy material which may maintain a low surcharge angle on the belt, a 45° troughing idler may be required.

For speeds less than 150 fpm the cost of the belt conveyor per ton of bulk material handled usually exceeds that of other types of conveyors suited to the particular operation. The normal speeds of 50–100 fpm for picking or sorting belts may be increased for coarse, comparatively clean material whose refuse is easily discernible. For belt speeds of 200–400 fpm a standard tripper is preferable, but for speeds exceeding 400 fpm a tripper chute is necessary. If plows are used, the belt speed should not exceed 200 fpm. For wet material the minimum speed should be about 300 fpm.

The maximum slope at which a belt conveyor will operate satisfactorily is governed by the belt speed and the type, condition, size, and shape of the material to be handled. Conditions which may allow large lumps of material to fall back down the belt, which is a hazard to workers and may damage the belt and equipment, are a slope exceeding 16°, excessive belt speed, and a partially loaded belt. To prevent wet or sloppy material from slipping on the belt, small slopes must be used. However, for damp material a steeper angle may be used. The slope of an inclined conveyor handling coal, ore, or stone preferably should not exceed 15° or 16°.

Spacing of idlers. The life of idlers as well as that of the belt is influenced by the idler spacing. For any given belt tension, the larger the idler spacing, the greater the load on the idlers and the greater the sag of the belt. If the sag is too great, the belt tends to flatten out between idlers, which permits some of the load to spread out on the belt and possibly spill over the edge. This shifted load is squeezed back as it passes over the next idler, which consumes extra power. If the sag is excessive, the idlers and belt are subjected to impact as lumps of conveyed material rise rapidly over each idler.

Spacing of idlers under the loading chute should be one-half of the regular spacing (Fig. 1-8). This spacing maintains a small enough sag under the loading skirts to prevent jamming of the conveyed material between the skirts and the belt. Locating the first idler 6 in. back of the point where the material first strikes the belt prevents damage to the belt and the idler. The troughing idler nearest the discharge end should be located close enough to prevent material from spilling as the belt flattens out to go over the head pulley but not so close as to cause excessive loading of the idler end rolls and overstressing of the belt edges. Return idlers should be placed close enough to cross supports to protect the return belt.

Belt life and conveyor capacity both depend on the manner in which the conveyor is loaded. Materials should be baffled to assure a gentle drop onto the belt and should be delivered to the center of the belt in the direction of travel and at as near the speed of the belt as possible to reduce wear of the belt due to slippage of the material. Because the stored material accelerates as it is loaded on the belt, skirtboards may be required to prevent spilling. Skirtboards are usually an extension of the loading chute. Feed should be controlled so that the cross section of load is uniform throughout the length of the conveyor. Material consisting of large lumps intermixed with fines should drop on a screen or grizzly to allow the fines to pass through to the belt to form a cushion for the large lumps. The rate of flow should be such that the conveyed material will be at rest on the belt in a distance of 6-9 ft from the point of loading. This will require skirt boards from 5-10 ft in length, depending on the belt width. Boards longer than necessary will cause undue belt wear.

The *catenary idler* (Fig. 1-9) consists of five steel shells covered with rubber and fabric and supported at each end by bearing on a stub shaft. The belt makes four gentle curves instead of the two sharp bends required for troughing idlers. The rollers trough and train a belt easily and provide excellent resistance to impact. Mechanically spliced belts run smoothly and quietly. The flanged cable belt in conjunction with this type of idler offers the following advantages: increased belt life, larger spacing of idlers, and the

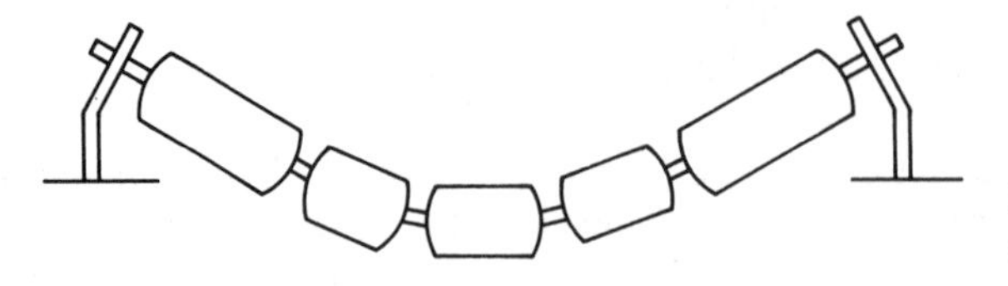

Figure 1-9 Catenary idler for belt conveyor.

possibility of lower power requirements. However, these advantages must be balanced against increased first cost.

Discharging. The simplest method of discharging material from a belt conveyor is to allow it to pass over a terminal pulley. A troughed conveyor belt must go through a transition section onto a flat pulley. To prevent spilling of the material over the edge of the flat portion, the travel time on the transition section should not exceed 1 sec.

Belt-cleaning devices. Unless a conveyor belt turns through 180° at the discharge end, a belt-cleaning device should be installed to maintain a clean belt surface to prevent carry-back of material on the return run. Temperature of the material and its size and grading must be considered in designing the cleaner. Because cleaners require constant maintenance and frequent adjustment, adequate space must be provided to allow inspection, repair, and maintenance.

A single- or multiple-blade rubber scraper located under the discharge pulley and held against the belt by springs or a counterweight is very effective for materials such as coal and stone. Rotary brushes with diameters up to 12 in. are suitable for both dry and damp materials. Speeds of 400–600 fpm are satisfactory for dry materials. Damp materials may require speeds of 1000–1500 fpm.

A brush operating at a right angle to the belt cleans as it travels across the surface. High-pressure air sprays and hydraulic sprays are used to remove damp or wet material. A thin steel wire stretched across and $\frac{1}{8}$ in. above the belt surface removes sticky materials. A secondary brush may be necessary to ensure a clean belt.

FEEDERS

1-9. Hopper Gates

The simplest feeder is a gate attached to the bin or hopper and controlled by an operator by means of a hand lever, motor, hydraulic piston, or air. Gates should be used only with free-flowing materials and are generally not suitable if automatic control of discharge is required. Typical gates are shown in Fig. 1–10: a horizontal gate in Fig. 1–10a, a vertical gate in Fig. 1–10b, a single-quadrant gate in Fig. 1–10c, a duplex-quadrant gate in Fig. 1–10d, a hinged chute in Fig. 1–10e, and a swinging gate in Fig. 1–10f. The first four are used mostly for fine and medium materials and the last two for coarse materials such as coal and ore. These gates can be modified in a variety of ways.[6-8]

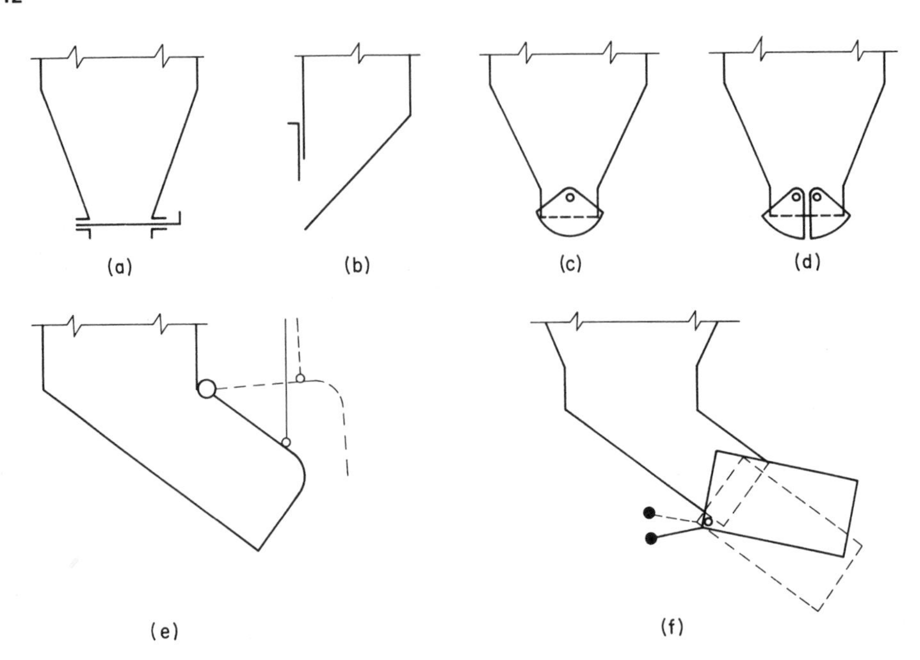

Figure 1-10 Hopper gates.

1-10. Apron Feeders[1,3]

Apron feeders are used for heavy, lumpy, or abrasive materials such as sand, gravel, stone, coal, and ore. They are designed to withstand the impact of heavy materials falling some distance and are recommended where impact or lump sizes prevent the use of belt feeders. They are also recommended for feeding hot materials which might damage other types of feeders.

The apron feeder is similar to the apron conveyor. It consists of overlapping pans attached to two or more strands of chain to form an endless apron. Sides may be attached to the pans to prevent spillage. Widths vary from 2 to 10 ft and lengths from 8 to 100 ft. Lengths greater than 10–15 ft are usually used for conveying rather than feeding. Although apron feeders are usually rated at 100–2000 tons/hr they are usually limited to about 1600 tons/hr at a speed of 10–50 fpm. They require 50–100% more power than belt feeders of equivalent capacity.

1-11. Belt Feeders[1,3]

The construction of belt feeders is similar to that of belt conveyors (Sec. 1-8). The feeder consists of an endless belt operating over closely spaced idler rolls which support the belt and material. The feeder operates smoothly and will handle materials which are not too hot or which do not contain very large lumps. A long, slotted hopper opening feeding along the length of the hopper assures economical feeding. Side skirting retains the material.

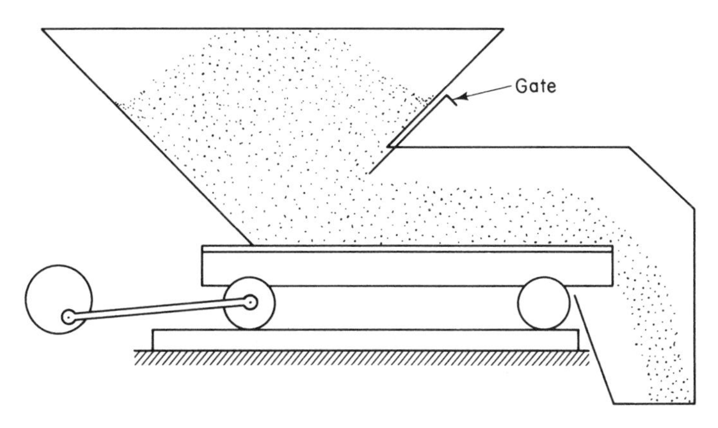

Figure 1-11 Reciprocating plate feeder.

Belt feeders are usually 2–6 ft wide and 5–10 ft long and operate at a capacity of between 5 and 1500 tons/hr. Feeders with greater capacity have been built using wider belts.

1-12. Reciprocating Plate Feeders[1,3]

The reciprocating plate feeder is one of the most reliable types of feeders for handling nonsticky or nonadhering materials such as stone, coal, ore, and sand. It can handle large quantities of materials as well as material composed primarily of large lumps.

The feeder consists of a horizontal or slightly inclined plate supported on rollers. The plate is subjected to a forward and backward motion by means of a connecting rod attached to a disc crank (Fig. 1–11). Stationary skirt plates attached to the bin or hopper contain the conveyed material. As the reciprocating plate moves forward, it carries the material forward while material from the hopper flows into the vacant space at the rear of the feeder. On the return stroke the plate slides beneath the conveyed material, which cannot move with the plate because of the flow of new material from the hopper, and is discharged. Thus material is fed from the bin or hopper on the forward stroke and discharged from the feeder on the return stroke. Flow rate can be controlled by moving the gate or by varying the stroke.

1-13. Vibrating Feeders*

The vibrating feeder, which can deliver several thousand tons of material per hour, consists of a feeder trough (or pan) and an excitation and mounting structure. The drive system may be either a brute-force or a tuned system.[1,3,9]

*See also Sec. 1–7.

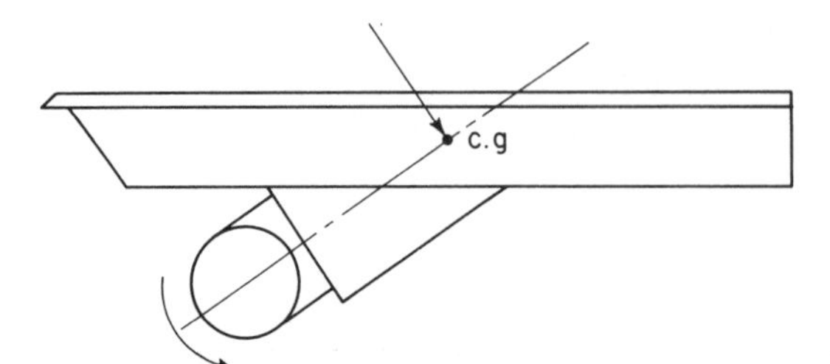

Figure 1–12 Vibrating feeder (brute-force system).

The *brute-force system* (also called direct-force system) consists of a rotary vibrator attached directly to the trough so that the driving force will be through the center of gravity of the trough (Fig. 1–12). The vibrator consists of eccentric weights which are either belt driven or connected to the shaft of the motor. The drive may consist of a single shaft or dual counter-rotating shafts. A dual exciter drive assures a straight-line transmission of forces.

The feeder of the *tuned system* (also called the two-mass system) consists of a trough mass and an exciter mass coupled by springs which select the natural frequency very near the operating frequency. Thus advantage is taken of the magnification of amplitude that occurs when a vibrating system is operated near resonance. The drive force may be supplied by either electromagnets or rotating eccentric weights. This system requires less power, causes very little force transmission to surrounding structures, and offers a controlled, variable rate of feed. If its operating frequency is somewhat lower than the natural frequency of the spring-coupled system, its output is relatively insensitive to variations of material head load.

1-14. Bar-Flight Feeders[1,3]

The bar-flight feeder consists of two or more strands of endless steel chain to which cross bars are attached at intervals to move coarse materials along a steel trough bottom to the point of discharge. The bar-flight feeder is usually enclosed and can be made dust-tight. To prevent buildup in the trough and conveyor chains, the material should be free-flowing, nonabrasive, and only mildly corrosive. Units of two or more strands may be placed in parallel.

These feeders are compact in design and often afford a solution where a minimum of headroom is available. Horizontal feeders and those inclined up to 10° can successfully handle a material bed whose height is several times the depth of flight.

1-15. Screw Feeders[1-3]

Screw feeders are designed to control the flow of material from bins, tanks, and storage hoppers at either a constant rate or a variable rate. They assure positive discharges at low output and can handle a wide variety of materials ranging from fines to a combination of fines and lumps. They are compact, simple in design, and economical to install, operate, and maintain, and when totally enclosed are dust-tight.

The feeder consists primarily of a helical screw rotating beneath the hopper outlet (Fig. 1-13). Feeders with uniform-diameter screws are used for handling fine, free-flowing materials. Because of the uniform diameter, the material is fed from the forepart of the inlet rather than over the entire length. This type is satisfactory and economical for hoppers which are to be completely emptied or where inert or dead areas of material over the inlet are not objectionable.

A double-flight screw (Fig. 1-5f) delivers a more uniform flow of material than a single-flight screw and because of its uniform discharge provides more accurate control.

Tapering-diameter screws (Fig. 1-5c) will handle materials containing a fair percentage of lumps. They assure uniform draw of the material across the entire length of the opening, thus eliminating inert or dead pockets in the storage bin.

Stepped-pitched screws are screws whose succeeding sectional flights increase progressively in pitch (Fig. 1-5e). The portion of the screw with the smaller pitch is placed under the inlet opening. This type of screw is sometimes used instead of one with a tapering diameter.

Sticky or gummy material will usually collect on a standard flight at the juncture of the flight and the screw. The open space between the flight and the shaft of the *ribbon-flight screw* (Fig. 1-5a) eliminates this problem. A double-ribbon flight (Fig. 1-5b) provides a more uniform flow of material and therefore a more even discharge.

Multiple-screw feeders, which consist of a series of parallel screws, may be used in flat-bottom bins to discharge materials which tend to pack or arch under pressure. The entire bottom may be provided with these feeders to carry material to conveyors.

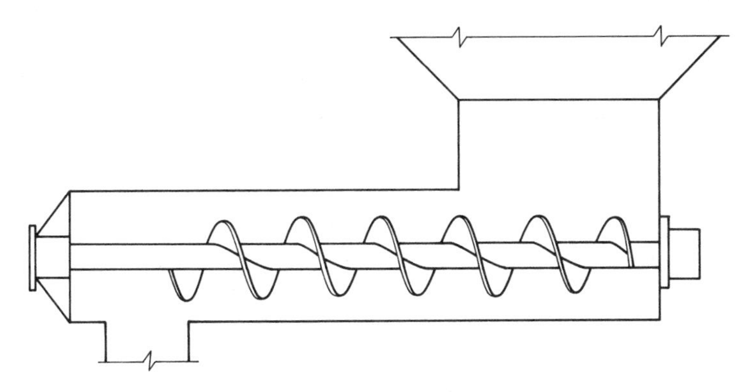

Figure 1-13 Screw feeder.

1-16. Rotary Table Feeders[3,7]

The rotary table feeder (Fig. 4–7b) is generally used for handling materials which tend to arch (damp sand, wood chips, etc.). It consists of a power-driven circular plate rotating directly below the opening of a round vertical bin. An adjustable feed collar, located immediately above the table, regulates the volume of material delivered.

If the lower edge of the feed collar is either helical or spiral, material will flow under the collar uniformly around the complete circle and spread over the table as it revolves. A steady stream of material is then plowed off into the discharge chute. The rate of feed is adjusted by repositioning the feed collar.

Objectionable features of the rotary table feeder are the tendency of certain cohesive materials to cling to the hopper walls, high installation costs, and the size of the machinery compared to the output of the feeder.

1-17. Rotary Plow Feeders[3,7]

Rotary plow feeders provide a dependable means of moving large quantities of bulk material from tunnels, from under stock piles, and from beneath long storage bins. They are used for reclaiming materials such as sinter, coal, potash, phosphate, limestone, and other materials which will not actually flush through the opening and off the horizontal shelf. The plowing mechanism consists of a self-propelled traveling carriage mounted on a track. One or two motor-driven rotors, each with four or more curved arms rotating in a horizontal plane, are mounted on the carriage. This plowing unit travels continuously back and forth between predetermined limits and sweeps the material from a long horizontal shelf into a chute from which it flows to the belt conveyor.

Materials which tend to arch or pack are readily handled by the rotary plow feeder. The feeder opening is continuous for the entire loading distance, and arching of material is minimized because the rotating arms undercut the arch support on all but two sides. This arrangement prevents the stoppage of material flow caused by the dome effect which often occurs when material is supported by the four sides of a rectangular opening.

1-18. Rotary Drum and Vane Feeders[7,8]

The rotary drum feeder is one of the simplest forms of feeder (Fig. 1–14a). It is inexpensive and easy to maintain but can be used only with materials with good flowability. The drum prevents the material from flowing out freely and discharges it by rotation. The adjustable gate helps control the rate of flow.

The rotary vane feeder is also a simple form of feeder. It consists essentially of a rotor with pockets, mounted in a housing. Regulating the speed of

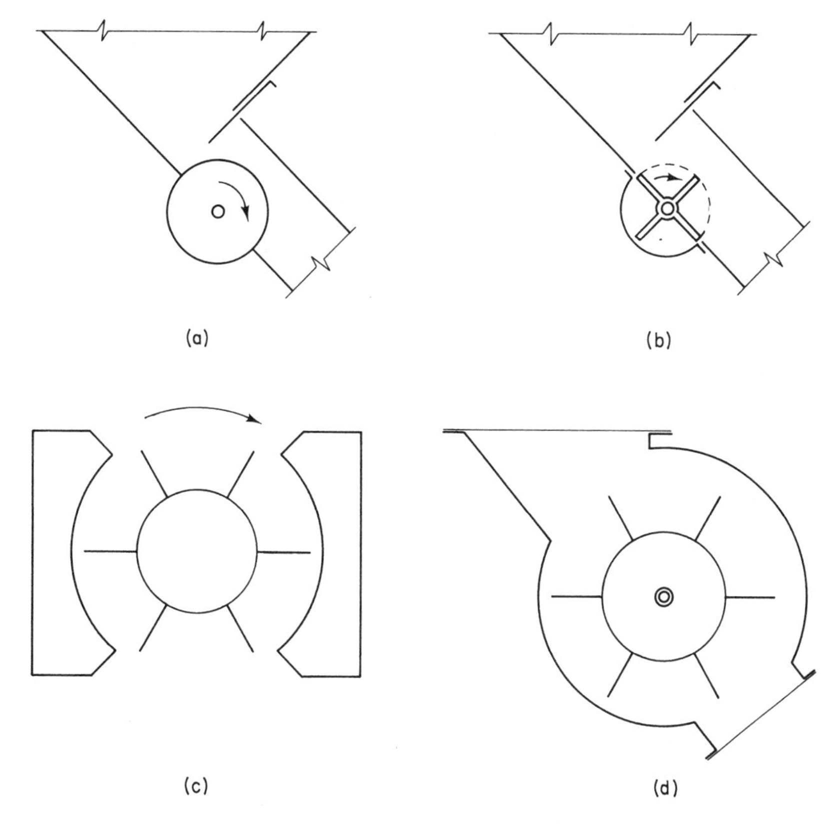

Figure 1-14 Rotary drum feeders.

the rotor controls the rate of flow. Many forms of rotor and housing are manufactured for a variety of materials and installation requirements. Only three are described here. The feeder of Fig. 1-14b is similar in operation to the drum feeder and is suitable for materials with good flowability. The feeder of Fig. 1-14c will handle fine, pulverized, or granular materials containing small friable lumps that will not jam between the close-fitting radial vanes and the housing. The material enters the feeder from above, and, as the rotor revolves, the material in the pocket is discharged through the opening in the bottom. Figure 1-14d shows a feeder which provides clearance between the vanes and the housing on the feeding side so that it can handle lumpy or nonuniform materials. It controls the flow without choking or jamming.

1-19. Selection of Feeder[8]

The size and properties of the material as well as the quantity to be delivered are of prime importance in selecting the type and size of feeder required. The ease with which materials of various sizes can be handled by some of the

TABLE 1-1. GUIDE TO SELECTION OF FEEDER[a]

Type of feeder	Maximum size of material[b]					
	Very coarse, > 12 in.	Coarse, < 8 in.	Medium, < 2 in.	Fine, $<\frac{1}{2}$ in.	Very fine, < 100 mesh	Concentrate, < 200 mesh
Apron	E	E	S	F		
Belt		F	S	E	E	S
Rotary plow		F	E	E		
Rotary table				S	E	S
Screw			F	S	E	E
Vibrating (brute force)	E	E	S	F		
Vibrating (tuned)	E	E	E	E	F	F

[a]From Ref. 8.
[b]E = excellent, F = fair (special consideration), S = satisfactory.

feeders discussed in this chapter is shown in Table 1-1. It will be noted that the apron and vibrating feeders are excellent choices for coarse and very coarse materials but only fair for fine materials, while the belt feeder is an excellent choice for fine and very fine materials but only fair for coarse materials. Also, the screw feeder is most acceptable for concentrates and fine powders, while the vibratory tuned feeder can handle a range of materials from fine to very coarse.

The horsepower required to deliver a ton of material by feeder is an important consideration. Power requirement usually decreases with increase in feeder capacity. The rotary table and the screw feeders are among the highest in power requirement, the rotary table ranging from about 8 hp/100 tons for a feeder with a capacity of 50 tons/hr to 2.5 hp/100 tons for one with a capacity of 300 tons/hr and the screw feeder ranging from about 9.5 hp/100 tons for a capacity of 50 tons/hr to about 6 hp/100 tons for a capacity of 300 tons/hr.

Of the rotary plow, apron, belt, and vibratory feeders, the rotary plow requires the greatest power (about 5 hp/100 tons for capacities ranging from 300 to 3000 tons/hr) because of the torque needed to rotate the plow on the shelf. The apron feeder is next at about 4 hp/100 tons for capacities from about 200 to 2000 tons/hr. The belt feeder delivers from 200 to 1500 tons/hr when operating at 1–3 hp/100 tons. The brute-force vibratory feeder delivers from 500 to 1500 tons/hr at about 1 hp/100 tons and the tuned vibratory feeder 600–3000 tons/hr at about 0.5 hp/100 tons.

FLOW INDUCERS

In cases where it is impracticable or impossible to design a bin or hopper that will be self-emptying, various devices, usually called *flow inducers*, can be used to promote flow. In many cases, a funnel-flow bin with a flow inducer may be a better choice than a mass-flow bin. Flow inducers can be classified according to type as mechanical dislodging, pneumatic, and vibratory.

1-20. Mechanical Dislodgers

The simplest mechanical dislodger is a sledgehammer, which can be used to pound the bin or hopper to break up walls of funnels, arches, or other blockages. Although this is an obviously unsophisticated technique, there are occasional layouts in which it may be adequate. To provide for this situation, the lower part of a hopper can be made with a flexible rubber cone, developed by the General Rubber Company and called a Flexicone, which can be pounded without destroying equipment.[7] In another mechanical dislodging scheme the bin and hopper are provided with holes through which rods can be inserted to loosen the material in cases of arching or funneling. Various vertical or horizontal revolving stirrers, rodding devices, rotating paddles, etc., have been developed. Most of them require frequent repair, especially if they are used with hard and/or large-size material, and are not generally satisfactory.[8]

1-21. Pneumatic Inducers

Air can be used to induce flow in three ways: (1) by aeration of the material by a uniform, gentle flow of air; (2) by jets of compressed air as needed; and (3) by inflation of cushions attached to the bin or hopper walls.[7,10,11]

Aeration is generally limited to use with powders. Blowing air into the lower part of a hopper reduces the contact among grains of the material and enables it to move at a constant rate. The entire cross section of a bin can be aerated, which allows the surface of the stored material to drop uniformly, so that mass flow is achieved. Various bin configurations and aeration elements have been developed. For example, porous ceramic tiles, woven cotton, and similar materials can be installed between the air-pressure chamber and the stored material.

Compressed-air jets are injected through tubes, called lances, which are installed in the bin or hopper walls in the areas where funneling or arching is likely to occur. The ends of the lances are bent so as to direct the air toward the outlet of the bin. Airflow is regulated by valves.

Inflatable panels or cushions installed in the bin and/or hopper walls can be used to destroy arches and restore flow. The cushions are inflated with compressed air, as needed, and deflated after flow is restored. They may be controlled automatically by sensing units near the hopper outlet. However, they are high in first cost and maintenance.

1-22. Vibrators[3,7]

Vibrators may be elements suspended in a bin or hopper or units mounted on the wall. Alternatively, the entire hopper may be designed to vibrate (Sec. 1-23).

Wall-mounted vibrators are of three types: piston, electromagnetic, and eccentric-rotor. Air-powered piston vibrators with capacities of 1–10 ft^3 are available for small bins handling lightweight materials. Models with capacities of 5–50 tons are available with either metal or rubber striking blocks. Rubber blocks are less noisy, which makes them ideal for confined areas with workers nearby. Models of 100–150 tons are ideal for large hoppers or chutes handling material such as coal and iron ore. Available frequencies range from about 1000 to 10,000 c/min.

The rotary vibrator imparts a nonlinear, slightly rocking action and operates at a low noise level. Available frequencies range from about 1000 to as much as 20,000 c/min. Vibrators with the higher frequencies can operate at small amplitudes and still improve flow, which is advantageous as regards stresses in the bin or hopper wall.

The electromagnetic vibrator generates a vibration perpendicular to the surface on which it is mounted. It is simpler in operation and has a longer life than the rotary vibrator. Available frequencies range from about 3000 to 6000 c/min. Many models are waterproof and dust-tight, while separate waterproof, dust-tight cases are available for other models which are to be used in dusty, dirty atmospheres or which will be exposed to the weather.

For maximum efficiency, vibrators must be located in the correct position with a mounting stiff enough to transmit vibrations directly to the contents of the hopper. For conical and rectangular hoppers the vibrator should be bolted at a distance of about 12–18 in., but not more than $x/4$, from the discharge end (Fig. 1–15a). If the hopper has a vertical face, the vibrator should be located at these same distances on the sloping face opposite the vertical one. If a second vibrator is required on a rectangular hopper, it should be bolted on the opposite face at a slightly higher elevation. The vibrator on a hopper with a sloping discharge chute should be mounted on the hopper centerline as close to the discharge end as possible; a second vibrator may be required on the chute (Fig. 1–15b). The vibrator should be located as shown in Fig. 1–15c on flat-bottomed rectangular or cylindrical bins with central discharge.[3]

The position of the vibrator on an unsymmetrical hopper, or on a symmetrical hopper with unsymmetrical filling, can influence the flow pattern. In one case, a vibrator was mounted as shown in Fig. 1–16a on a hopper with

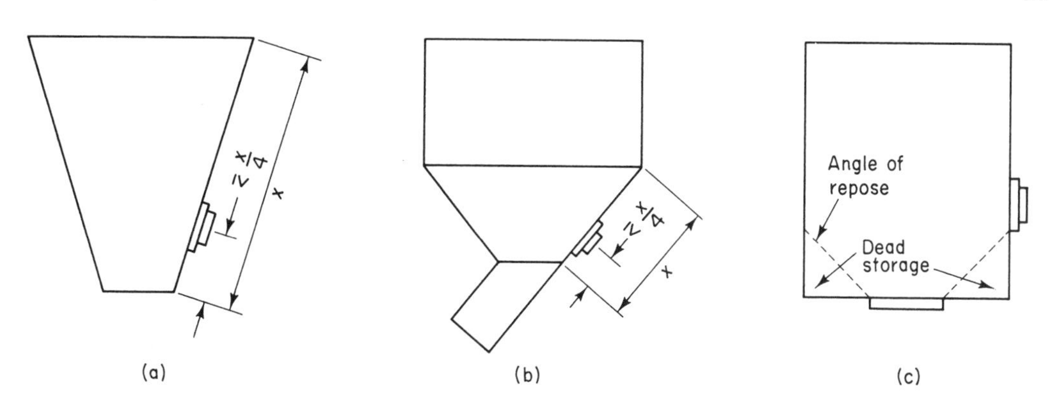

Figure 1-15 Recommended positions of wall-mounted vibrators.

a centrally located screen. When, to increase capacity, a second screen was located alongside (Fig. 1-16b), the resulting unsymmetrical loading made it difficult to avoid overflowing the hopper. Moving the vibrator to the position shown in Fig. 1-16c solved the problem.[7]

A steel plate, with vibrator attached, suspended in the center of a bin is shown in Fig. 1-17a. Such a unit can operate at a much larger amplitude of vibration than a wall-mounted vibrator because very little of the vibration is transmitted to the walls, which considerably reduces the possibility of damage to the walls. This vibrating plate was found to be much more effective than vibrating walls.[12] Therefore, this device is useful with bulk materials which require a very large amplitude of vibration for adequate flow. A vibrator attached to a suspended casing which moves with a slow swinging motion produces a similar vibration (Fig. 1-17b). A suspended steel grating activated by a vibrator mounted outside the hopper wall has given good re-

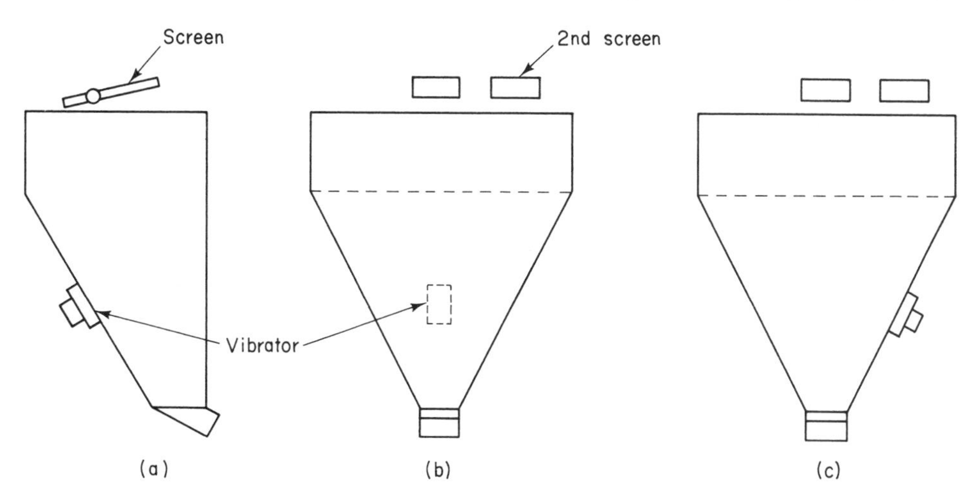

Figure 1-16

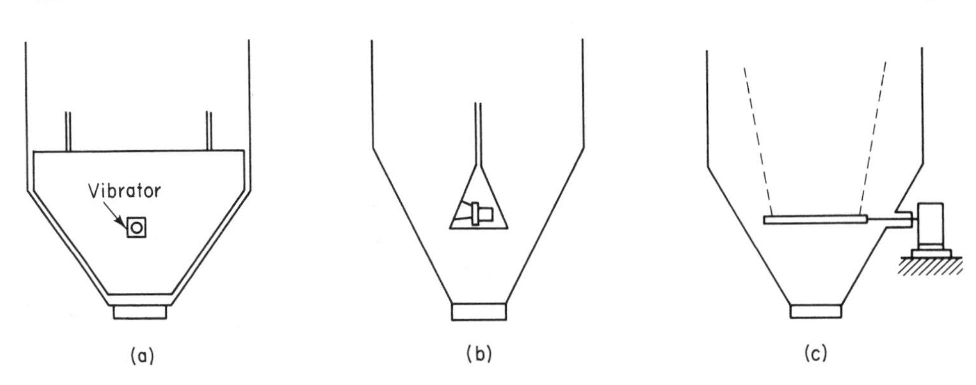

Figure 1-17 Suspended vibrators.

sults with materials such as salt, chalk, sulfur, quartz, rubber, sawdust, etc. (Fig. 1-17c).[8]

Misuse of vibrators may result in extreme compaction of the material. Vibrators also increase maintenance and its associated costs.

1-23. Vibrating Hoppers[13,14]

There are two principal types of vibrating hoppers. One, called a gyrated hopper, utilizes an exciter mechanism consisting of one or more motor-operated eccentric weights which produce a vibration perpendicular to the flow channel (Fig. 1-18a). The other, called a whirlpool hopper, utilizes two vibrating motors 180° apart and mounted so that their eccentric weights balance each other. In operation it exerts a lift-twist action on the material (Fig. 1-18b). Both types are equipped with internal pressure cones or baffles whose function is to force the material to converge at the outlet and to help produce a larger flow from the upper part of the bin. The cone for a gyrated hopper usually has a clearance of one discharge diameter and is positioned

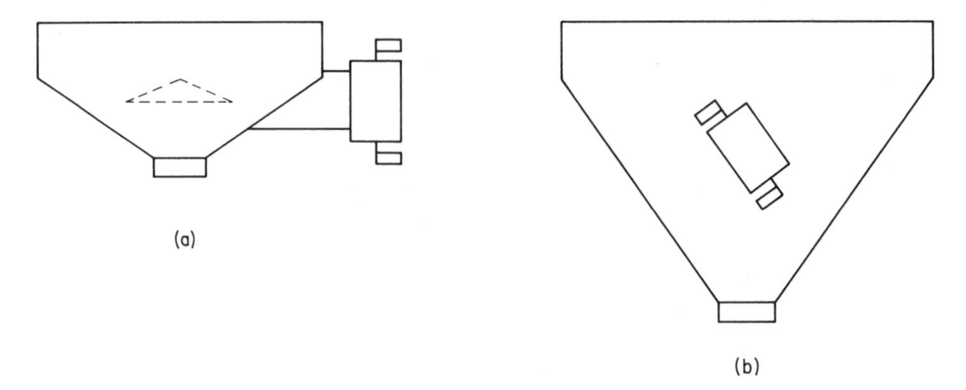

Figure 1-18 Vibratory hoppers: (a) gyrated, (b) whirlpool.

one discharge diameter above the outlet. The cone for a whirlpool hopper usually has a clearance of one-half the discharge diameter and is positioned $1\frac{1}{2}$ discharge diameters above the outlet.

The operating frequency of most vibratory hoppers ranges from 900 to 1800 c/min at amplitudes of $\frac{3}{8}-\frac{1}{8}$ in., respectively. It has been found that lower frequencies at large amplitude penetrate bulk materials more effectively. However, such vibrations are more difficult to isolate.

The whirlpool hopper is more expensive than the gyrated hopper, but it will handle a larger variety of materials at wider extremes of moisture.

A vibrating hopper must be attached to the bin with a flexible connection to avoid structural damage. A relatively thin rubber strip held to both bin and hopper by drawbands is commonly used for this purpose.

LEVEL-INDICATING DEVICES

The two basic methods of measuring the depth or amount of material in a bin are (1) a continuous material-level indicator and (2) a limit-level indicator.

1-24. Continuous Level Indicator

A continuous level indicator is required if the depth or amount of material must be determined at any time. This device consists of a weighted probe supported by a steel cable and operated from a control console. The probe is lowered into the bin, and a counter in the control console provides a reading of the amount or depth of material. When the probe reaches the material level, a trigger reverses the motor, stops the counter, and returns the probe to its original position. The counter reading remains on the control console until another reading is required or until the reading is manually erased. The control console can be located close to or remote from the bin, or it can be set up in a production control area to monitor many units.

Several types of probes are available. A cast aluminum plumb bob is recommended for granular or larger-particle materials. Probes recommended for other materials are a stainless inverted cone for materials with a bulk density from 5 to 20 pcf and a cast epoxy plumb bob for applications requiring a nonmetallic probe.

1-25. Paddle Unit

A paddle unit consists of a motor-driven sensing paddle. It is usually installed through the deck or through the wall at the middle or near the bottom of the bin. The unit must be shielded in low-level mounting. The sensing paddle is kept immobile by the material stored around it. As the material level lowers during withdrawal, the paddle is freed and rotates to trigger an alarm system such as a horn, bell, light, or other device. The motor operating the signal device is in an off position until the material level is below the paddle. This unit operates efficiently on all types of material ranging from aerated flour to grains and aggregates.

1-26. Pressure-Sensitive Diaphragm Sensor

A pressure-sensitive diaphragm sensor performs the same function as the paddle unit. Pressure exerted by the stored material against a diaphragm generates a signal. As the material is removed from the bin and clears the diaphragm, a different signal is generated. The sensitivity of the unit is adjustable.

1-27. Level Control by Conductivity

Level control by conductivity can be used only for stored material which has electrical conductivity. A transducer consisting of several partially insulated probes of varying length is installed inside the bin and connected to a control device. The bin wall forms the other pole of the circuit. During filling the stored material rises until it contacts a probe which closes the circuit and causes the controller to activate a signal.

ELEVATORS

1-28. Screw Elevator

The screw elevator is a very compact, efficient, and economical means of elevating all free-flowing materials as well as many sluggish ones. The elevator consists of a vertical housing which encloses a helicoid screw conveyor, an inlet, a top or bottom drive, and a discharge housing.

The screw flighting for the conveyor is rolled from steel and stretch-mounted on a section of straight pipe. Housing for the screw elevator is constructed of steel sheet. For unusual applications the screw, housing, and accessories may be formed from alloy materials.

1-29. Bucket Elevators

The simplest and most dependable unit for lifting material vertically is the bucket elevator, which is available in a wide range of capacities. It can be designed to lift practically any bulk material such as cement, coal, crushed stone, grain, lime, ore, sand, etc. It may be enclosed or operate in the open.

A bucket elevator consists of a series of buckets which are connected at uniform intervals to two strands of chain or a belt. The buckets, which are spaced to prevent interference in loading or discharging, scoop the material from a boot, pit, or pile located at the foot pulley. If the belt speed is high enough, the buckets are emptied by centrifugal force as they pass over the head pulley. This type, which is called a centrifugal-discharge elevator, may be positioned either vertically or at an angle. Because discharge from a vertical elevator depends entirely on centrifugal force, the belt must travel at a relatively high speed. If the elevator is inclined, the discharge chute can be located partly under the head pulley so that the discharge is not entirely dependent on the centrifugal force and the elevator may run at a lower speed. This is the most common type.

For sticky materials or those which tend to pack, the buckets are mounted on two strands of chain. As the bucket passes over the head sprocket, it is snubbed back and inverted for positive discharge. If the buckets do not completely empty, knockers which hit the buckets at the discharge point may be installed. Because this type of unit operates at a slower speed, larger or more closely spaced buckets are required to equal the capacity of the centrifugal-discharge unit.

Materials composed of large lumps which are difficult to handle may require a continuous-bucket elevator. The buckets for this type are closely spaced, with the back of the preceding bucket serving as a discharge chute for the bucket which is dumping as it rounds the head pulley. Because of the close bucket spacing, this type of elevator can run at a reduced speed and maintain a capacity comparable with the spaced-bucket elevator. The reduced speed produces a more gentle discharge, which prevents excessive degradation and assures effective handling of finely pulverized or fluffy materials.

REFERENCES

1. Link-Belt Co. (now FMC Corporation), *Materials Handling and Processing Equipment,* Chicago, Il., 1959.
2. C. K. Andrews, The Performance of Helical Screw Equipment for the Handling of Solids, *ASME Publ. 68-MH-38,* American Society of Mechanical Engineers, New York, 1968.
3. FMC Corporation, Materials Handling Equipment Div., Product literature, Homer City, Pa.
4. W. L. Hickerson, Vibrating Conveyor and Feeding Systems, ASME Vibrational Conference, January 3, 1967, Boston, Mass.
5. H. Colijn, Vibratory Conveyors, *Bulk Materials Handling,* Vol. VI, University of Pittsburgh, Pittsburgh, 1978.
6. W. G. Bauer, Factors Affecting the Movement of Material Through Bins, Gates, and Feeders, *Pit and Quarry,* Aug. 1945.
7. C. A. Lee, Solids Feeders, *Chem. Eng.,* Feb. 1964.

8. W. Reisner and M. v. Eisenhart Rothe, *Bins and Bunkers for Handling Bulk Materials,* Trans Tech Publications, Clausthal-Zellerfeld, West Germany, 1971.
9. P. J. Carroll, Vibratory Feeders, *Chem. Eng. Prog.* June 1970.
10. T. A. Engh, Effect of Injected Air on the Rate of Flow of Solids, *ASME Publ. 68-MH-18,* American Society of Mechanical Engineers, New York, 1968.
11. S. P. J. Ellis, Design of Silos with Aeration-Assisted Discharge, Symposium on Practical Bunker Design, Institute of Mechanical Engineers, Feb. 1969, London.
12. A. D. Sinden, Getting Difficult Materials out of Bins, *ASME Publ. 61-BSH-8,* American Society of Mechanical Engineers, New York, 1968.
13. J. L. Myers, Vibrating Hoppers, *Mech. Eng.,* March 1970.
14. G. D. Dumbaugh, Concepts of Vibratory Hoppers, *ASME Publ. 65-MH-11,* American Society of Mechanical Engineers, New York, 1965.

2

MATERIALS OF CONSTRUCTION

2-1. Mechanical Properties of Metals

The most important factor affecting mechanical properties of the steels is the chemical composition. Some of the other factors are the total reduction from ingot to finished product, finishing temperatures, and rate of cooling. Because thin plates involve larger reduction in the ingot than thick plates, which requires more passes through the rolls, they have a higher yield strength. However, yield strength may be kept independent of thickness by varying the chemical composition with the thickness of the plate to be rolled. This is done with A36 steel for thicknesses to 8 in.

The mechanical properties of structural steels are determined by tests prescribed by ASTM specifications A6, A20, and A370. Tensile strength, yield point or yield strength, and elongation are determined from a tensile test.

Ductility implies a large capacity for inelastic deformation before rupture. It is usually measured by the elongation at fracture of a 2-in. or 8-in. gage length or by the reduction, in percent, of the cross-sectional area. Because ductility is an extremely important property, ASTM specifies minimum elongations which are subject to their modification.

Brittle fractures are usually catastrophic, low-ductility fractures that originate at a point of concentration of stress such as a defect or a geometrical stress raiser. Failures usually occur at low temperatures and in welded structures. The Charpy V-notch test (ASTM E23) is commonly used to evaluate the susceptibility of steels to brittle fracture.

The composition of the steel is of major importance as it affects its brittle-fracture behavior; the lower the carbon content, the greater the notch toughness and energy-absorbing capacity. The toughness of any given steel decreases with increase in thickness because thicker plates require less rolling and cool more slowly, which results in a coarser-grained material. Residual stresses also affect brittle-fracture behavior. Low-temperature, low-stress brittle fractures have usually occurred in situations where there were large residual stresses, usually because of welding, in addition to sharp notches and low temperatures. Cold working of steels is also detrimental because of the resultant lowering of ductility.

Fine-grain steels have greater toughness than coarse-grain steels. Heat-treated fine-grain steels are tougher than as-rolled fine-grain steels. Unless otherwise specified, the producer may furnish either coarse-grain or fine-grain steel.

Good design details and good fabrication and erection are very important in protecting against brittle fracture. A well designed and fabricated detail in a low-ductility steel may be as brittle-fracture-resistant as a poorly executed detail in a notch-tough steel. For example, improper welding can result in flaws that may lead to brittle fracture because of the resulting stress concentrations. Also, corners in openings should be made with as large a radius as possible to reduce stress concentrations at these points. Other precautions are discussed in Ref. 1.

Plate steels as finished off the hot mill usually have sufficient toughness for use in bin construction, provided design, fabrication, and erection are all of good quality.

2-2. Steels for Bins, Tanks, and Hoppers

The steels used in the construction of bins, tanks, and hoppers and their supporting structures are (1) hot-rolled plates and shapes, (2) sheet and strip steels, and (3) pipe and tubing. They are listed in Table 2-1. Minimum tensile properties are given in Tables 2-2, 2-3, and 2-4.) Some of these steels are used only occasionally for bins. The following should be noted.

The most commonly used plates are A283 Grades A, B, C, and D. A36, A131 in Grades A, B, and C, and A573 Grade 58 are also in common use. Structural shapes are usually A36.

For A36, copper can be specified as an alloying element in an amount which doubles its resistance to corrosion.

A285 may have slightly better weldable properties than A283 but, for this purpose, is not likely to be worth the premium involved.

A588 is in seven grades, all with the same yield point, each produced by a different company.

For A606 sheet and strip, copper may be specified as an alloying element in an amount which increases its resistance to corrosion to twice that of plain carbon steel. It is also available in Type 2, which is twice as resistant to corrosion as plain carbon steel, and in Type 4, which is four times as resistant.

A446 is a galvanized sheet produced with eight classes of hot-dip zinc coatings so that sheets with coating consistent with expected service life are available.

Steel pipe and tubing are often used as structural members. A53 is allowed by the AISC specification only in Grade B, which is electric-resistance-welded.

2-3. Clad Steels

Clad steel is a composite plate formed by mill-rolling under pressure a sheet of cladding metal and a sheet or plate of base metal until they bond integrally over their entire surface. Cladding metals include stainless-steel alloys, nickel and nickel alloys, and copper and copper alloys. Typical base metals for bins and hoppers are A36 and A283 carbon steels and A514 structural alloy steel. The clad steels include A263 corrosion-resisting chromium steel clad plate or sheet, A264 stainless chromium-nickel steel clad plate or sheet, and A265 nickel and nickel-base alloy clad steel plate. The cladding thickness is usually specified as a percentage of the total thickness of the composite plate. The percentage is determined by the requirements of the application. The most common thicknesses are 10 and 20%, but any thickness from 5 to 50% is obtainable.

Because of its corrosion and abrasion-resistant surface, stainless-clad steel has proven economical for coal bunkers and hoppers. The stainless surface is not affected by the sulfur in the coal, and because of the abrasive action of the coal, the surface maintains a highly polished finish. The smooth surface helps to eliminate ratholing and arching. A thickness of 20% is commonly used.

2-4. Non-ASTM Steels

Steels not covered by an ASTM standard may be used provided such material conforms to the chemical and mechanical process and heat-treatment requirements of one of the listed standards or other published standard which establishes its properties and suitability and provided it is subjected by either the producer or the purchaser to analyses, tests, and other controls to the extent and in the manner prescribed by one of the listed standards.

2-5. Structural Bolts

The two commonly used types of bolts are the unfinished bolt (A307) and the high-strength bolt (A325 and A490).

The A307 bolt is furnished in Grades A and B, the former for general

TABLE 2-1. STEELS FOR BINS, TANKS, AND HOPPERS

ASTM designation	Product	Use
A36	Carbon-steel shapes, plates, and bars	Welded, riveted, and bolted construction; general structural purposes
A53	Welded or seamless pipe, black or galvanized	Welded, riveted, and bolted construction
A131	Structural steel shapes, plates and bars	
A167	Stainless and heat-resisting chromium nickel steel plate, sheet, and strip	
A242	High-strength, low-alloy shapes, plates, and bars	Welded and bolted construction; general structural purposes; atmospheric-corrosion resistance about four times that of carbon steel; a weathering steel
A283	Low- and intermediate tensile strength carbon-steel plates, shapes, and bars	General application
A285	Low- and intermediate tensile strength carbon steel	Fusion-welded pressure vessels
A446	Zinc-coated (galvanized) sheets in coils or cut lengths	Welded, cold-riveted, bolted, and metal-screw construction
A500	Cold-formed welded or seamless carbon-steel structural tubing	Welded, riveted, or bolted construction
A501	Hot-formed welded or seamless carbon-steel structural tubing	Welded, riveted, or bolted construction
A514	Quenched and tempered alloy-steel plates of structural quality	Welded structures
A516	Carbon-steel plates for moderate- and lower-temperature service	Welded pressure vessels for service at temperatures requiring improved notch toughness

TABLE 2-1. *(continued)*

ASTM designation	Product	Use
A537	Heat-treated carbon-manganese-silicon steel plate	Fusion-welded pressure vessels and structures
A570	Hot-rolled carbon-steel sheets and strip in coils or cut lengths	Welded, cold-riveted, bolted, and metal-screw construction
A572	High-strength, low-alloy columbium-vanadium steel shapes, plates, and bars	Welded and bolted construction
A573	Structural-quality carbon-manganese-silicon steel plates	Primarily for service at atmospheric temperatures where improved notch toughness is important; for fusion welding
A588	High-strength low-alloy steel shapes, plates, and bars	A weathering steel; atmospheric-corrosion resistance about four times that of carbon steel; welded or bolted construction
A606	High-strength, low-alloy hot- and cold-rolled sheet and strip	Where greater strength, savings in weight, and enhanced atmospheric corrosion resistance are important
A607	High-strength, low-alloy columbium and/or vanadium hot- and cold-rolled sheet and strip	Where greater strength, savings in weight, and enhanced atmospheric corrosion resistance are important
A611	Cold-rolled carbon structural steel sheet	
A633	High-strength, low-alloy structural steel shapes, plates and bars	Welded and bolted construction; tion; serviceable at ambient temperatures of -50°F
A678	Quenched and tempered carbon-steel plates for structural applications	Welded and bolted construction

TABLE 2-2. MINIMUM TENSILE PROPERTIES OF STRUCTURAL STEELS FOR BINS, TANKS, AND HOPPERS

ASTM designation	Yield (ksi)	Strength (ksi)	Elongation (% in 8 in. unless noted)	Electrode filler metal
Carbon steels				
A36	36	58-80	20	E60XX, E70XX, unless restricted
A131	34	58-71	21	E60XX, E70XX, unless restricted
A283				E60XX, E70XX, unless restricted
Grade A	24	45-55	27	
Grade B	27	50-60	25	
Grade C	30	55-65	22	
Grade D	33	60-72	20	
A285				E60XX, E70XX, unless restricted
Grade A	24	45-65	27	
Grade B	27	50-70	25	
Grade C	30	55-75	23	
A516				E7018, E7028
Grade 55	30	55-75	23	
Grade 60	32	60-80	21	
Grade 65	35	65-85	19	
Grade 70	38	70-90	17	
A573				
Grade 58	32	58-71	21	
Grade 65	35	65-77	20	
Grade 70	42	70-90	18	
High-strength steels				
A242				E70XX
To $\frac{3}{4}$ in. incl., and Group 1 shapes	46	67	18	
Over $\frac{3}{4}$ to $1\frac{1}{2}$ in. and Group 2 shapes	50	70	18	
Over $1\frac{1}{2}$ to 4 in. and Group 3 shapes	42	63	11	
A572				E70XX
Grade 42 to 6 in. incl., all groups of shapes	42	60	20	
Grade 50 to 2 in. incl., all groups of shapes	50	65	18	
Grade 60 to $1\frac{1}{4}$ in. incl., Groups 1 and 2 shapes	60	75	16	
Grade 65 to $1\frac{1}{4}$ in. incl., Group 1 shapes	65	80	15	

TABLE 2-2. *(continued)*

ASTM designation	Yield (ksi)	Strength (ksi)	Elongation (% in 8 in. unless noted)	Electrode filler metal
A588 to 4 in. thick	50	70	18	
A633				
Grade A to 4 in.	42	63-83	18	
Grades C and D				
To $2\frac{1}{2}$ in. incl.	50	70-90	18	
Over $2\frac{1}{2}$ to 4 in.	46	65-85	18	
Grade E to 4 in.	60	80-100	18	
Quenched and tempered steels				
A514				E110XX[a]
To $2\frac{1}{2}$ in.	100	110-120	18 in 2 in.	
Over $2\frac{1}{2}$ to 6 in.	90	100-130	16 in 2 in.	
A678				
Grade A to $1\frac{1}{2}$ in.	50	70-90	22	
Grade B to $2\frac{1}{2}$ in.	60	80-100	22	
Grade C to $\frac{3}{4}$ in.	75	95-115	19	
Over $\frac{3}{4}$ to $1\frac{1}{2}$ in.	70	90-110	19	
Over $1\frac{1}{2}$ to 2 in.	65	85-105	19	
Heat-treated steels				
A537				
Class 1 normalized				
To $2\frac{1}{2}$ in.	50	70-90	18	
Over $2\frac{1}{2}$ to 4 in.	45	65-85	18	
Class 2 quenched and tempered				
To $2\frac{1}{2}$ in.	60	80-100	–	
Over $2\frac{1}{2}$ to 4 in.	55	75-95	–	
Stainless and heat-resisting chromium-nickel steel				
A167[b] 40 in 2 in.	25-30	70-75	40 in 2 in.	

[a]Low-hydrogen classification.

[b]26 types depending on chemistry.

TABLE 2-3. MINIMUM TENSILE PROPERTIES OF SHEET AND STRIP STEELS FOR BINS, TANKS, AND HOPPERS

ASTM designation	Yield (ksi)	Strength (ksi)	Elongation (% in 8 in. unless noted)	Electrode filler metal
Carbon steels				
A570				E60XX, E70XX, unless restricted
Grade 30	30	49	17–19[a]	
Grade 33	33	52	16–18[a]	
Grade 36	36	53	15–17[a]	
Grade 40	40	55	14–16[a]	
Grade 45	45	60	12–14[a]	
Grade 50	50	65	10–12[a]	
A611				
Grade A	25	42	26 in 2 in.	
Grade B	30	45	24 in 2 in.	
Grade C	33	48	22 in 2 in.	
Grade D	40	52	20 in 2 in.	
Grade E	80	82	[b]	
Low-alloy steels				
A606				E8018G
Hot-rolled, cut length	50	70	22 in 2 in.	
Hot-rolled, annealed	45	65	22 in 2 in.	
Cold-rolled	45	65	22 in 2 in.	
A607				E8018G
Grade 45	45	60	22,[c] 25[d]	
Grade 50	50	65	20,[c] 22[d]	
Grade 55	55	70	18,[c] 20[d]	
Grade 60	60	75	16,[c] 18[d]	
Grade 65	65	80	15,[c] 16[d]	
Grade 70	70	85	14,[c] 14[d]	
Zinc-coated (galvanized) steels				
A446				EXX10, 11, 12, 13
Grade A	33	45	20	
Grade B	37	52	18	
Grade C	40	55	16	
Grade D	50	65	12	
Grade E	80	82	[e]	
Grade F	50	70	12	
Stainless and heat-resisting chromium-nickel steel				
A167[f]	25–30	70–75	40 in 2 in.	

[a]Depends on thickness.

[b]Full-hard product, no specified minimum.

[c]Hot-rolled, in 2 in.

[d]Cold-rolled, in 2 in.

[e]Not specified.

[f]26 types depending on chemistry.

TABLE 2-4. MINIMUM TENSILE PROPERTIES OF STRUCTURAL PIPE AND TUBING

ASTM designation	Yield (ksi)	Strength (ksi)	Elongation (% in 2 in.)
Welded and seamless pipe			
A53			
Grade A	30	48	–
Grade B	35	60	–
Round structural tubing			
Grade A	35	45	25
Grade B	42	58	23
Grade C	46	62	21
Shaped structural tubing			
A500 (cold-formed welded seamless)			
Grade A	39	45	25
Grade B	46	58	23
Grade C	50	62	21
A501 (hot-formed welded seamless)	36	58	23

purposes and the latter for joints in pipe systems. They are made of low-carbon steel with a minimum tensile strength of 60 ksi.

The A325 bolt is made of medium-carbon steel. Its tensile strength decreases with increase in diameter of the bolt, so that two ranges of diameter are specified (Table 2-5). The A490 bolt is made of alloy steel in one tensile-strength grade. Tensile properties for both are based on the *stress area,* which is larger than the section at the root of the thread but smaller than the unthreaded area. The tensile strength of the bolt, based on this area, is about the same as the coupon strength.

A193 covers alloy- and stainless-steel bolting material for high-temperature service. Eight grades are available: four ferritic steels and four austenitic stainless steels. Tensile strengths range from 90 to 125 ksi and yield strengths from 75 to 105 ksi, both on the stress area. Elongations in 2 in. range from 15 to 18%.

A320 is an alloy-steel bolting material for low-temperature service. Ten grades are available, including both ferritic and austenitic steels. The austenitic steels may be strain-hardened for increased tensile properties. Tensile strengths range from 75 to 125 ksi and yield strengths from 30 to 105 ksi, both on the stress area. Elongations in 2 in. range from 12 to 35%.

2-6. Welding Electrodes

Electrodes are classified on the basis of the mechanical properties of the weld metal, the welding position, the type of coating, and the type of current

TABLE 2-5. TENSILE PROPERTIES OF STRUCTURAL BOLTS

ASTM designation	Diameter (in.)	Tensile strength on stress area[a] (ksi)	Proof load on stress area[a] (ksi)
Low-carbon steel			
A307			
Grade A	All	60	
Grade B	All	60–100	
High-strength structural bolts			
Medium-carbon steel			
A325	$\frac{1}{2}$-1	120	85
	$1\frac{1}{8}$-$1\frac{1}{2}$	105	74
Alloy steel			
A490	$\frac{1}{2}$-$1\frac{1}{2}$	150–170	120

[a]Stress area = $0.785(D - 0.9743/n)^2$, where D = nominal diameter and n = threads per inch.

required. Electrodes for shielded metal arc welding of carbon and low-alloy steels are covered by AWS A5.1 (mild steel electrodes) and AWS A5 (low-alloy steel electrodes). Mild steel electrodes are identified by a code number E60XX or E70XX, where E stands for electrode and the first two digits indicate the tensile strength, in kips per square inch, of the deposited metal. The first X represents a digit denoting the positions in which the electrode can be used, the number 1 meaning all positions, the number 2 flat and horizontal fillet welds, and the number 3 flat welding only. The last number denotes the type of covering and the type of current (ac or dc). Each low-alloy steel electrode is identified by a code number E70XX-X, E80XX-X, etc. The last X represents a suffix A1, B1, etc., and designates the chemical composition of the deposited metal.

REFERENCE

1. W. H. Munse, Fatigue and Brittle Fracture, Sec. 4 in E. H. Gaylord and C. N. Gaylord (eds.), *Structural Engineering Handbook,* 2nd ed., McGraw-Hill, New York, 1979.

3

PROPERTIES OF BULK SOLIDS

Bulk solids such as ore, concentrate, grain, flour, feed, chemicals, rock, sand, powders, and plastic pellets are assemblies of discrete solid particles. The properties of a solid refer to the assembly of its particles.

3-1. Pressure-Density Relations

Bulk density. Bulk density γ is the weight of a bulk material per unit volume. *Loose bulk density* (also called aerated bulk density) is determined by weighing a sample in a noncompacted condition. *Packed* or *compacted bulk density* is determined by weighing a sample that has been compacted by vibrating it in a container[1] or by measuring volume change under pressure in a test cell similar to the one shown in Fig. 3-1. A typical variation of bulk density with consolidating pressure is shown in Fig. 3-8b. *Working bulk density* is given by

$$\gamma_w = \frac{(\gamma_p - \gamma_a)^2}{\gamma_p} + \gamma_a$$

where γ_a = aerated bulk density and γ_p = packed bulk density.

The aerated bulk density should be used to compute bin and hopper capacities. The working bulk density should be used to determine rate of feed.

Compressibility. Compressibility is a measure of the change in volume of a bulk solid caused by a change in the stress system acting on it. Carr[1]

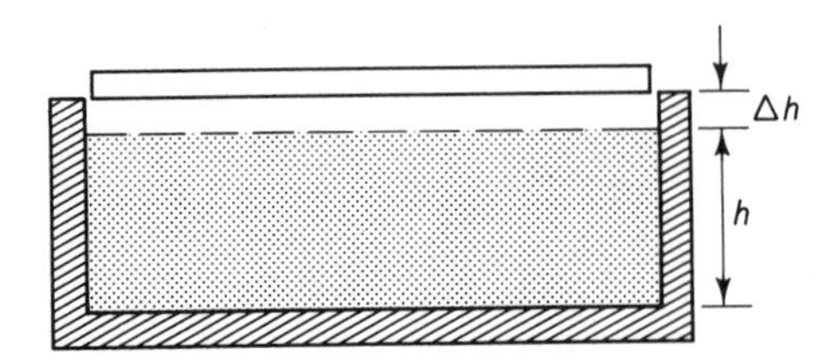

Figure 3-1 Pressure test cell.

defines the compressibility coefficient of a bulk solid as

$$\frac{\gamma_p - \gamma_a}{\gamma_p} = 1 - \frac{\gamma_a}{\gamma_p} \tag{3-1}$$

The compressibility coefficient gives an indication of the flowability of a bulk solid (Sec. 3-7, Table 3-3). Carr recommends that the packed density be measured by weighing a loose sample after it has been vibrated for 5 min.

3-2. Particle Size

Granular materials are generally noncohesive and free-flowing, while powders are not. The heavier the material is in bulk density, the finer its powder particles will be. Therefore, the powder fraction of a material is defined in terms of its bulk density and a screen mesh size. Table 3-1 gives a classification by Carr. The powder fraction of a material is that portion which passes the mesh size specified in this table, and the granule fraction is the portion retained.

3-3. Angle of Repose

When an unconsolidated bulk solid falls freely to a horizontal surface from a height low enough to minimize impact, the particles of the solid roll down the pile. The angle of repose α is the slope of the cone formed by dropping material through a funnel onto a horizontal surface. At the start of the test the funnel is held close to the horizontal surface. It is then gradually raised as the height of the pile increases, maintaining about a 1-in. clearance. The funnel nozzle should have a diameter of four to six times the maximum par-

TABLE 3-1. POWDER SCREEN MESH[a]

Bulk density (pcf)	Mesh size (mm)
18	100
25-55	200
60-90	325

[a]From Ref. 2.

ticle size but not less than $\frac{1}{2}$ in. The volume of the test sample must be large enough so that the distance from the bottom of the funnel to the horizontal surface on which the completed pile rests is at least six times the maximum particle size but not less than 4 in.

For free-flowing solids with a narrow range of particle size the angle of repose equals the angle of internal friction (Sec. 3-5).

3-4. External Angle of Friction

Pressures within a solid in a bin are dependent on the coefficient of friction μ' between the solid and the wall of the bin. The coefficient of friction is given by

$$\mu' = \tan \phi' \tag{3-2}$$

where ϕ' is the external angle of friction.

To determine ϕ', a sample of the solid contained in a ring is placed on a sample of the wall material (Fig. 3-2). The height of the ring should be at least 1.5 times the largest particle size but not less than $\frac{1}{2}$ in., and the inside diameter should be 6 times the height. A cover with an attached bracket is placed on the solid. The pin shown in the figure is at midheight of the ring, and the force S is applied at the plane of contact of the solid with the wall sample. A weight W is placed on the cover, and the force S to cause sliding is measured. The ring is then rotated and raised slightly to ensure that the load W is transferred to the wall material by the solid. A part of the weight W is then removed, and a second value of S is determined. The process is repeated until the weight W is zero. A plot of the unit frictional force q/A and the unit compression $p = (W + w)/A$, where w is the weight of the ring, cover, and contained solid, is called a wall yield locus. The locus may be a straight line through the origin, in which case ϕ' is a constant, but it is often convex upward and ϕ' decreases with increasing p (Fig. 3-3). Although it would appear that some account should be taken of this difference, since smaller values of ϕ' give larger bin-wall pressures and smaller vertical compression in the wall, this is not generally done.

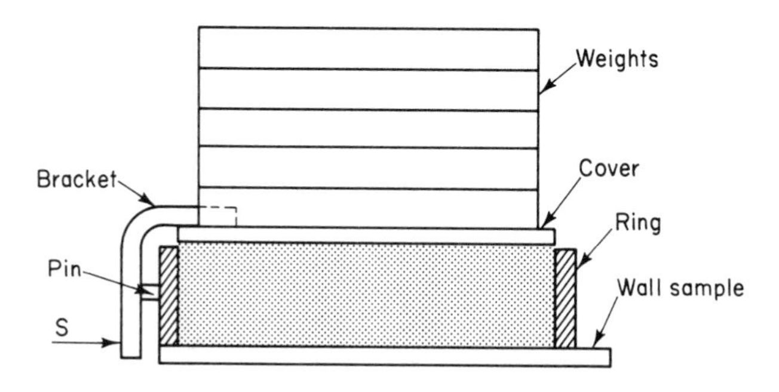

Figure 3-2 Shear cell. (From Ref. 3.)

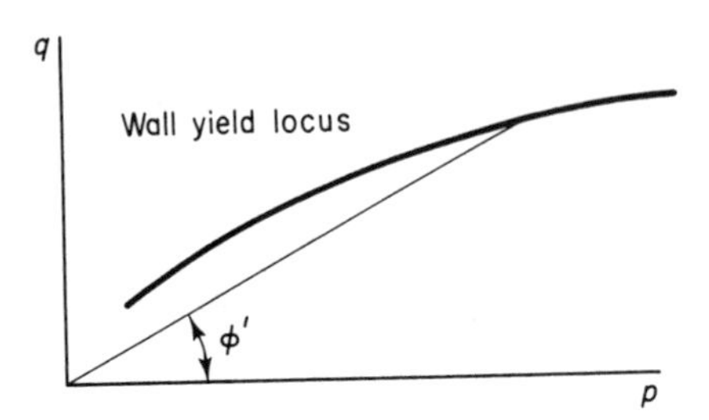

Figure 3–3 Wall yield locus.

3-5. Internal Angles of Friction

Unconsolidated granular material has no strength, but it gains strength by consolidation when it is stored in a bin. Materials do not gain the same strength under equal pressures; for example, gravel and dry sand, which are cohesionless, gain little if any strength within the range of pressures acting in a bin. Such materials shear when $q = p \tan \phi$, where q = unit shear, p = unit pressure, and ϕ = angle of internal friction (Fig. 3–4a). The line $q = p \tan \phi$ in this figure is called a yield locus. A typical yield locus for a cohesive material is shown in Fig. 3–4b. Cohesive materials gain strength with consolidation and have a yield locus for each consolidating pressure. In general, these loci form a family of convex-upward lines, so that the angle of friction varies with both the yield locus and the normal pressure p. The curvature of the yield locus is more pronounced at low values of p.

The relationship between the principal pressures in a bulk solid of semi-infinite mass is given by

$$\frac{p_1}{p_2} \leqslant \frac{1 + \sin \phi}{1 - \sin \phi} + \frac{2c}{p_2} \frac{\cos \phi}{1 - \sin \phi} \tag{3-3}$$

where c = cohesion, which varies with the degree of consolidation. The inequality represents an elastic state of pressure and the equality a plastic state.

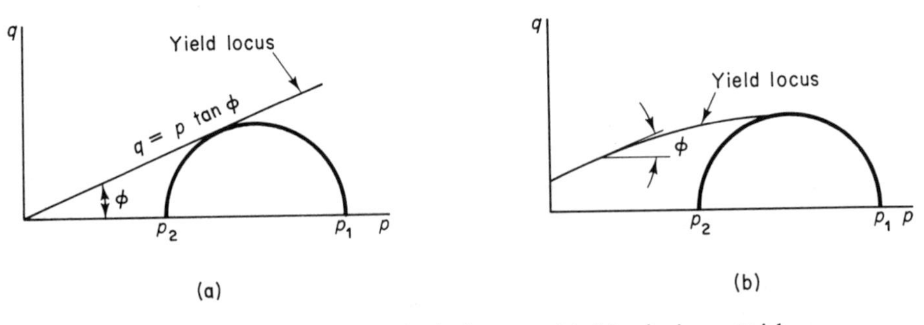

Figure 3–4 Yield loci: (a) cohesionless material, (b) cohesive material.

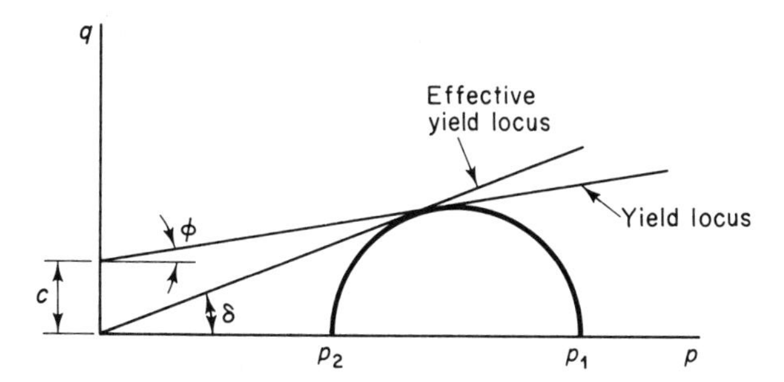

Figure 3-5 Yield locus and effective yield locus.

Measurement of principal stresses during flow show that the ratio p_1/p_2 is practically constant for wide ranges of pressures and is given by

$$\frac{p_1}{p_2} = \frac{1 + \sin \delta}{1 - \sin \delta} \tag{3-4}$$

where δ is an *effective* angle of friction. A line through the origin at an angle δ with the p axis and tangent to a Mohr circle is called the effective yield locus (Fig. 3-5).

Tests to determine the effective yield locus are made in the shear cell shown in Fig. 3-6. A mold is fitted on the ring of the cell and the mold and ring placed in an offset position on the base. Layers of the sample are placed in the cell and lightly packed with the fingers until the mold is filled. The material is scraped off level with the top of the mold and a twisting top placed in position. A number of oscillating twists are applied to the cover while it sustains a vertical force W. This preconsolidation assures a uniform specimen. The load W, the twisting top, and the mold are then removed and the solid scraped level with the top of the ring. The cover (Fig. 3-2) is then placed in position on the solid and loaded with a force W_1. A shear force is applied and increased until the specimen shears at a value S_1. This gives one

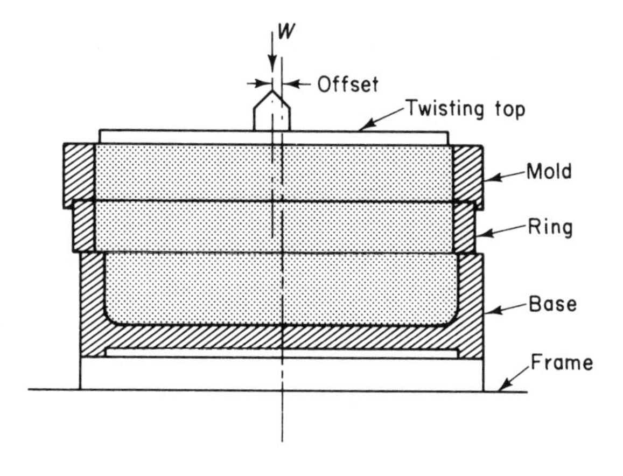

Figure 3-6 Shear cell. (From Ref. 3.)

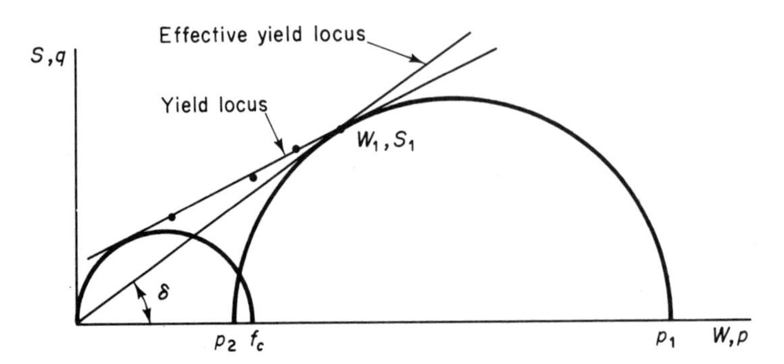

Figure 3-7 Determination of yield locus.

point on the yield locus (Fig. 3-7). Two other points are determined at the same preconsolidation, one for a value of about $\frac{2}{3}W_1$ and one for a smaller value. A Mohr circle tangent to the resulting yield locus is drawn through the origin. This circle determines the unconfined yield strength f_c of the solid, since it gives $p = 0$ on one principal plane and $p = f_c$ on the other. As a check, a third point on the locus is established to the right of the tangent point. A second Mohr circle is then drawn tangent to the yield locus at W_1, S_1, which gives the consolidating pressures p_1 and p_2 of the solid. A straight line through the origin tangent to this circle determines the effective angle of friction δ.

If the flow of a solid is interrupted for a period of time, the solid may consolidate under static pressure. The strength of some solids increases significantly with this consolidation in which case their flowability is reduced. The time yield locus of such solids is determined as already described except that the load on the test sample is maintained for a specified interval of time before the sample is sheared.

3-6. Flow Function

Some solids will not flow freely through the outlet of a bin because of the formation of an arch in the solid or a pipe through it. It is reasonable to assume that when such an obstruction collapses the failure starts at an exposed surface of the obstruction.[3] Since there are no stresses on the exposed surface, the stress parallel to it is a principal stress whose maximum value is the unconfined yield strength f_c discussed in Sec. 3-5. A plot of f_c against the major consolidating pressure p_1 is called a flow function. These curves are developed from yield loci. For example, the values of f_c and p_1 in Fig. 3-7 determine one point on a flow function. Tests at other consolidating pressures determine additional pairs of values. Typical flow functions are shown in Fig. 3-8c. The yield loci used to determine the flow function also

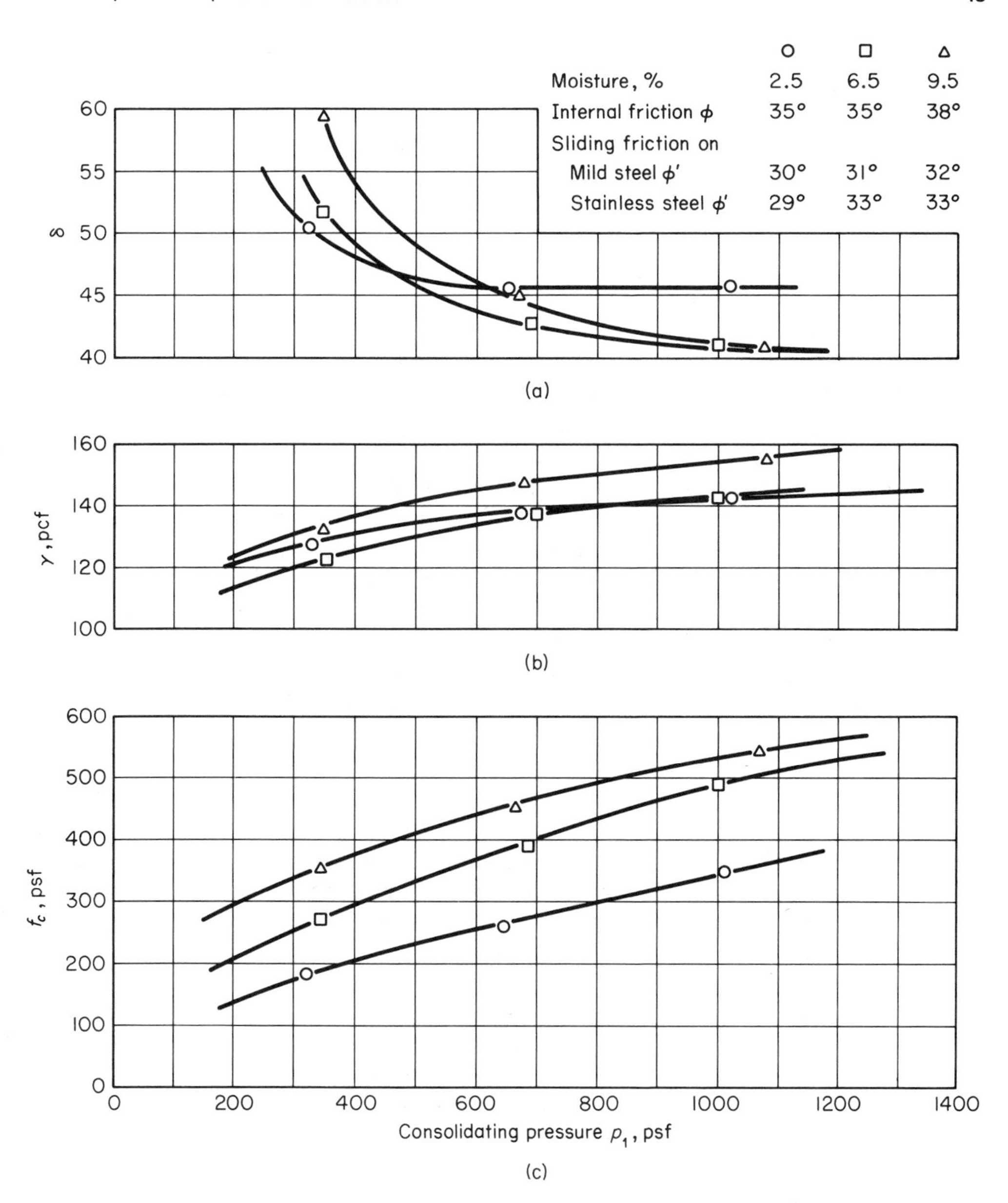

Figure 3–8 Flow properties of taconite concentrate. (From Ref. 6.6.)

give corresponding values of the effective angle of friction, for which typical plots are shown in Fig. 3–8a. Bulk density γ is also a function of consolidation (Fig. 3–8b).

Flow properties of solids which consolidate under static pressures during periods of interruption of flow are given by flow functions constructed from the time yield loci discussed in Sec. 3–5. These are sometimes called time flow functions.[3]

3-7. Flow Properties

A solid for which $f_c = 0$ (dry, round sand, for example) is perfectly free-flowing. According to Jenike,[3] solids which do not contain particles smaller than, say, 0.01 in. are usually free-flowing; coal and most ores are examples. However, there are exceptions to this rule. For instance, grain ferments under adverse moisture and atmospheric conditions and may develop sufficient strength to cease being free-flowing. Again, soybean meal contains oil which under certain conditions of high moisture and temperature binds the particles into a nonflowing mass. Also, flaky or stringy materials such as wood shavings, mica, and asbestos interlock to form obstructions to flow. The flowability of a solid containing a range of sieve sizes including both fine and coarse particles is invariably governed by the flow properties of the fine particles, because during flow the solid shears across the fine particles. The size of the coarse particles affects the tendency of the solid to interlock at the outlet, and the impact of these heavy, coarse particles charged into a container may cause compaction of the solid along the trajectory of the falling stream.

Materials with a relatively low compressibility, such as fluid-like powders and granular powders, may form weak or momentary arches at the outlet during flow. Only a small amount of energy is required to cause a weak arch to fall through the hopper. This phenomenon is called *hanging up.*

According to Carr[2] the angle of repose is a direct indication of the potential flowability of a material. An evaluation of the results of tests on more than 3500 samples of dry materials is shown in Table 3–2.

Tests also show that very fluid powders have a very unstable angle of repose of about 35°. The angle for less fluid powders will range from 40° to 55°. The smaller the angle of repose, the more flowable or floodable a material will be. Floodable flow is an unstable, gushing, discontinuous flow caused by the fluidizing of a mass of particles by air. Floodable materials may hang up or form a weak arch which breaks suddenly and causes the particles to be fluidized. Starch, clay, flour, talc, and sulfur are examples of floodable materials.

The effect of compressibility on the flow of solids is shown in Table 3–3 from Carr.[4] The compressibility coefficient in this table is defined in Sec. 3–1.

TABLE 3-2. FLOWABILITY OF MATERIALS[a]

Angle of repose (deg.)	Flowability
25-30	Very free-flowing granules
30-38	Free-flowing granules
38-45	Powdered granules, fair to passable flow
45-55	Cohesive powders, may require special agitation
55-70	Very cohesive powers require special agitation

[a]From Ref. 2.

Carr has developed a system of classifying masses of dry particles of granular, powdered, laminar, and fibrous form (Table 3-4). From this information one may evaluate the potential flow of a given material and predict whether it will flow freely, arch, flood, or tend to be fluidized.

The relationship between flowability and compressibility and the angle of repose of dry materials is shown in Fig. 3-9. The relationship is not exactly linear as shown, and the values at which the flow changes are approximate.

3-8. Conditions Affecting Flow Properties

Materials which flow freely from a bin under conditions of continuous flow may gain strength and obstruct flow after storage at rest. The results of stor-

TABLE 3-3.
COMPRESSIBILITY VERSUS CLASS OF SOLID AND TYPE OF FLOW[a]

Compressibility coefficient	Class of dry solid	Flow
0.05 to 0.15	Free-flowing granules	Excellent
0.12 to 0.18	Free-flowing, powdered granules	Good
0.18 to 0.22	Flowable, powdered granules	Fair to passable
0.22 to 0.28	Very fluid powders	Poor, unstable
0.28 to 0.33	Fluid, cohesive powders	Poor
0.33 to 0.38	Cohesive powders	Very poor
0.38 to > 0.40	Very cohesive powders	Very, very poor

[a]Excerpted by special permission from *Chemical Engineering* (Oct. 13, 1969) © 1969, by McGraw-Hill, Inc., New York, N.Y. 10020.

TABLE 3-4. RELATIVE FLOW PROPERTIES OF SOLIDS[a]

Description of class	Uniformity coefficient	Powder fraction percent	Flowability[b]	Floodability[b]	Archability[c]
CORPUSCULAR (roundish three-dimensional)					
Granules					
Uniform (sand, salt, sulfur)	1-10	< 5	70-100	0	1
Nonuniform (salt, sulfur, coal)	15-30	< 5	60-75	0	12
Very nonuniform (salt, sulfur, gravel)	30+	<5	50-70	0	11
Soft or sticky (rubber, detergent, carbon black)	–	< 5	40-75	0	12
Powdered granules					
Uniform (salt, sulfur, perlite)	1-12	< 30	70-80	0	1
Less uniform (salt, coal, perlite)	8-18	< 30	55-70	0	5
Nonuniform (salt, coal, gravel)	15-30	< 30	50-65	0	2 or 8
Very nonuniform (salt, coal, gravel)	30+	< 30	30-60	0	8
Soft or sticky (carbon black, rubber, wax)	–	< 30	50-70	0	5
Fluid powders					
Very fluid, granular (some)	–	60-90	55-70	70-90	7
Very fluid, granular and powder	–	30-60	50-65	30-70	6
Very fluid, powder	–	95-100	45-65	80-95	4
Fluid					
Granular (some)	–	60-90	45-55	50-75	2 or 4
Granular and powder	–	30-60	40-60	20-60	2 or 4
Powder	–	95-100	35-50	45-75	2 or 4
Fluid cohesive, granular (some)	–	60-90	20-45	25-60	2
Fluid cohesive, granular and powder	–	29-60	20-40	10-35	2
Fluid cohesive, powder	–	95-100	10-40	15-45	2 or 3
Cohesive powders					
Powder	–	95-100	5-25	0-20	13
Granular and powder	–	30-60	5-40	0-20	2 or 3

TABLE 3-4. *(continued)*

Description of class	Uniformity coefficient	Powder fraction percent	Flowability[b]	Floodability[b]	Archability[c]
LAMINAR (micaceous, two-dimensional)	Average size range				
Micaceous					
Thin	$\frac{1}{2}$ to 3 in.		1-15	0	10
Thick	$\frac{1}{4}$ to 1 in.		10-20	0	8
Powdered	-200 mesh		30-50	35-60	4
Film, very thin	$\frac{1}{16}$ to $\frac{1}{2}$ in.		1-10	0	10
Chips					
Fine, uniform	+100 to +10 mesh		70-80	0	1
Fine, nonuniform	-60 to +10 mesh		60-70	0	12
Large, nonuniform	-10 mesh to 3 in.		30-40	0	8
Flakes					
Thin	-100 mesh to 1 in.		1-20	0	10
Fine, uniform	-20 mesh to $\frac{3}{8}$ in.		50-80	0	1,9
Fine, nonuniform	-40 mesh to $\frac{3}{4}$ in.		45-50	0	11
Powdered	-200 mesh to $\frac{1}{4}$ in.		50-70	0	11
FIBRILLAR (fibrous, one-dimensional)					
Stems					
Very short	$\frac{1}{4}$ to $\frac{3}{8}$ in.		50-60	0	11
Short	$\frac{1}{4}$ to 1 in.		10-20	0	3
Long	2 to 3 in.		1-10	0	10
Fibrous					
Bunches	$\frac{1}{2}$ to 2 in.		5-20	0	10
Fine bunches	-100 mesh		20-30	0	8
Powder, coarse	-200 to +10 mesh		40-50	0	4
Acicular					
Fine	+100 to +40 mesh		65-75	0	9
Fine, nonuniform	-100 to +10 mesh		55-70	0	11
Medium	+20 mesh to $\frac{7}{8}$ in.		30-40	0	8
Rare	-200 mesh		35-50	40-60	4

[a]From Refs. 2 and 5.
[b]See Table 3-5.
[c]See the following descriptions:

1. No
2. Yes
3. Very greatly
4. If packed
5. May if packed
6. If excessively packed
7. If very excessively packed
8. Hang up
9. May hang up
10. Hang up very greatly
11. Hang up if packed
12. May hang up if packed
13. Very greatly if packed

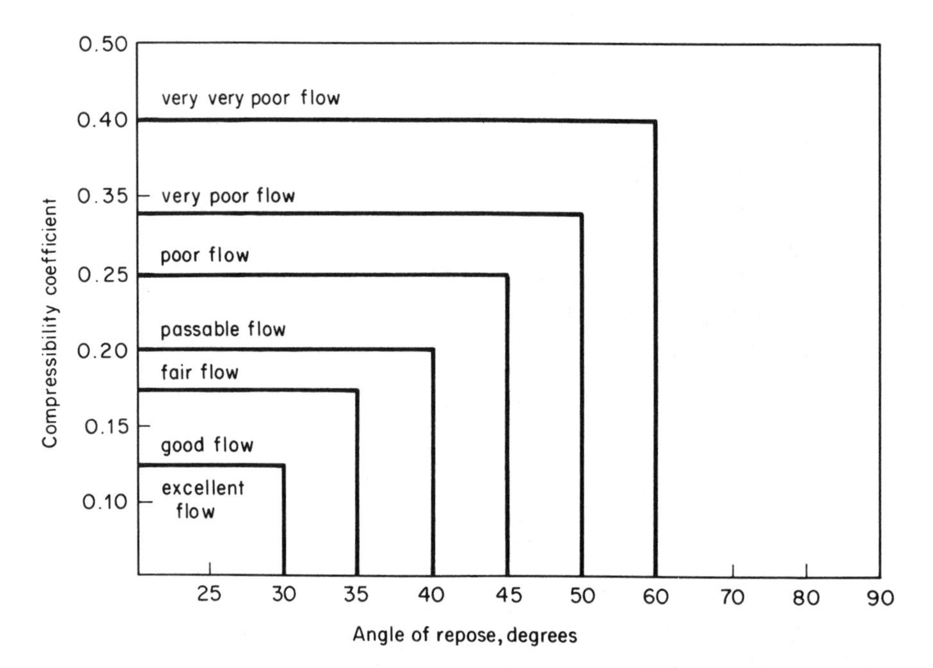

Figure 3-9 Approximate points at which flowability changes. (From Ref. 2.)

age time at rest may be due to any one or a combination of the following:

(a) During charging, powders entrain air which escapes with time and results in a more intimate contact between particles of the mass and increases in its density, the areas of contact between particles, and the cohesive forces.

(b) Physical instability due primarily to changes in surface moisture and temperature.

(c) Chemical instability due to chemical changes occurring at the surface of the particles.

Moisture content. Water in a bulk solid may occur in free, solvent, or chemically bound form. Free and solvent water may be in the form of surface or pore water.

Insoluble substances, such as metallic ores, are affected only by surface water which through capillary action (surface tension) increases the cohesive forces betwen the particles. Capillary action is absent in a dry solid as well as in a water-saturated solid. Cohesive forces due to capillary action reach a maximum for a range of moisture content between 70 and 90% of saturation and are not significantly affected by small changes of temperature. A saturated solid allowed to drain has a tendency to reach a moisture content corresponding to maximum cohesive forces. In porous solids, such as activated carbon, free water first fills the pores, and the surfaces may appear dry even

TABLE 3-5. FLOWABILITY AND FLOODABILITY SCORES FOR TABLE 3-4

Flowability	Score	Floodability	Score
Excellent	90–100	Very floodable	80–100
Good	80–89	Floodable	60–79
Fair	70–79	Likely to flood	40–59
Passable	60–69	Might flood	25–39
Poor	40–59	Will not flood	0–24
Very poor	20–39		
Very, very poor	0–19		

though the material contains a substantial percentage of water in the pores. Cohesive forces due to capillary action may be absent under these conditions.

Water-soluble substances such as crystalline salts (sodium chloride, ammonium sulfate, and fertilizers containing mixture of salts), organic materials (starch and sugar), and more complex materials (cocoa, coffee solids, bakery mixes, and powdered milk) adsorb water on the surface in the presence of water vapor and produce a saturated solution. The vapor pressure over this solution, at a given temperature, corresponds to a definite value of relative humidity. If the relative humidity is above the equilibrium pressure of the solution, vapor condenses with a progressive dissolution of the solid. On the other hand, if the relative humidity is below the equilibrium pressure, water evaporates from the surface of the crystals, solute precipitates, and links form between the particles, leading to caking.

Some freshly produced materials which have not been thoroughly dried prior to storage contain pore-water vapor which diffuses to the surface. This dissolves the solid at the surface, and if the relative humidity is low, the water evaporates and cakes the solid.

Temperature. The effects of temperature changes depend on the types of stored solid, which may be categorized as follows:

(a) Soluble solids containing moisture. Under constant temperature a solution of solids may remain in equilibrium with no change in the properties of the solid. If the temperature changes, the equilibrium pressure of the solution changes, resulting in dissolution or precipitation which usually lead to caking. Cyclic changes in temperature alternately produce dissolution and precipitation; this leads to severe caking. Similar effects may occur even with an insoluble solid if the surface water contains a solute.

(b) Solids subject to a phase change or a significant change in hardness with a change in temperature (plastics, for example). These solids are placed in storage at a temperature of 70 to 400°F. If the temperature remains constant during storage at rest, the solids usually retain their initial properties, but if the temperature drops, they cake.

(c) Frozen or partly frozen solids. A frozen solid stored at a temperature below 32°F retains its initial properties. However, if only a part of the solid is frozen, heat is exchanged, and the free water freezes to bind the mass.

In all the preceding categories temperature changes disturb the equilibrium, which causes surface bonds to develop with time. These effects are minimized by keeping the temperature of the stored solid constant and by breaking the bonds as they develop. Insulating the bin helps keep the temperature of the solid constant. A continuous or periodic relative motion of the particles within the stored solid will prevent bonds from developing, which can be accomplished by continuous or periodic circulation of the solid in a mass-flow hopper, a hopper equipped with a vibrating discharge mechanism, or a hopper equipped with multiple openings feeding a common collection point. When normal operation of the bin does not assure continuous or periodic and frequent flow, a conveying system can be used to circulate the solid.

Gradation. An indication of the particle-size content of a solid is determined by passing a sample of the solid through a set of screens of certain standard sizes and recording the percent of each size range passing a given screen size and retained on the next smaller screen. The *uniformity coefficient* is defined as the ratio of the width of the sieve opening through which 60% of the sample passes to the width of the sieve opening through which only 10% passes.[1] The more uniform the particles are in size and shape, the more flowable the material will be.

Segregation. Solids with a wide range of particle sizes, and blends containing particles of a wide range of density, size, or shape, tend to segregate as they are charged into a bin. The larger, heavier, and more nearly spherical particles roll to the periphery of the bin, while the finer, lighter, and flakier particles congregate along the trajectory of the falling stream. The greater the height of free fall, the greater the segregation. Additional segregation occurs within the falling stream if the stream is not vertical.

Degradation. A degradable solid is one whose lumps or particles may be broken and reduced in size as a result of impact, agitation, or attrition. The change in size may affect the behavior of the solid in the bin. Degradation usually occurs during charging, although it may occur during discharge.

TABLE 3-6. MOH'S HARDNESS SCALE

Talc:	1	Feldspar orthoclase:	6
Gypsum:	2	Quartz:	7
Calcite:	3	Topaz:	8
Fluorite:	4	Sapphire:	9
Apatite:	5	Diamond:	10

Corrosiveness. Corrosive solids are those which chemically attack the confining surfaces with which they come in contact. Corrosion is frequently promoted by the acid or alkaline quality of the solid as denoted by its pH value. Solids with pH values from 1 to 7 are acidic. The lower numbers represent the more acidic solids, and 7 is neutral. Solids with pH values from 7 to 14 are alkaline, the higher numbers representing more alkalinity. Two materials of a given pH value may not have the same corrosive action. Also, some materials of neutral pH may be corrosive to metal surfaces.

Abrasion. Movement of a solid during charging or discharging abrades the walls and bottom of the bin. The abrasive character of the solid depends on the hardness, size, and shape of its particles and on its bulk density.

No method of indicating material hardness is applicable to all materials. One of the earliest measures of hardness, Moh's scale (Table 3-6), is widely used for ores and minerals. Any material of a given Moh's hardness number can scratch any material of a lower hardness. Iron and steel range in hardness from 4 to over 8.5, depending on the alloy content and heat treatment.

Funnel-flow bins are advantageous for the storage of hard, abrasive, lumpy solids because there is little wear of the hopper walls.

3-9. Tables of Properties

The properties of bulk solids that influence their behavior in bins depend on so many variables, e.g., chemical composition, moisture content, consolidation pressure, storage time, temperature, particle size, and gradation, that tabulated values should be used with caution.

Figure 3-8 shows the variation of effective angle of friction δ, the bulk density γ, and the unconfined yield strength f_c with the principal consolidating pressure p_1 for a taconite concentrate at three moisture contents. The tests were conducted under conditions simulating the behavior of the material as it flows in a bin and hopper. Table 3-7 gives ranges in values of these properties determined from corresponding curves for a number of other materials. Since each is from a specific source, the tabulated values should not be used for all materials of similar chemical composition. Material properties from other sources are given in Tables 3-8 and 3-9.

TABLE 3-7. TYPICAL PROPERTIES OF SELECTED BULK MATERIALS[a]

Material	Moisture (%)	p_1 (psf)	f_c (psf)	γ (pcf)	δ (deg)	ϕ (deg)	ϕ' on steel (deg)			
							New	Rusted	Stainless	Ni-Hard
Taconite with bentonite	9.6	200-900	185-560	145-175	62-56	39-42	–	39	30	29
Taconite concentrate	10.1	200-1000	180-500	152-182	63-50	32-40	–	36	27	29-33
Cherty taconite	4	200-1200	95-360	116-125	51-52	45	–	–	25	25
Slaty taconite	4	200-1200	70-200	116-128	55	50	–	–	25	25
Lac Jeannine concentrate	3	200-1100	80-160	175-195	43-38	33	23	29	23	–
Mesabi ore	13.5	200-1300	115-650	110-140	56	43	21	34	28	–
Alpheus coal	3	200-1100	100-250	48-51	50	41	24	–	22	–
Stoker coal	12	200-1000	40-130	44-50	50	46	–	33	15	22
Limestone	4	200-1100	80-420	97-108	52	43	–	33	29	31
Dolomite	6.8	200-1000	90-380	100-110	48-43	35	26	31	22	–
Gypsum	7.6	200-1200	230-560	96-109	64-52	40	–	37	23	26
Shale	8	200-1100	80-360	86-94	52-53	45	–	30	20	20
Fluorspar	4.5	300-1400	140-500	101-114	55-52	43	26	33	23	–
Sand	11	100-1100	60-120	97-110	53-42	38	–	32	22	25
Starch	6	100-700	30-195	40-42	52	46	–	–	22	28
Bentonite	6	200-900	60-120	54-62	41	36	–	–	26	24

[a]Adapted from Ref. 6.

TABLE 3-8. TYPICAL PROPERTIES OF SELECTED BULK MATERIALS

Material	γ (pcf)	α (deg)	ϕ' on steel (deg)	$K = \frac{1 - \sin \phi}{1 + \sin \phi}$ [a]
Ashes				
Dry, compact	45	40	25	0.217
Loose	40	30	20	0.333
Wet, saturated	70	20	15	0.490
Coal, anthracite				
Dry, broken	55	27	22	0.376
Pulverized, aerated	40	20	15	0.490
Pulverized, compact	60	25	20	0.406
Coal, bituminous				
Dry, broken	50	35	25	0.271
Pulverized, aerated	35	20	15	0.490
Pulverized, compact	55	25	20	0.406
Coke, dry, loose	30	30	20	0.333
Gravel and sand	100–110	25–30	15	0.406–0.333
Sand				
Dry, loose	90–100	30–35	20	0.333–0.271
Saturated	120	25–30	15	0.406–0.333
Shale	90	45	20	0.172
Stone, crushed	100–110	32–39	–	0.307–0.228

[a] Assuming $\phi = \alpha$.

TABLE 3-9. TYPICAL PROPERTIES OF GRAINS

Grain	γ (pcf)	α (deg)	ϕ' on steel	$K = \frac{1 - \sin \phi}{1 + \sin \phi}$ [a]
Barley	39	27	21	0.472
Beans	46	32	20	0.490
Corn	44	27	20	0.490
Flaxseed	41	25	19	0.509
Oats	28	28	22	0.376
Peas	50	25	15	0.490
Wheat	49	29	22	0.376

[a] Assuming $\phi = \alpha$.

REFERENCES

1. Ralph L. Carr, Jr., Evaluating Flow Properties of Solids, *Chem. Eng.,* Jan. 18, 1965.
2. Ralph L. Carr, Jr., Particle Behavior, Storage, and Flow, *ASME Publ. 68-MH-6,* American Society of Mechanical Engineers, New York, Oct. 1968.
3. Andrew W. Jenike, Storage and Flow of Solids, *Bull. No. 123,* Utah Engineering Experiment Station, University of Utah, Salt Lake City, Nov. 1964.
4. Ralph L. Carr, Jr., Properties of Solids, *Chem. Eng.,* Oct. 13, 1969.
5. Ralph L. Carr, Jr., Classifying Flow Properties of Solids, *Chem. Eng.,* Feb. 1, 1965.
6. J. R. Johanson, Properties of Various Raw Materials Related to Gravity Flow, *Report 30.016-005 (8),* U.S. Steel Applied Research Laboratory, Monroeville, Pa., March 30, 1965.

4

FUNCTIONAL DESIGN OF BINS

Some of the factors which must be considered in selecting a type and size of vessel for the storage of bulk solids are the following: type and maximum quantity of material to be stored, segregation tolerance, blending needs, storage duration, and discharge rates. Feeding and discharge rates and the maximum quantity of material to be stored determine the size of a bin, and the flow properties of the solid may influence its shape.

4-1. Funnel-Flow Bins

Funnel-flow bins (Sec. 1-2 and Fig. 4-1) are useful for storing free-flowing nonperishable materials and for hard, abrasive, lumpy materials. They are also useful for many other solids but may require accessories such as flow-promoting devices, air locks, etc. They also provide more storage within a fixed space than mass-flow bins.

Funnel flow has the following characteristics, some of which are disadvantageous for some solids but can be offset by using devices such as those already noted:

(a) Only a portion of the stored solid is in motion during draw.

(b) A last-in, first-out flow sequence prevails. The solid surrounding the flow channel at the bottom of the bin remains at rest until the flow channel is completely empty. This may lead to consolidation, caking, deterioration, spontaneous combustion of coal, and oxidation of ores.

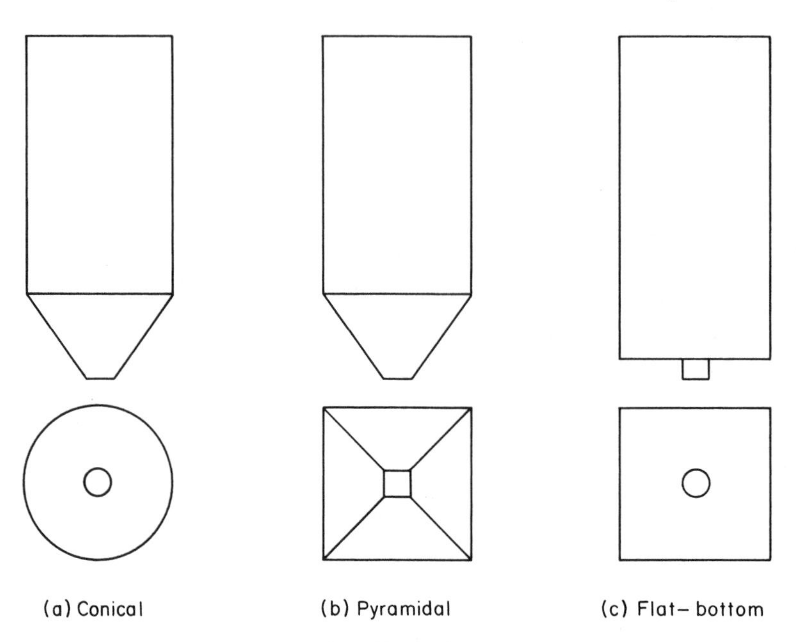

Figure 4–1 Funnel-flow bins.

(c) Solids which remain at rest under pressure for a long period of time may gain strength and obstruct flow. This may lead to piping (ratholing).

(d) When a pipe collapses, impact of the falling mass may aerate a fine solid and cause flooding.

(e) The bulk density of the solid in the bin may vary. Some of the old solid which slides into the channel may flow in consolidated chunks, while the freshly charged unconsolidated solid may flow straight through the channel. This leads to surging flow and may cause flushing and flooding.

(f) Segregation usually occurs because no remixing takes place in the hopper.

(g) Circulation of the solid in the bin produces little useful effect.

(h) Low-level controllers and indicators are unreliable if they become embedded in nonflowing solid.

(i) The bin is useless as a degasifier or a gas seal. Air locks are needed to prevent the flooding of powders and to separate regions of different gas pressure.

4–2. Mass-Flow Bins

Mass-flow bins (Sec. 1–2 and Fig. 4–2) have the following characteristics:

(a) The total volume of stored solid is available for process by gravity.

(b) Channeling, hang-ups, surging, and flooding do not occur, provided

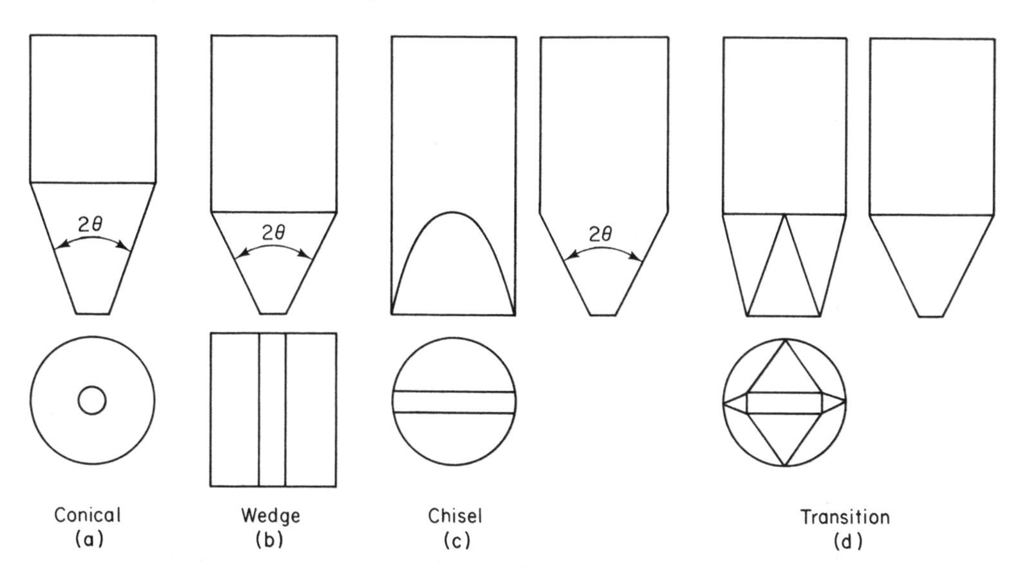

Figure 4-2 Mass-flow bins.

in the case of powder no attempt is made to withdraw it faster than its inherent rate of discharge.

(c) The bulk density of the drawn solid is constant and practically independent of the head of the stored solid. This is advantageous in all cases of controlled flow rate and essential when the rate is controlled volumetrically.

(d) A first-in, first-out flow pattern, which is desirable for solids that deteriorate during storage and essential in chemical reactors, is readily obtained. This flow pattern is also desirable when segregation must be taken into account.

(e) Since no dead regions occur within the vessel, there is a minimum of degradation, spoilage, spontaneous combustion of coal, oxidation of ore, and consolidation at rest.

(f) A mixture may be blended by circulating it through and around a suitable mass-flow vessel.

(g) Pressures throughout the solid are relatively low, which results in low consolidation and attrition of the solid. Also, the pressures are relatively uniform over a horizontal cross section of the hopper, which gives uniform consolidation and permeability.

(h) Flow is relatively uniform, so that steady-state flow can be closely approached.

(i) Low-level controllers and indicators work reliably.

(j) Powders degasify and settle out if sufficient storage time is attained.

(k) Air locks (rotary vane feeders) can often be eliminated as long as a minimum level of solid is retained to provide a seal.

Disadvantages of mass-flow bins are the following:

(a) The steep hopper wall required to attain mass flow necessitates a relatively tall bin for the amount of material stored.

(b) Flow pressures on the hopper wall at the juncture with the bin are relatively high and may require a thicker wall than a funnel-flow bin.

(c) The flow of abrasive materials results in much more wear on the walls than in a funnel-flow bin.

4-3. Expanded-Flow Bins

The disadvantages of the funnel-flow bin can be obviated by attaching to it a mass-flow hopper (Fig. 4-3). The hopper forces the flow channel to expand to a size large enough to eliminate the possibility of ratholing, to reduce segregation to an acceptable level, and to ensure deaeration and smooth flow. This type of bin is useful for the storage of large quantities of nondegrading solids such as ores. A low-level indicator can be placed on the mass-flow hopper. Several mass-flow hopper-feeder units can be placed under one large funnel-flow bin (Fig. 4-3c).

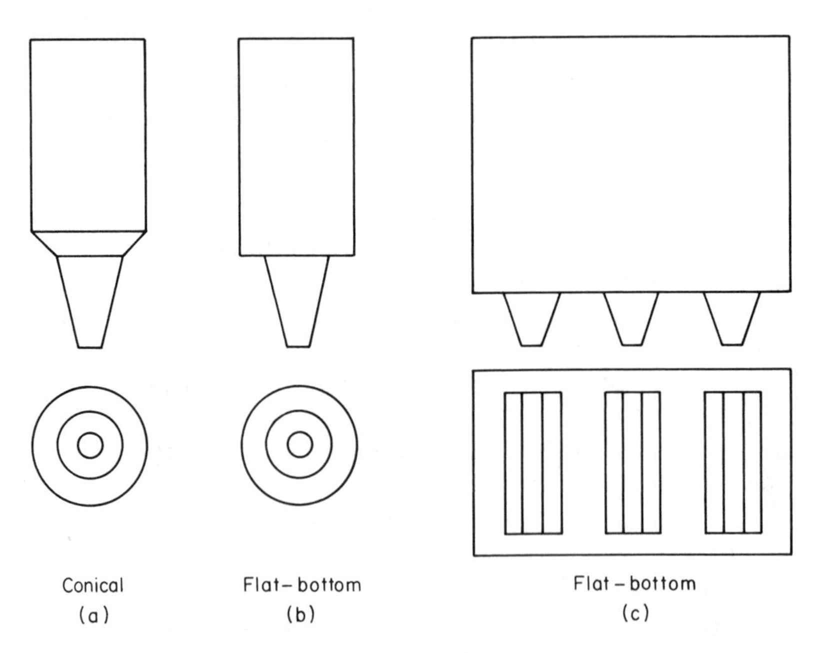

Figure 4-3 Expanded-flow bins.

4-4. Type of Material

Perishable solids. Perishable solids should not be stored for a long enough period of time to cause decay. Moisture in grain migrates from warm grain to cold grain, and considerable migration occurs when the temperature differential is large. This excess moisture causes heating and spoilage due to molds, germination, or oxidation of starches in grain due to respiration. Migration of moisture may be prevented by the movement of air through the grain to produce a more uniform temperature. Unless a bin storing perishable solids is completely emptied at suitable intervals, the bin should be of the mass-flow type or have other provisions for complete movement of the material on a first-in, first-out basis.

Powders. Powders must degasify after being charged into a bin and remain degasified to obtain a controlled draw. This requires a minimum storage time prior to discharge. A mass-flow bin in which a minimum level of powder is always retained produces uniform feed without flooding. The last-in, first-out flow pattern of a funnel-flow bin may lead to inadequate degasification and flushing of fluidized solid.

Caking. Many solids which are relatively free-flowing under conditions of continuous flow cake if left at rest under pressure. These solids develop their full strength in the nonflowing regions surrounding the flow channel of a funnel-flow bin. Flow may be restored by vibrating the solid or by installing inflatable panels. Caking in mass-flow bins from which regular withdrawals are made may be less severe than in funnel-flow bins since dead areas are eliminated. If the solid cakes and obstructs flow, external vibrators more readily break up the domes and restore flow in mass-flow bins than in funnel-flow bins.

Gradation. Solids which contain fine particles as well as heavy coarse lumps are capable of developing the strength to form obstructions to flow and of packing hard as the lumps fall on the deposited fines. Mass-flow bins satisfactorily handle these materials provided a minimum level of solid is retained to prevent packing in the region of the outlet. A baffle or impact breaker placed in the stream of charged solid spreads the stream and eliminates high local packing.

Nonperishable, nonflooding, stable materials, such as sand and gravel, can be stored equally well in funnel-flow and mass-flow bins. A nonfibrous stable solid containing no fines is free-flowing and can be stored in funnel-flow bins.

Mine-run ores. Mine-run ores which may contain rock as long as 4–8 ft should be stored in piles or in funnel-flow bins. Large rock causes severe wedging in mass-flow bins as well as rapid wearing of the walls because of the large impact pressures. The grade of a run-of-mine ore may be evened out by charging a series of bins in sequence and drawing them simultaneously.

4-5. Segregation

Mass-flow bins which are operated with a level of solids above the hopper enforce a first-in, first-out flow pattern so that material which segregates during charging automatically remixes in the hopper while discharging. Segregation is reduced in a multiple-outlet bin by discharging through all outlets simultaneously and blending the draw.

Funnel-flow bins do not remix. In a funnel-flow bin the feed from an outlet located under the trajectory is mostly fine when the level of solid in the bin is rising (Fig. 4–4a), mostly coarse when the level is falling (Fig. 4–4b), and the same as the charge when the level is constant (Fig. 4–4c). The last case produces a true last-in, first-out sequence.

4-6. Blending

A solid composed of very small particles or of particles of uniform size, shape, and density can be blended in a mass-flow bin by recirculating the solid in a closed circuit about the bin. The height-to-diameter ratio of the bin should not exceed one-half, and the hopper walls should be very steep to assure a substantial velocity gradient in all horizontal cross sections of the

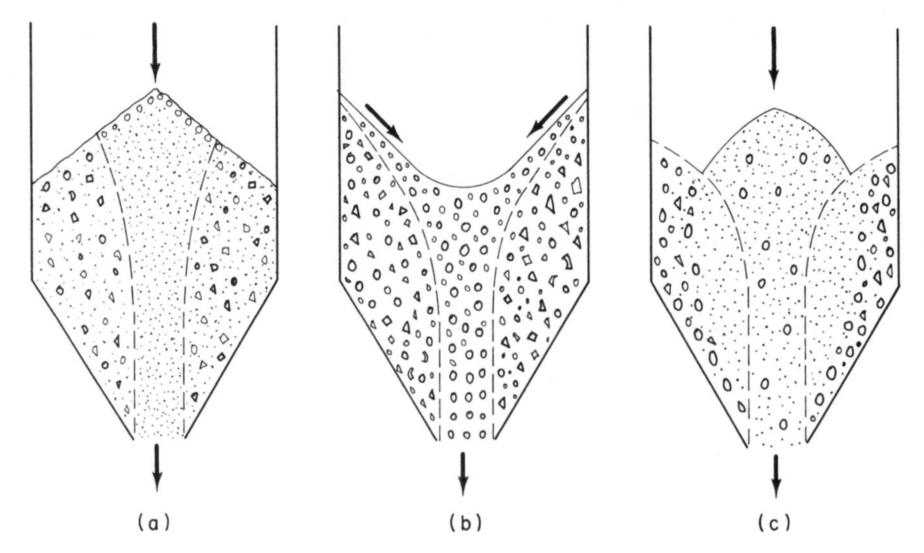

Figure 4–4 Funnel-flow patterns.

flow channel to prevent a first-in, first-out flow. A larger height-to-diameter ratio would induce first-in, first-out flow similar to that of a nonsegregating bin. If the hopper walls are not sufficiently steep, dead regions will develop at the transition.

4-7. Hoppers

The success of bin operation depends largely on the design of the hopper. A hopper changes the direction of flow of the stored material and forces it to converge and flow through the smaller opening of the hopper. The ease with which the material flows and converges toward the opening depends almost entirely on the shape of the hopper and the smoothness of its walls.

The shape of the upper part of the hopper is usually that of the cross section of the bin. The symmetrical, conical hopper attached to a bin of circular cross section is commonly used (Fig. 4–1a). It performs satisfactorily with free-flowing materials.

A pyramidal hopper which may be attached to a square or rectangular bin is shown in Fig. 4–1b. Fillets should be placed in the corners and valleys of the hopper to prevent a hang-up of the solid resulting in funnel flow. Wedge and chisel hoppers (Figs. 4–2b and c) are preferable; the slotted outlets prevent funneling and assure mass flow, but a feeder over the full length of the slot is necessary to remove the solid. Round, square, and rectangular bins in which cohesive and nonflowing materials are stored may require an eccentric hopper (Fig. 4–5). Stable arches are less likely to form in these hoppers because the solid tends to slide down the vertical wall.

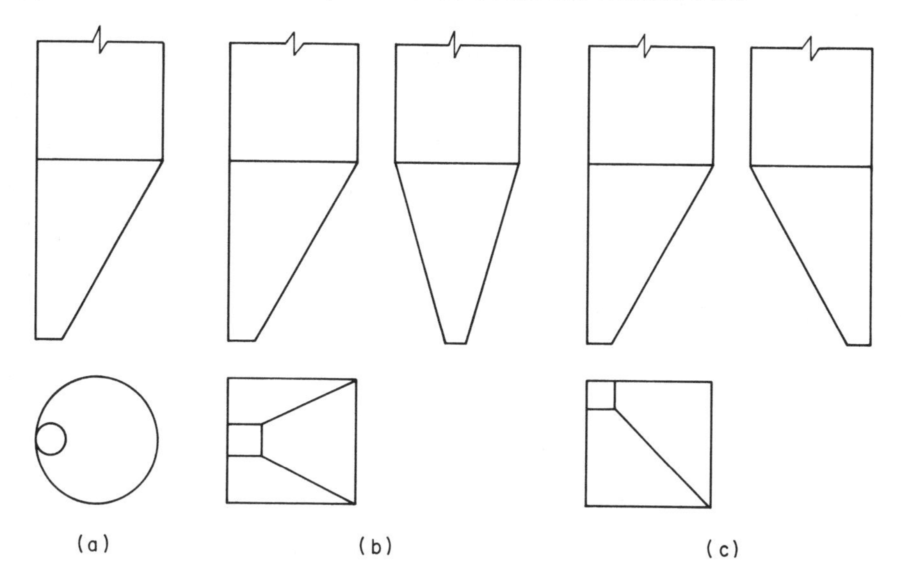

Figure 4–5 Eccentric hoppers.

Round or rectangular gravity-flow bins with flat bottoms and multiple circular openings are commonly used to store granular solids. Such bins are low in cost and occupy less vertical space than bins with a single hopper outlet. To maximize the live storage, the openings should be so spaced as to permit the flow patterns above the openings to intersect. Important dimensions of the flow channel are discussed in Sec. 4–14.

4-8. Outlets

The outlet of a hopper must be large enough to assure unobstructed flow at the required rate. Factors other than outlet dimensions that affect the rate of flow are cohesive strength of the solid, average particle size, and gas-pressure gradient. Unless the outlet for granular materials is larger than several particle sizes, flow may be obstructed by interlocking of large particles. Furthermore, certain minimum dimensions are necessary to prevent cohesive doming (Fig. 4–6a) and piping (Fig. 4–6b).

It is important to distinguish between the actual area of an outlet of a hopper, bin, or storage pile and the effective area, which is the area through which the solid flows. In many cases, particularly in the case of rectangular outlets, the effective area may be only a small part of the actual area. For

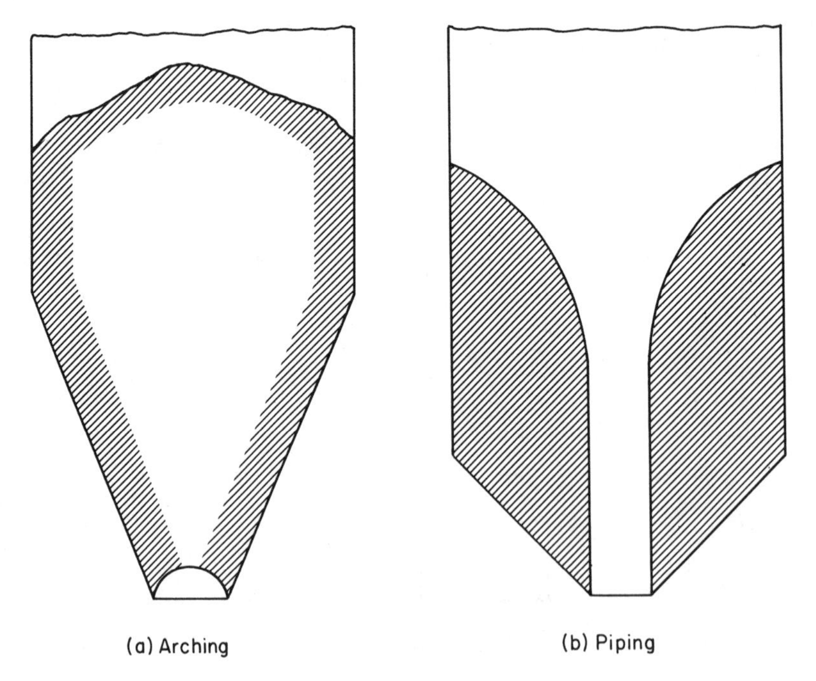

Figure 4–6

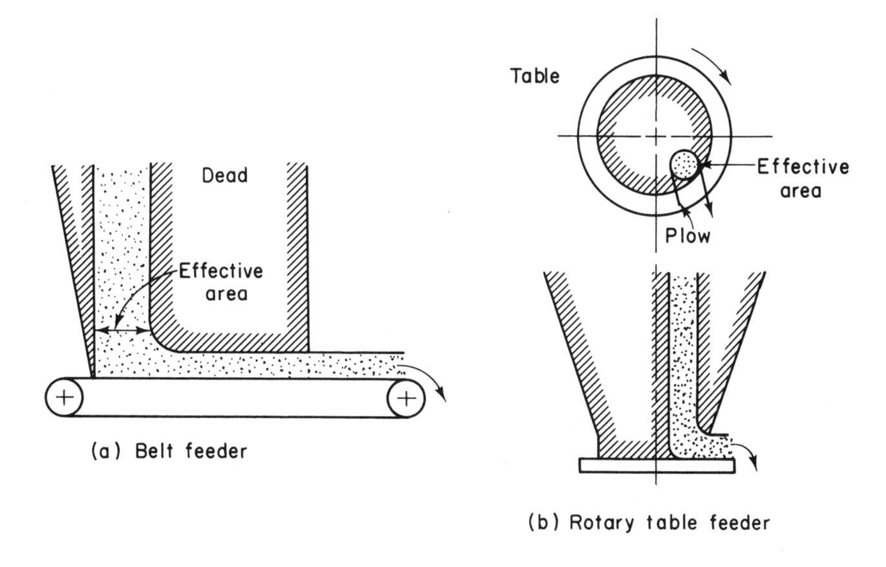

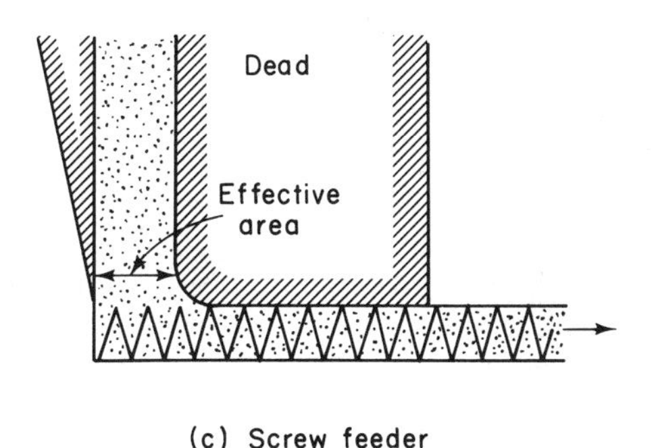

(c) Screw feeder

Figure 4–7 Effective area of outlets.

example, if the length of the outlet to a belt feeder of constant width exceeds, say, twice the width of the outlet, the solid is almost certain to feed at one end of the outlet (Fig. 4–7a). It is impossible to say which end since this depends on the friction that develops between the solid and the surface of the feeder. There is a similar situation in a rotary table feeder when the skirts seal the solid around the table (Fig. 4–7b). In this case the solid flows through a channel over the plow but remains stationary outside the channel. Finally, if the length of the outlet exceeds about twice the diameter of a screw feeder of constant diameter and pitch, the solid fills the screw at the end opposite the discharge, and the screw runs full over the remainder of its length and cannot accept more material (Fig. 4–7c).

Stationary solid that slides over a feeder packs hard and develops large frictional forces on the feeder. This increases power consumption and accelerates wear. For a feeder to be effective along the full length of an outlet the capacity must increase in the direction of flow. This can be done with a tapered screw (Fig. 1–5c) or by an increasing-pitch screw (Fig. 1–5e). A tapered outlet can be used with a belt feeder. Alternatively, the skirts may be raised on a belt feeder or a rotary table feeder to permit side flow of the solid.[1]

4–9. Size of Outlet

In general, outlets should be sized to prevent arching of the solid or piping through it.

Arching. Figure 4–8a shows an arch or dome in a flow channel. An arch forms if the channel is wedge-shaped and a dome if it is conical. If the blockage is assumed to consist of a stack of self-supporting domes or arches so that the upper and lower boundaries are free surfaces, the minor principal stress (normal to the free surface) is zero at any section, and the corresponding major principal stress is tangent to the arch. The maximum shears for this stress system are on planes at 45° with the principal directions and are equal to one-half the major principal stress. The maximum span of arch which can be self-supporting under this stress system is attained when the shears on the vertical sections at the abutments reach their maximum values. This condition is realized when the major principal stress at the abutments equals the unconfined yield strength f_c of the solid and acts at 45° with the horizontal (Fig. 4–8b). The corresponding shear on the vertical section is $f_c/2$. Equilibrium of vertical forces for an arch of unit vertical thickness forming over a rectangular opening of width b and length l gives

$$\frac{2lf_c}{2} = \gamma bl$$

from which

$$b = \frac{f_c}{\gamma} \tag{4-1}$$

Similarly, for a dome forming over a circular opening of diameter d,

$$\frac{\pi d f_c}{2} = \frac{\gamma \pi d^2}{4}$$

$$d = \frac{2f_c}{\gamma} \tag{4-2}$$

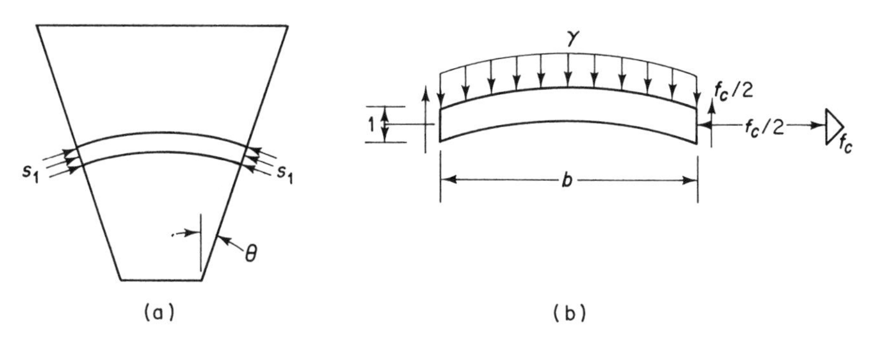

Figure 4-8 Arching.

According to Jenike,[1] the required outlet diameter is also a function of the hopper half angle θ (Fig. 4-8a), and Eqs. (4-1) and (4-2) are correct only for $\theta = 0$. His solution gives values of b for wedge hoppers varying essentially linearly from f_c/γ at $\theta = 0$ to $1.3f_c/\gamma$ at $\theta = 60°$. Values of d for conical hoppers range from $2f_c/\gamma$ at $\theta = 0$ to $2.6f_c/\gamma$ at $\theta = 40°$.

In a later report Jenike and Johanson[2] suggest the following outlet dimensions: For circular outlets,

$$d \geqslant \frac{2.2f_c}{\gamma} \tag{4-3}$$

For rectangular outlets,

$$b \geqslant \frac{1.3f_c}{\gamma} \tag{4-4}$$

The value of f_c to be used in Eqs. (4-1)-(4-4) is determined as follows. Since s_1 and p_1 are zero at the vertex of a hopper and are linear functions of the hopper diameter (or width), their ratio is constant for a given hopper (Fig. 4-9). The ratio p_1/s_1 is called the flow factor and is denoted by *ff.* With *ff* known, the critical value of f_c is determined by drawing the straight line $ff = p_1/s_1$ on a flow-function plot *FF* (Fig. 4-10a). To the right of the intersection of *ff* with *FF*, f_c is less than s_1 and an arch cannot be supported, while to the left the opposite is true. The two regions are called the *flow* and *no-flow* regions, respectively, and the intersection itself gives the critical value of f_c.

Flow function FF. Construction of the flow function is discussed in Sec. 3-6. Flow functions are usually convex upward, as in Fig. 4-10a, or straight lines through the origin. In the latter case, if *ff* lies above *FF*, f_c is always less than s_1, and an arch cannot be supported. Therefore, the size of the opening

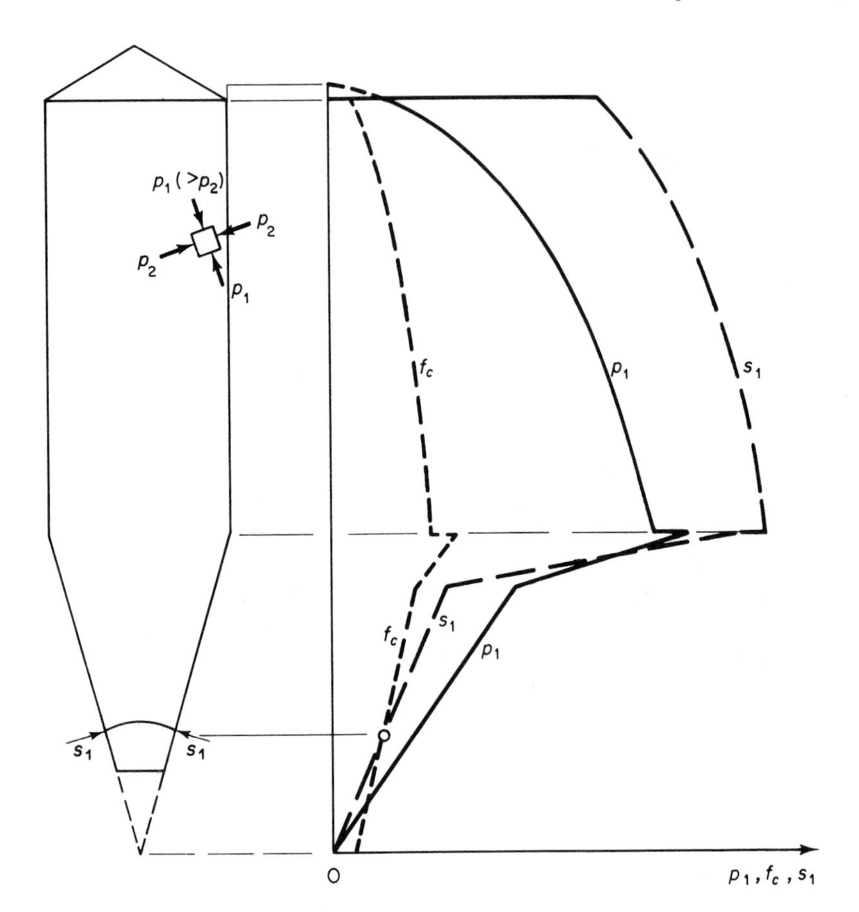

Figure 4-9 Arching and consolidating flow pressures.

is dictated by particle size* and the required rate of discharge (Sec. 4–15). If *ff* lies below *FF*, the material will not flow. In this case, if *FF* is a time flow function, a reduction in the planned storage time at rest may change it enough to lie below *ff*, while if it is an instantaneous function, it may be possible to reduce *ff*, such as by changing the slope of a mass-flow hopper or by reducing the moisture and/or temperature of the solid at which the operation was planned.

Flow functions for some solids are convex downward (Fig. 4–10b), and the flow, no-flow regions are opposite to those for the convex-upward flow function. In this case material may flow from the lower region of the hopper while an arch forms farther up. Therefore, the dimension of the outlet by

*An empirical rule for a material with no fines is that the opening should be not less than five times the largest particle size.

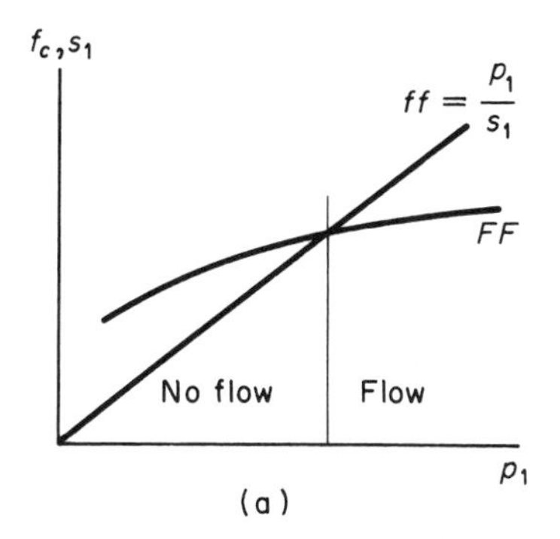

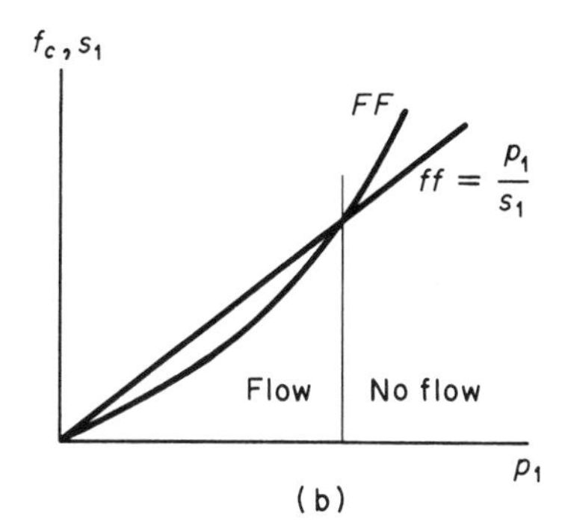

Figure 4-10 Flow, no-flow criteria.

Eqs. (4–1)–(4–4) is the maximum diameter, or width, of the bin if vibrators or other flow inducers are not to be used.

The flow function should be determined for the most severe conditions of storage. They are usually the highest moisture content, the highest temperature, and the longest period of storage.[3] In general, it is difficult to predict changes in properties over an extended period of time. Some materials show no increase in strength above the instantaneous strength even after a week or more of consolidation. Other materials, such as coal, may continue to increase in strength for years because of slow chemical reactions that take place. Some fine powders may have practically no strength at room temperature but become very strong after a few hours of consolidation at higher temperatures. Flow inducers may be required for a material which gains considerable strength if it is left in a bin for some time, because the opening size to prevent arching would be too large.

Flow factor ff. The flow factor for arching in mass-flow hoppers is a function of the effective angle of friction δ, the angle of friction ϕ' of the solid on the hopper wall, the slope of the wall, and the type of flow. Two types of flow have been investigated: plane flow (flow between two plane converging walls, theoretically only if they are infinitely long) and conical flow, which occurs in conical and square-pyramid hoppers. According to Jenike,[1] plane flow is approached closely in a rectangular outlet whose length is three or more times its width.

Flow-factor charts for arching in mass flow in conical channels and plane-flow channels have been prepared by Jenike.[1] Colijn[4] gives charts, in different form, for three values of the half-apex angle θ (Figs. 4–11, 4–12, and 4–13). The Jenike charts, one of which is shown in Fig. 4–14, cover a wider range of θ. However, the $10°$–$30°$ range of Figs. 4–11, 4–12, and 4–13 is sufficient to cover the usual range for mass-flow conical hoppers and can be extrapolated a few degrees on either side of the limiting values. These fig-

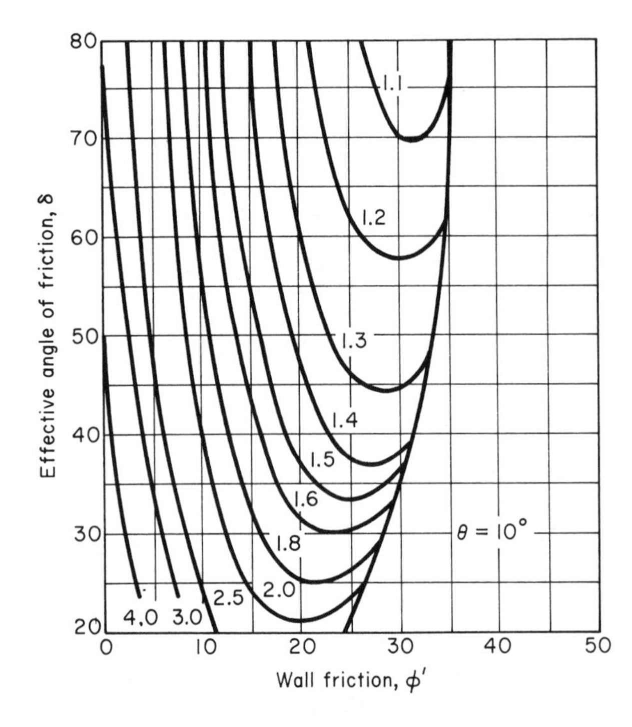

Figure 4-11 Flow factors for mass-flow conical hoppers, $\theta = 10°$. (From Ref. 4.)

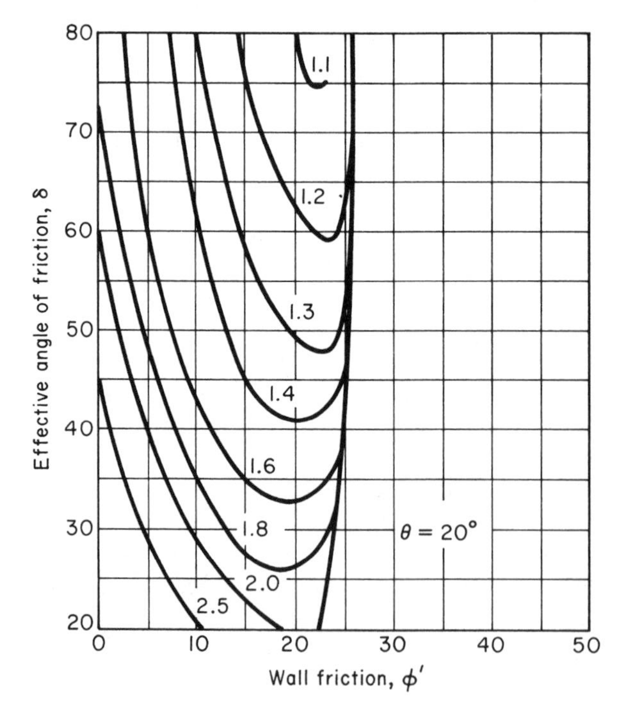

Figure 4-12 Flow factors for mass-flow conical hoppers, $\theta = 20°$. (From Ref. 4.)

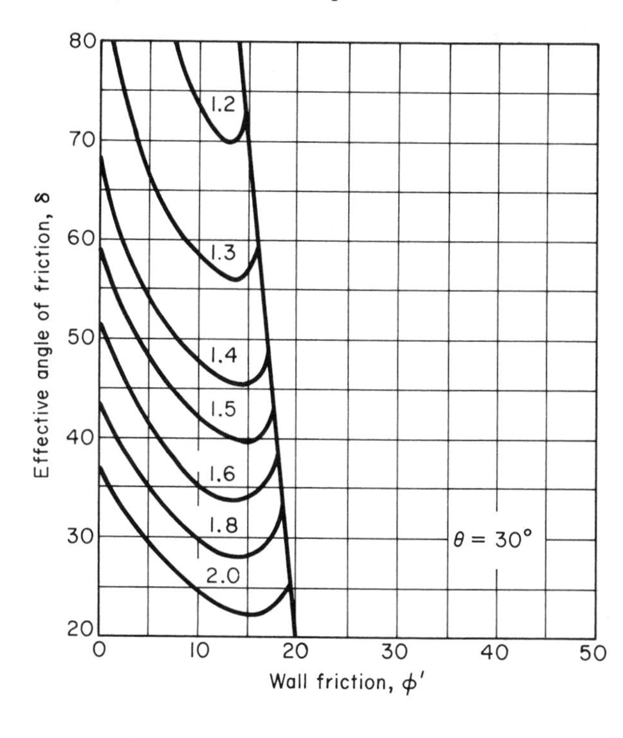

Figure 4-13 Flow factors for mass-flow conical hoppers, $\theta = 30°$. (From Ref. 4.)

ures can also be used with good accuracy for wedge-shaped hoppers, since flow factors for plane flow are only slightly larger than for conical flow.

Walker's theory (Sec. 5–13) yields the following formulas for the conical-flow factor[5]:

$$ff = \frac{1 + \sin\delta}{\Delta} \qquad \theta + \phi' \geqslant 45° \tag{4-5a}$$

$$ff = \frac{1 + \sin\delta}{\Delta} \sin 2(\theta + \phi') \qquad \theta + \phi' < 45° \tag{4-5b}$$

where

$$\Delta = 2 \sin\delta \sin 2(\theta + \epsilon) - \tan\theta [1 - \sin\delta \cos 2(\theta + \epsilon)]$$

$$\epsilon = \frac{1}{2}\left(\phi' + \sin^{-1}\frac{\sin\phi'}{\sin\delta}\right)$$

These flow factors are to be used with Eqs. (4–1) and (4–2). Walker does not give flow factors for plane flow but says that they are only slightly larger than the conical-flow factors.

Flow factors by Eq. (4–5) with $\delta = 50°$ are compared in Fig. 4–14 with the corresponding Jenike chart. According to Walker's formula, the smallest

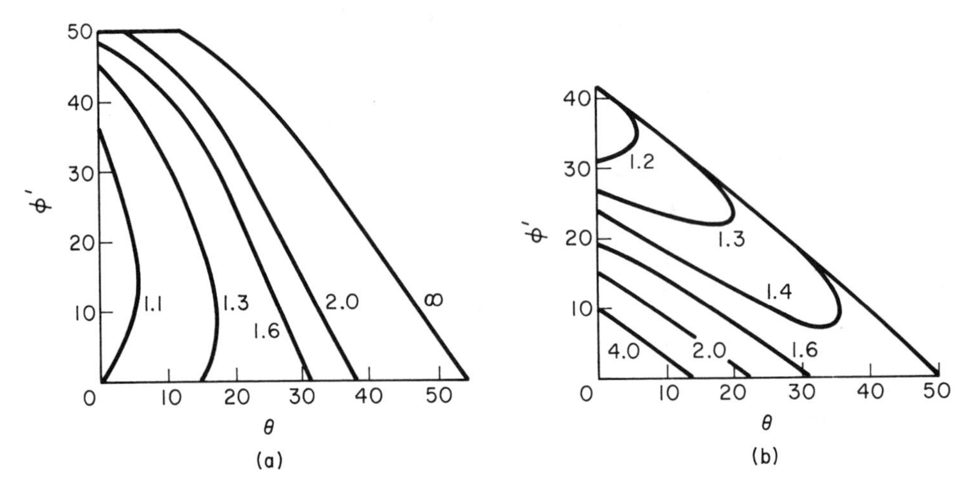

Figure 4-14 Flow factors for $\delta = 50°$: (a) Walker, (b) Jenike. (From Ref. 5.)

flow factors are for the steepest, smoothest hoppers; that is, the steeper and smoother the hopper, the better its handling characteristics. On the other hand, Jenike's flow factors show an optimum hopper slope for a given angle of friction, and a slope steeper than the optimum, or one less steep, decreases the flowability. The difference is mainly in the factor sin $2(\theta + \phi')$ in Eq. (4-5b). This factor arises from Walker's assumption that failure of an arch is by slip along the wall rather than by shear in the abutting solid if the walls are smooth and steep, while Jenike assumes that failure is through the solid no matter how steep and smooth the wall. The conclusion from Walker's flow factors that steep, smooth hopper walls give better handling characteristics is in agreement with Walker's tests and tests by others.[5]

Tests to assess the validity of the Jenike design method are reported in Ref. 6.

Flat-bottom bins. The flow factor for flat-bottom bins is a function of the effective angle of friction δ and the internal angle of friction ϕ (Fig. 4-15).

Funnel-flow bins need to be checked for arching only in determining the width of a rectangular outlet, using the flow factor for flat-bottom bins. The length of a rectangular outlet and the diameter of a circular outlet are determined by piping rather than by arching (Sec. 4-10).

Arching under initial pressures. The flow factors that have been discussed are based on pressures during steady flow. Johanson has investigated the likelihood of arching under the larger static pressures that exist after filling.[7] He

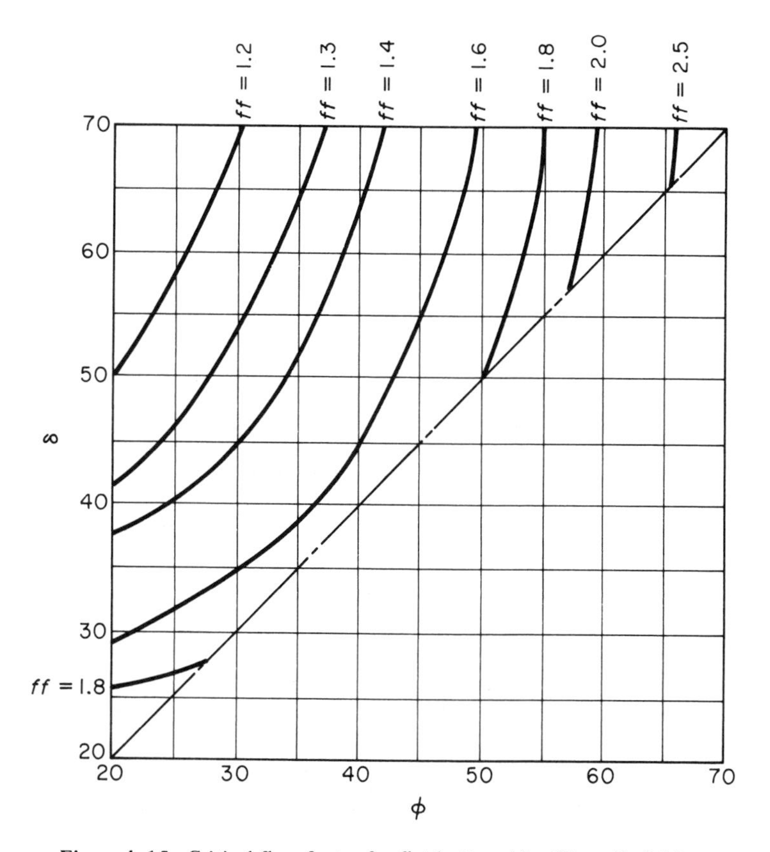

Figure 4-15 Critical flow factor for flat-bottom bin. (From Ref. 3.)

shows that for a mass-flow bin the largest flow factor *ff* likely to develop is less than unity. Since flow factors during flow exceed unity, stable arches in mass-flow bins are unlikely to develop under initial (filling) pressures.

4-10. Outlets to Prevent Piping

Piping can develop in bins which are not designed for mass flow. An analysis similar to that for arching gives the diameter d to make a pipe unstable so that a stoppage will not develop. The result is

$$d = \frac{f_c}{\gamma} G(\phi) \tag{4-6}$$

where the piping factor $G(\phi)$ is a function of the material's internal angle of friction (Fig. 4-16). The flow factor for determining f_c is a function of the material's internal angle of friction and effective angle of friction (Fig. 4-17).

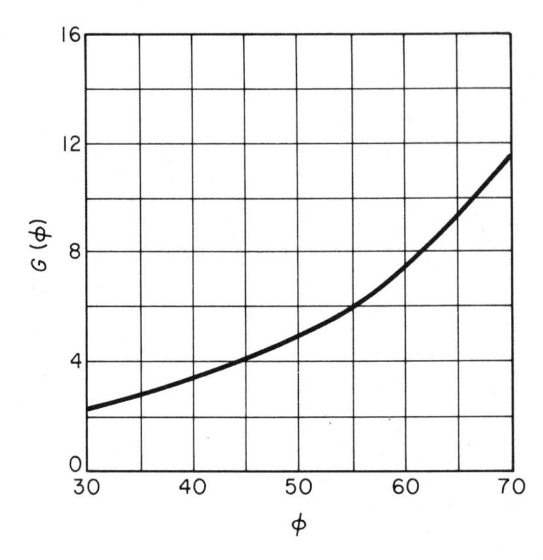

Figure 4-16 Piping factor versus angle of friction. (From Ref. 1.)

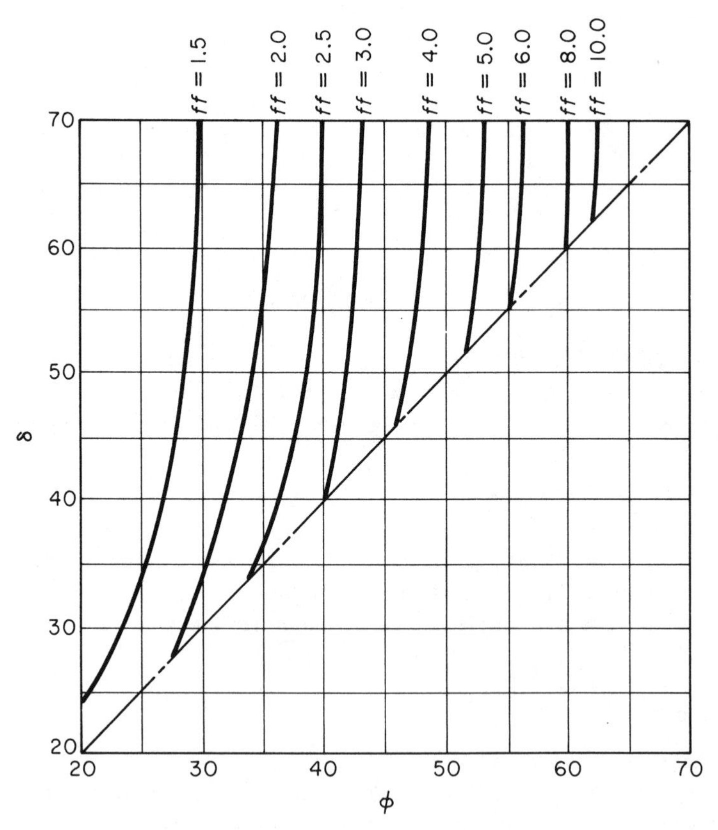

Figure 4-17 Critical flow factor for piping. (From Ref. 3.)

Piping under initial pressures. Johanson has investigated piping under static filling pressures.[7] The diameter d is given by Eq. (4–6) with f_c based on the following flow factors:

For a flat-bottom cylindrical bin (circular or prismatic),

$$ff = \frac{p_v}{d} G(\phi) \tag{4-7}$$

where p_v is the vertical filling pressure and $G(\phi)$ is given by Fig. 4–16. Johanson suggests that p_v be evaluated by the Janssen formula in deep bins but taken equal to γH, where H = height of bin, for shallow bins (say $0 < H/D < 1$).

For a hopper without surcharge,

$$ff = FG(\phi) \tag{4-8}$$

where F is given by Figs. 4–18 and 4–19. The parameter k on which F depends is defined in Ref. 7. The value 0.8 is typical and is suggested for general use.

For a bin with a hopper, ff is the smaller of the values by Eq. (4–8) and

$$ff = \left(\frac{p_{vt}}{\gamma d} + \frac{D - d}{2d \tan \theta}\right) G(\phi) \tag{4-9}$$

where p_{vt} = vertical pressure in bin at transition
D = diameter of cylindrical bin or width B of rectangular bin

According to Colijn,[4] an increase of about 30% in the piping flow factor $G(\phi)$ will generally account for the higher initial pressures.

These flow factors are for pressures from filling with no outflow. If material is withdrawn during filling, the flow factors for steady flow apply. Flow factors for filling with no outflow may be several times those for steady flow.

4-11. Impact Pressure Due to Filling

The consolidating pressure caused by free fall of a solid may exceed both the initial consolidating pressure and the consolidating pressure during flow. The impact pressure can be estimated by[3]

$$p_1 = \frac{Q}{gA} \sqrt{2gh} = 0.25 \frac{Q}{A} \sqrt{h} \tag{4-10a}$$

where Q = weight-flow rate into bin, lb/sec
A = area of impact, ft^2
h = height of fall, ft
g = acceleration of gravity, ft/sec^2

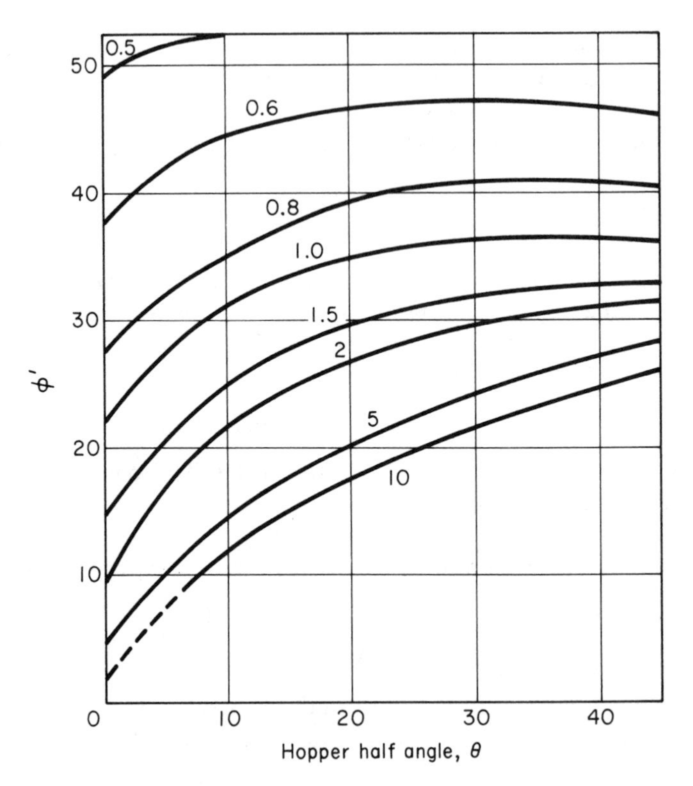

Figure 4-18 Consolidating pressure factor F for conical hoppers, $k = 0.8$. (From Ref. 7.)

The flow rate for a belt conveyor is given by $Q = av\gamma$, where a = cross-sectional area of solid on belt (ft^2), v = belt velocity (ft/sec), and γ = bulk density (pcf). The area of impact A can be approximated by $A = 2a$ for a belt conveyor. Substituting these values of Q and A into Eq. (4-10a) gives

$$p_1 = 0.125 v\gamma\sqrt{h} \tag{4-10b}$$

The value of f_c corresponding to p_1 from Eqs. (4-10) should be read from an instantaneous flow-function curve, and should be used to determine the opening size if it is larger than f_c for steady flow in mass-flow bins or for both steady flow and initial pressure in funnel-flow and flat-bottom bins.

Impact pressure is not likely to give the critical value of f_c except for a material whose instantaneous flow function determines f_c for both flow pressure and filling pressure, that is, for materials which do not experience an increase in consolidation with time.

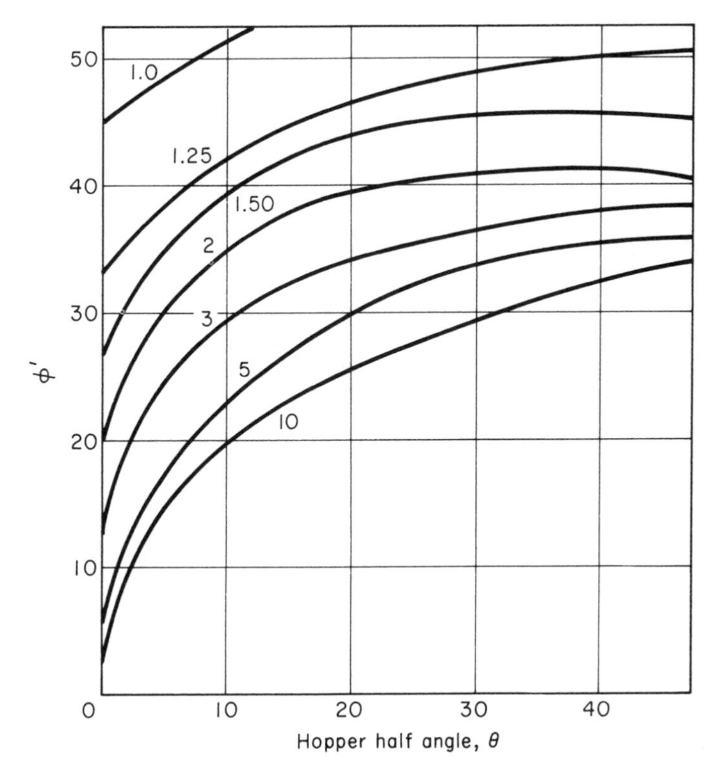

Figure 4-19 Consolidating pressure factor F for wedge-shaped hoppers, $k = 0.8$. (From Ref. 7.)

4-12. Hoppers for Mass-Flow Bins

For mass flow to occur, the hopper walls must be sufficiently steep and smooth to assure sliding of the solid along the wall. The largest included angle 2θ (Fig. 4-20) for which mass flow will occur depends on the angle of friction ϕ' between the solid and the wall and on the shape of the horizontal cross section of the hopper. In general, ϕ' is not a constant but decreases with increasing pressure and therefore assumes its largest value at the outlet of the hopper where the pressures between the solid and the walls are smallest during flow. Since the required value of θ decreases as ϕ' increases, the slope of the hopper walls is usually determined by the value of ϕ' at the outlet.

Conical hoppers. Mass flow will occur in a conical hopper if point (θ, ϕ') lies below the shaded area shown in Fig. 4-20. If the point is within the

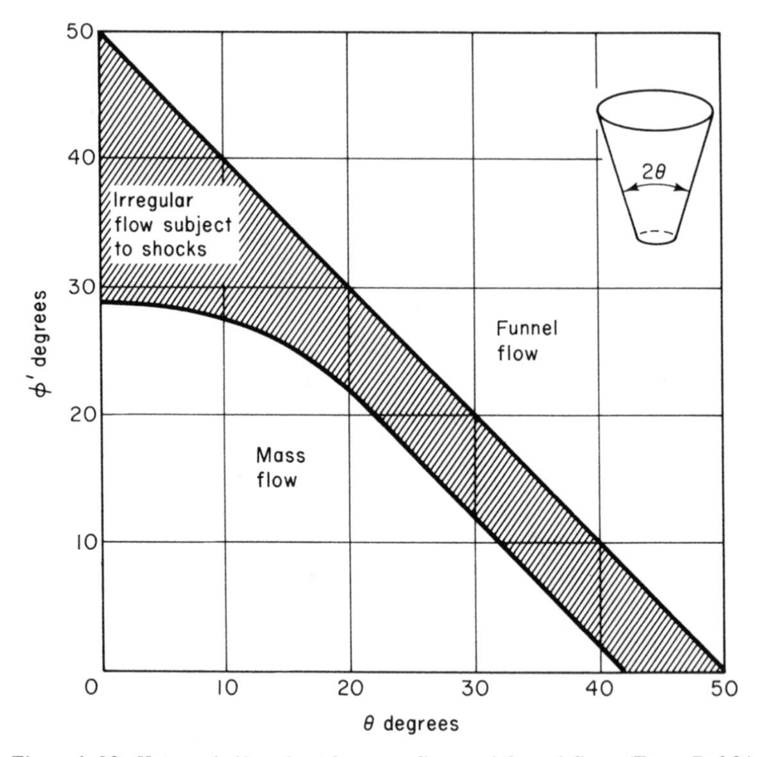

Figure 4-20 Hopper half angle θ for mass flow and funnel flow. (From Ref.2.)

shaded area, mass flow is likely to alternate with funnel flow with resultant shocks.

Walker's formula does not give limiting values of θ directly. However, Fig. 4-14a shows that the flow factor rapidly approaches infinity after it exceeds about 2. This sets practical limits on the wall angle.

Pyramidal hoppers. The angle θ_v of the hopper valley with the vertical is determined as for a conical hopper. The adjoining-wall angles θ_1 and θ_2 with the vertical are related to θ_v by

$$\tan^2 \theta_v = \tan^2 \theta_1 + \tan^2 \theta_2 \tag{4-11}$$

Wedge and chisel hoppers. Mass flow will occur in wedge and chisel hoppers if

$$\theta \leqslant 60^\circ - 1.33\phi' \tag{4-12a}$$

$$\phi' \leqslant 0.9\delta \tag{4-12b}$$

$$l \geqslant 6b \tag{4-12c}$$

Transition hoppers. For transition mass-flow hoppers with rectangular outlets the end slopes are determined as for conical hoppers and the side slopes as for wedge hoppers. The length-to-width ratio of the outlet should not be less than 3.

Surface finish. It is essential that the completed hopper have a surface which is at least as smooth as the sample of the material used in the determination of the angle of friction ϕ'. It is unwise to assume that the hopper surface will polish during discharge. If mass-flow conditions are not met for the first filling of the bin, the solid will flow through a central channel instead of at the walls so that the polishing cannot occur. On the contrary, if corrosion of the walls is possible, their roughness will increase and perpetuate the funnel-flow pattern.

Example 4-1

Determine the wall slope and minimum opening size of a stainless-steel mass-flow conical hopper for the dolomite whose properties are given in Fig. 4-21.

1. From Fig. 4-20, $\theta \leqslant 16°$ for $\phi' = 25°$. Use 15°.
2. From Fig. 4-21a assume $\delta = 55°$ as a first approximation. Interpolating between Figs. 4-11 and 4-12 with $\delta = 55°$ and $\phi' = 25°$ gives $ff = 1.28$. The line $ff = p_1/f_c = 1.28$ intersects the curve in Fig. 4-21c at $f_c = 120$ psf. The corresponding value of δ is 57°, which gives $ff \approx 1.25$. The resulting changes in f_c and δ are negligible.
3. Figure 4-21b gives $\gamma = 96$ pcf for $ff = 1.28$. Then from Eq. (4-3)

$$d \geqslant \frac{2.2 f_c}{\gamma} \geqslant \frac{2.2 \times 120}{96} \geqslant 2.75 \text{ ft}$$

Walker's Eq. (4-5b) gives $ff = 1.30$ for $\theta = 15°$, $\phi' = 25°$, and $\delta = 55°$. This is only slightly larger than the Jenike value and so would give the same value of f_c. Equation (4-1) gives $d \geqslant 2 \times 120/96 = 2.50$ ft.

Example 4-2

Determine the wall slope of a square pyramidal hopper for the dolomite of Example 4-1.

1. The valley angle is chosen as for a conical hopper. Therefore, use $\theta_v = 15°$.
2. The wall slope is given by Eq. (4-11) with $\theta_1 = \theta_2$. Therefore, $\sqrt{2} \tan \theta_1 = \tan 15°$ and $\theta_1 = 10.7°$.

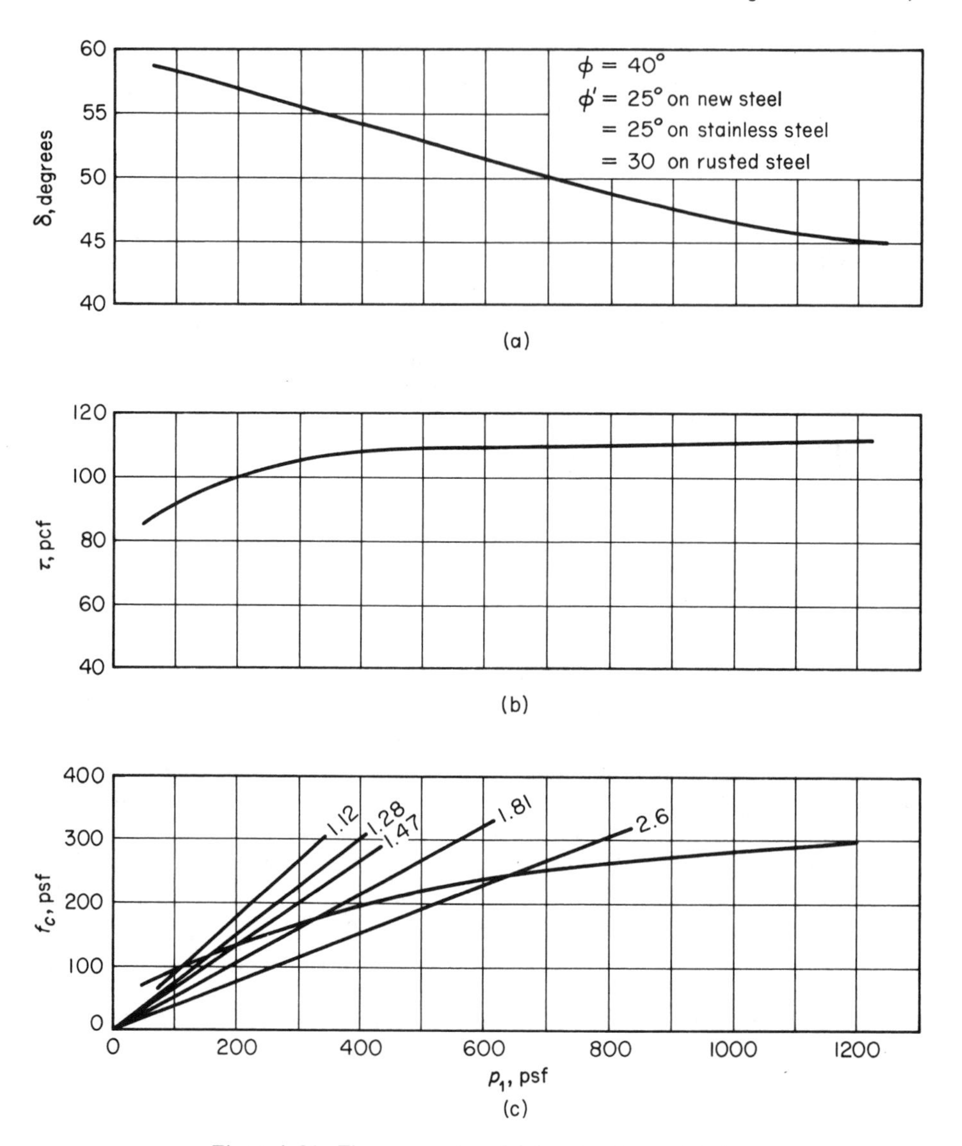

Figure 4-21 Flow properties of dolomite. (From Ref. 8.)

Example 4-3

Determine the wall slope and minimum width of the outlet for a stainless-steel mass-flow wedge hopper for the dolomite of Example 4-1.

1. Assume $\delta = 55°$. From Eqs. (4-12), $\theta \leqslant 60° - 1.33\phi' = 60° - 1.33 \times 25° = 27°$. Use $25°$.
   $0.9\delta = 0.9 \times 55° = 50°$. Since this is greater than ϕ', mass flow will occur.

2. Assume $\delta = 55^\circ$. Although flow in a wedge hopper is more nearly plane flow, flow factors for conical flow can be used with good approximation (Sec. 4-9). However, the point $(\phi', \delta) = (25^\circ, 55^\circ)$ is outside the range of the flow factors of Fig. 4-13. Walker's Eq. (4-5a) gives $ff = 1.81$ for which Fig. 4-21 gives $f_c = 180$ psf, $\gamma = 106$ pcf and $\delta = 55^\circ$. Equation (4-1) gives $b \geqslant 180/106 \geqslant 1.70$ ft.

Jenike's plane-flow charts in Ref. 1 give $ff \approx 1.12$, and from Fig. 4-21 $f_c = 100$ psf, $\gamma = 93$ pcf, and $\delta = 58^\circ$. The change in ff for $\delta = 58^\circ$ is insignificant. Therefore, from Eq. (4-4), $b = 1.3 f_c/\gamma \geqslant 1.3 \times 100/93 \geqslant 1.40$ ft.

4-13. Hoppers for Funnel-Flow Bins

The slope and surface condition of a funnel-flow hopper do not affect the flow pattern. Therefore, design is limited to determining the dimensions of the outlet.

Example 4-4

Determine the half angle θ and minimum size of opening of a steel funnel-flow hopper for the dolomite whose properties are given in Fig. 4-21.

1. From Fig. 4-21, $\phi' = 25^\circ$ for new steel. Then from Fig. 4-20, $\theta \geqslant 25^\circ$. A value larger than the minimum is to be preferred to guard against irregular flow.
2. From Fig. 4-21, $\phi = 40^\circ$. Use $\delta = 55^\circ$ as a first approximation. Then from Fig. 4-17, $ff = 2.6$. The line $ff = p_1/f_c = 2.6$ intersects the f_c curve in Fig. 4-21c at $f_c = 230$ psf. The corresponding values of γ and δ are 110 pcf and 52°, respectively. This value of δ is close enough to the assumed value.
3. From Fig. 4-16, $G(\phi) = 3.4$ for $\phi = 40^\circ$.
4. From Eq. (4-6),

$$d = \frac{f_c}{\gamma} G(\phi) = \frac{230}{110} \times 3.4 = 7.1 \text{ ft}$$

   This value of d is the minimum diameter of a circular opening or the minimum diagonal of a rectangular opening. If the opening is circular or square, doming or piping will not occur with this diameter. If the opening is rectangular, however, the short side b must be checked for doming as follows.
5. For $\delta = 55^\circ$ and $\phi = 40^\circ$, $ff = 1.47$ (Fig. 4-15). The line $ff = p_1/f_c = 1.47$ intersects the f_c curve in Fig. 4-21c at $f_c = 130$ psf. The corresponding values of γ and δ are 100 pcf and 57°, respectively. This value of δ is close enough to the assumed value, 55°. Then from Eq. (4-4), $b \geqslant 1.3 \times 130/100 = 1.7$ ft. Any of the following outlet dimensions satisfy this requirement and have a diagonal length not less than the required 7.1 ft determined in step 4: 2 × 7 ft, 3 × $6\frac{1}{2}$ ft, 4 × 6 ft, 5 × 5 ft.

The hopper diameter should also be checked for piping under initial pressures (Sec. 4-10).

4-14. Live Storage

An important consideration in the design of funnel-flow bins is the live storage that can be realized. Of course, the capacity of a bin which clears completely is easy to determine. A funnel-flow bin with a steep smooth pyramidal or conical hopper will clear by gravity. Stainless steel, ceramic tile, and resin coatings can be used to eliminate corrosion during storage so that the hopper will retain its smooth surface. Unless the solid is very free-flowing, the slope $\theta \leqslant 65° - \phi'$ is suggested by Jenike.[1] If the walls are steep but not designed to remain smooth, the bin can be cleared completely by vibration, aeration, or manual rodding and air lancing.

A formula for predicting the flow boundary in flat-bottomed bins with round openings has been developed by Giunta.[9] The boundary dimensions, defined in Fig. 4-22, are related by

$$d_f = d + \frac{2H - Md}{M + \cot \theta'} \tag{4-13}$$

where M is a coefficient and θ' is the boundary slope from the outlet, both given in Fig. 4-23. This equation is valid only if the opening is large enough

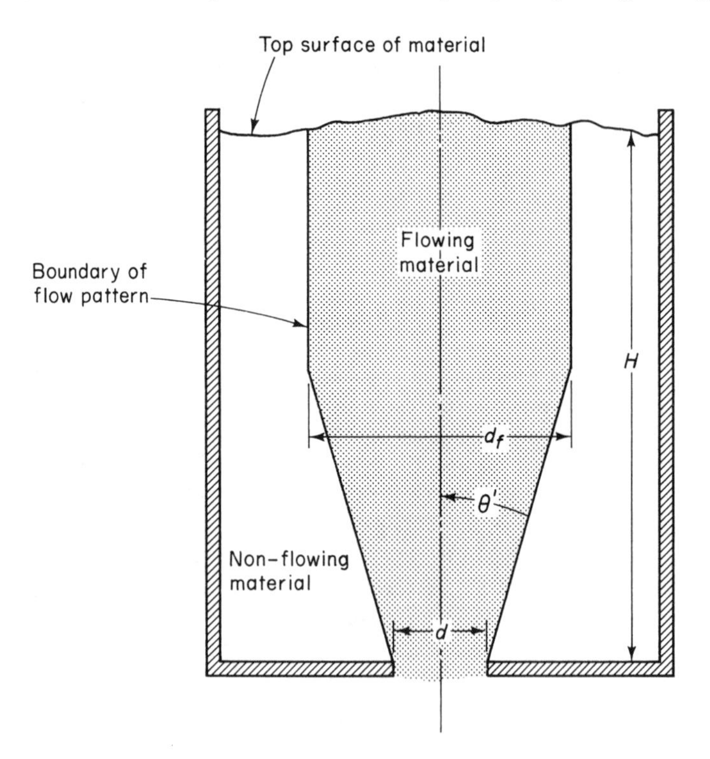

Figure 4-22 Flow boundary in flat-bottom bin. (From Ref. 9.)

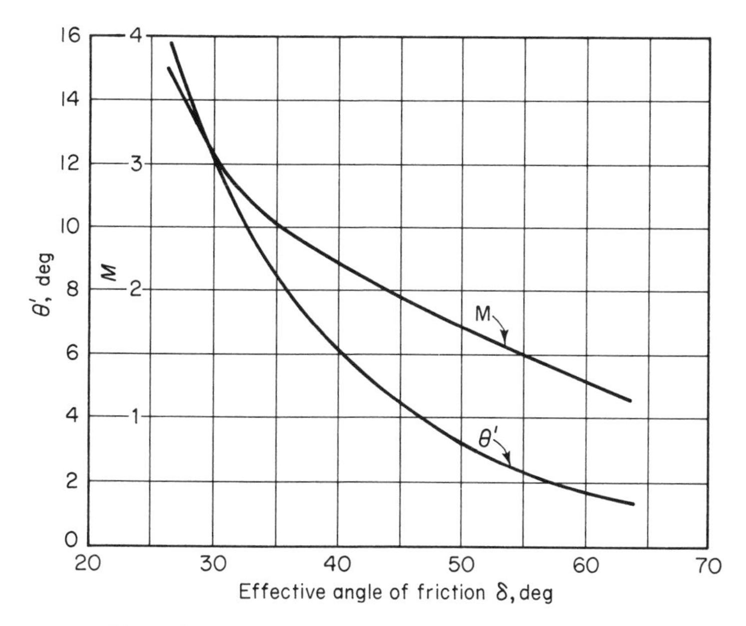

Figure 4-23 Coefficients for Eq. (4-13). (From Ref. 9.)

to prevent arching and ratholing and if $H \geqslant Md/2$. If $H \leqslant Md/2$, the diameter d_f of the flow boundary equals d.

Equation (4-13) has been verified by tests on model bins 18 in. in diameter by 24 in. high with round openings 1 to 8 in. in diameter. With one exception, good agreement with tests on starch, pulverized coal, and iron-ore concentrate was obtained. The maximum variation between the experimental and theoretical diameters of the flow channel at a height of 20 in. was 10% for coal with a 4-in. opening; the smallest was 2% for iron ore with a 6-in. opening. The exception was for iron ore with a 1-in. opening for which the measured diameter at a height of 20 in. was 34% more than the predicted value.

Example 4-5

Determine the flow channel for a flat-bottom bin 15 ft in diameter by 80 ft high for the dolomite whose properties are given in Fig. 4-21.

1. In Example 4-4 the critical dimension d for doming and piping was determined to be 7.1 ft for a funnel-flow bin. The same limit applies to a flat-bottom bin. For any opening smaller than this but large enough to prevent doming, a rathole will develop, and the maximum amount of material that can be expected to flow out of the bin is the material directly above the opening.
2. Assume the bin to have a circular opening 8 ft in diameter. This is larger than the critical diameter for piping. By using $\delta = 55^\circ$ from Example 4-4, Fig. 4-23 gives

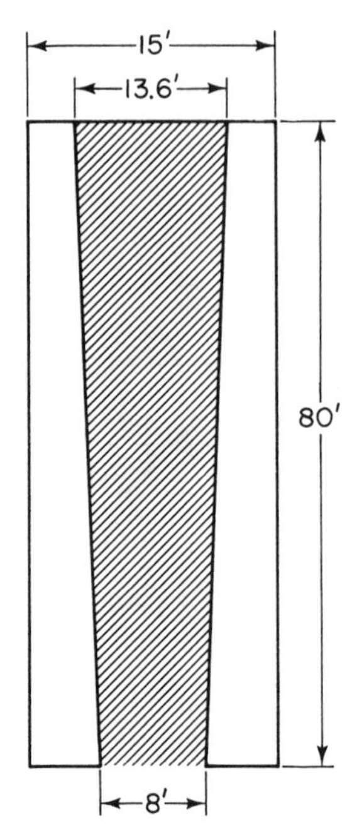

Figure 4-24

$M = 1.5$ and $\theta' = 2.3°$. Then from Eq. (4-13),

$$d_f = 8 + \frac{2 \times 80 - 1.5 \times 8}{1.5 + \cot 2.3°} = 13.6 \text{ ft}$$

The resulting flow channel is shown in Fig. 4-24.

4-15. Rate of Discharge

It has been shown that the shape of the bin cross section influences the flow channel but not the rate of discharge. The following formulas for calculating the rate of discharge of frictional, cohesive, granular solids are derived in Ref. 10. It is assumed that the material at the hopper opening forms a continuously failing arch that is in dynamic equilibrium. This assumption requires the principal stress s_1 in the arch to be equal to the unconfined compressive strength f_c of the material. The unconfined compressive strength depends in turn on the major consolidating pressure p_1 at the opening.

The consolidating pressure for a wedge hopper with width b is given by

$$p_1 = \gamma b(ff) \tag{4-14a}$$

and for a conical hopper with opening diameter d by

$$p_1 = \frac{\gamma d}{2}(ff) \tag{4-14b}$$

With the value of p_1, f_c is determined from a material properties curve. The maximum discharge rate Q_m for a wedge hopper is computed by

$$Q_m = \gamma A \sqrt{\frac{bg}{2 \tan \theta'} \left(1 - \frac{f_c}{p_1} ff\right)} \tag{4-15a}$$

and for a conical hopper by

$$Q_m = \gamma A \sqrt{\frac{dg}{4 \tan \theta'} \left(1 - \frac{f_c}{p_1} ff\right)} \tag{4-15b}$$

where g = gravitational acceleration. This value of Q_m is based on the steady-state velocity of the discharging material. Therefore, the rate of discharge will be overestimated if the gate is open for a short period of time. The average flow rate Q_a for the time the gate is open is given by

$$Q_a = Q_m \left(1 - 1.39 \frac{t}{T}\right) \tag{4-16}$$

where

$$t = \frac{Q_m}{2gA\gamma} \frac{1}{1 - ff(f_c/p_1)}$$

and T = time gate is opened. When $T > 4t$, the error in Eq. (4-16) is less than 1%.

Equations (4-15) were verified by laboratory tests on slotted hoppers with $\theta = 10°$, $25°$, and $45°$ and on conical hoppers with $\theta = 15°$ and $d = 0.14$, 0.46, and 0.70 ft. Lac Jeannine concentrate and Saxonburg sinter fines ($< \frac{1}{4}$ in.) were tested at various moisture contents. Full-scale tests were conducted at a sintering plant with Lac Jeannine concentrate, Venezuelan ore fines, and dolomite. In all cases the experimental results were in good agreement with predicted values. It should be noted, however, that the experimental verification was only for bins with flow along the hopper walls.

Example 4-6

Determine the rate of discharge of the material through a 3-ft round opening in the hopper of Example 4-1. The hopper slope $\theta = 15°$, the bulk density $\gamma = 96$ pcf, and the flow factor $ff = 1.28$ were determined in that example.

1. From Eq. (4-14b),

$$p_1 = \frac{96 \times 3}{2} \times 1.28 = 184 \text{ psf}$$

2. From Fig. 4-21c, $f_c = 130$ psf for $p_1 = 184$ psf. From Eq. (4-15b),

$$Q_m = 96 \times 7.07 \sqrt{\frac{3 \times 32.2}{4 \tan 15°} \left(1 - \frac{130 \times 1.28}{184}\right)} = 1992 \text{ lb/sec}$$

3. Assume the gate to be opened for 10 sec. The average rate of discharge during this time is determined by Eq. (4–16):

$$t = \frac{1992}{2 \times 32.2 \times 7.07 \times 96} \, \frac{1}{1 - 1.28 \times 130/184} = 0.48 \text{ sec}$$

$$Q_a = 1992(1 - 1.39 \times 0.48/10) = 1860 \text{ lb/sec}$$

REFERENCES

1. A. W. Jenike, Storage and Flow of Solids, *Bull. No. 123,* Utah Engineering Experiment Station, University of Utah, Salt Lake City, Nov. 1964.
2. A. W. Jenike and J. R. Johanson, Annual Report, Project No. 126—Fourth Part, Mass-Flow Bins, and Project 126A—First Part, Funnel-Flow Bins, American Iron and Steel Institute, Washington, D.C., 1971, unpublished.
3. J. R. Johanson and H. Colijn, New Design Criteria for Hoppers and Bins, *Iron and Steel Eng.,* Oct. 1964.
4. H. Colijn, *Weighing and Proportioning of Bulk Solids,* Trans Tech Publications, Clausthal-Zellerfeld, West Germany, 1975.
5. D. M. Walker, An Approximate Theory for Pressure and Arching in Hoppers, *Chem. Eng. Sci.,* Vol. 21, 1966.
6. H. Wright, An Evaluation of the Jenike Bunker Design Method, *Trans. ASME, J. Eng. for Ind.,* Feb. 1973.
7. J. R. Johanson, Effect of Initial Pressures on Flowability of Bins, *Trans. ASME, J. Eng. for Ind.,* May 1969.
8. Properties of Various Raw Materials Related to Gravity Flow, *Report 30.016-005 (8),* U.S. Steel Applied Research Laboratory, Monroeville, Pa., March 1965.
9. J. S. Giunta, Flow Patterns of Granular Materials in Flat-Bottom Bins, *Trans. ASME, J. Eng. for Ind.,* May 1969.
10. J. R. Johanson, Method of Calculating Rate of Discharge from Hoppers and Bins, *Trans. Society of Mining Engineers,* March 1965.

5

LOADS

5-1. Introduction

Early designers of vessels for the storage of bulk solids assumed that the stored materials behaved like liquids and designed the vessels for equivalent fluid pressures. Experiments by Roberts[1,2] on models and full-size silos showed that this is incorrect because some of the weight of the stored materials is transferred to the walls by friction. Janssen confirmed this conclusion and in 1895 published a theory that accounts for wall friction.[3] Soon after, Airy proposed another method of computing wall pressures.[4] A considerable amount of work by various experimenters followed, and Ketchum summarized the state of the art in a book published in 1909.[5]

It was noted as early as 1896 that pressures during discharge are larger than the initial pressures. Tests show that flow pressures can be two to four times as large as initial pressures even in bins with central discharge and that large peak pressures can occur in both the bin and the hopper. The idea that pressures during flow are larger because the initial (charging) pressure field approaches the active state but may change during discharge to one approaching the passive state is now generally accepted.[6,7]

5-2. Stresses in a Solid

A mass is said to be in a state of plastic equilibrium if it is on the verge of failing at every point. The states of plastic equilibrium in a semi-infinite mass acted upon by gravity alone were investigated by Rankine. Let *AB* in Fig.

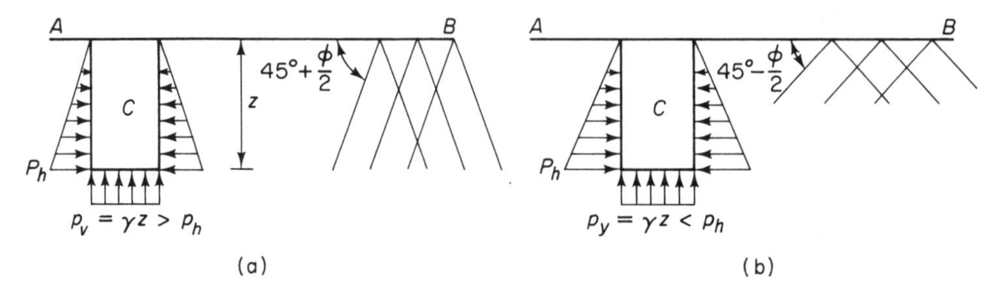

Figure 5-1

5-1a be the horizontal surface of such a mass of unit weight γ. Because of symmetry there can be no shear on the bottom surface of the element C in the figure. Therefore, the horizontal and vertical pressures p_h and p_v at every point in the mass are principal pressures either of which can be the major pressure. There are two corresponding states of plastic equilibrium: The mass is said to be in the *active Rankine state* if p_v is the major principal stress and in the *passive Rankine state* if it is the minor principal stress. The corresponding horizontal stresses are given by

$$p_h = K_a p_v \qquad p_h = K_p p_v \tag{5-1}$$

where K_a and K_p are the active and passive pressure coefficients, respectively. If the mass is cohesionless, these coefficients are given by

$$K_a = \frac{1 - \sin\phi}{1 + \sin\phi} = \tan^2\left(45 - \frac{\phi}{2}\right) \tag{5-2a}$$

$$K_p = \frac{1 + \sin\phi}{1 - \sin\phi} = \tan^2\left(45 + \frac{\phi}{2}\right) \tag{5-2b}$$

where ϕ is the angle of internal friction. If the mass is cohesive, formulas based on Eq. (3-3) can be used, but it is simpler to use Eqs. (5-2) with the effective angle of friction δ (Sec. 3-5) substituted for ϕ. Thus,

$$K_a = \frac{1 - \sin\delta}{1 + \sin\delta} = \tan^2\left(45 - \frac{\delta}{2}\right) \tag{5-3a}$$

$$K_p = \frac{1 + \sin\delta}{1 - \sin\delta} = \tan^2\left(45 + \frac{\delta}{2}\right) \tag{5-3b}$$

A granular mass will usually be in a state of elastic equilibrium which is intermediate between the two limiting plastic states, in which case $p_h = Kp_v$ where $K_a < K < K_p$. This value of K cannot change unless the mass expands or contracts laterally. Such a movement does not change p_v because the weight of the solid above any horizontal section is not changed, but p_h is reduced if the solid expands laterally and increased if it contracts. If the

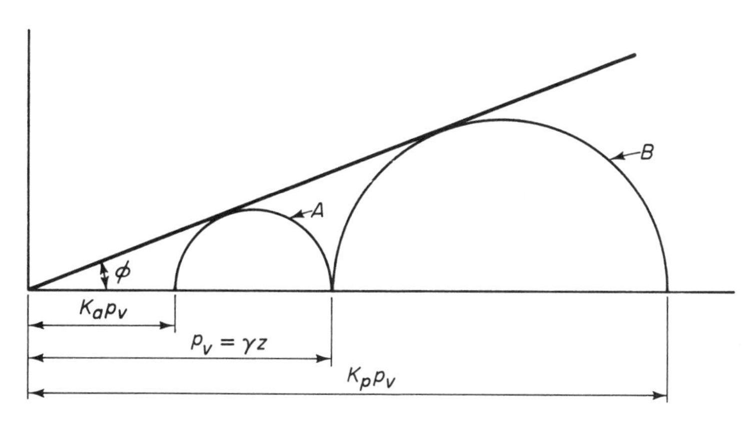

Figure 5-2 Rankine stress relationships.

solid expands, K decreases until it becomes equal to K_a. The solid is now in the active Rankine state. In this state it can slide, with no change in p_h, along the two sets of plane surfaces shown in Fig. 5-1a. These surfaces make the angle $45 + \phi/2$ with the horizontal. Similarly, if the solid contracts laterally, K increases until it becomes equal to K_p. The mass is now in the passive Rankine state and can slide along two sets of plane surfaces making the angle $45 - \phi/2$ with the horizontal (Fig. 5-1b).

The Mohr circles shown in Fig. 5-2 give the stresses on every plane in the solid in the two plastic states. Circle A represents the active Rankine state and circle B the passive. A line which makes the angle ϕ with the horizontal and is tangent to the circles is called a rupture line or yield line. This line intersects the origin if the solid is cohesionless (Fig. 5-2), while for a cohesive solid it intercepts the ordinate c on the shear axis (Fig. 3-5).

5-3. States of Stress in Bins

When a granular mass is deposited in a bin, the vertical pressure on an element tends to be the major principal pressure. Therefore, the mass is likely to assume an elastic state of equilibrium that approaches the active plastic state. Because of friction on the walls, however, the distribution of stress will not be as simple as in a semi-infinite mass, and the Rankine states described in Sec. 5-2 will be altered.

Assume the stresses on an element at the wall to be p_v, $p_h = Kp_v$, and $q = p_h \tan \phi'$, where ϕ' is the angle of wall friction (Fig. 5-3a). If $p_v > p_h$, this state of stress gives points A and B on the Mohr-circle coordinates shown in Fig. 5-4. Furthermore, if the element is in a plastic state, the circle $CADEB$ is tangent to the yield line. The corresponding principal pressures are given by OC and OE; they act in the directions α and $90° + \alpha$ with the horizontal (Fig. 5-3b), where α is half the angle AJE in Fig. 5-4.

Because of symmetry, there will be no shear on an element at the axis

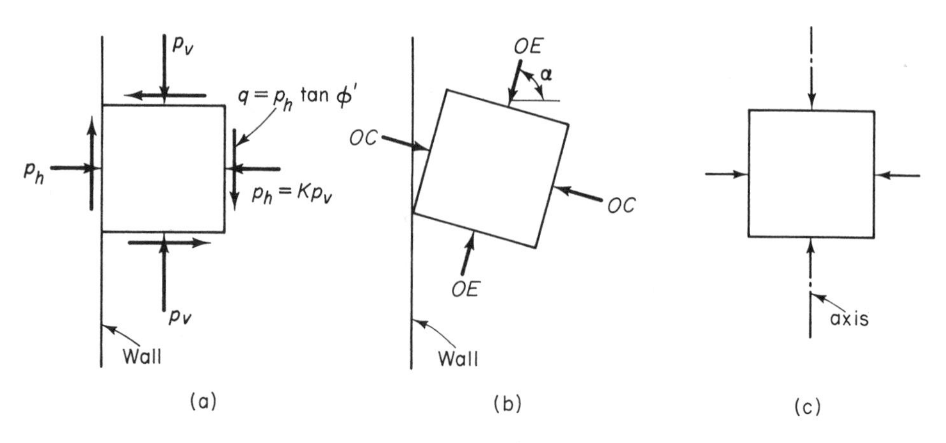

Figure 5-3

of the bin (Fig. 5-3c). Therefore, the pressures p_h and p_v on this element are principal pressures which plot on the p axis. If we assume p_h on this element to be equal to the horizontal pressure at the wall, the plastic condition for the element will be given by the Mohr circle FGH in Fig. 5-4. Thus, the vertical pressure OH at the axis of the bin is greater than the vertical pressure OE at the wall. Furthermore, if we assume the horizontal pressure to be constant over a horizontal cross section of the solid, the states of stress on all

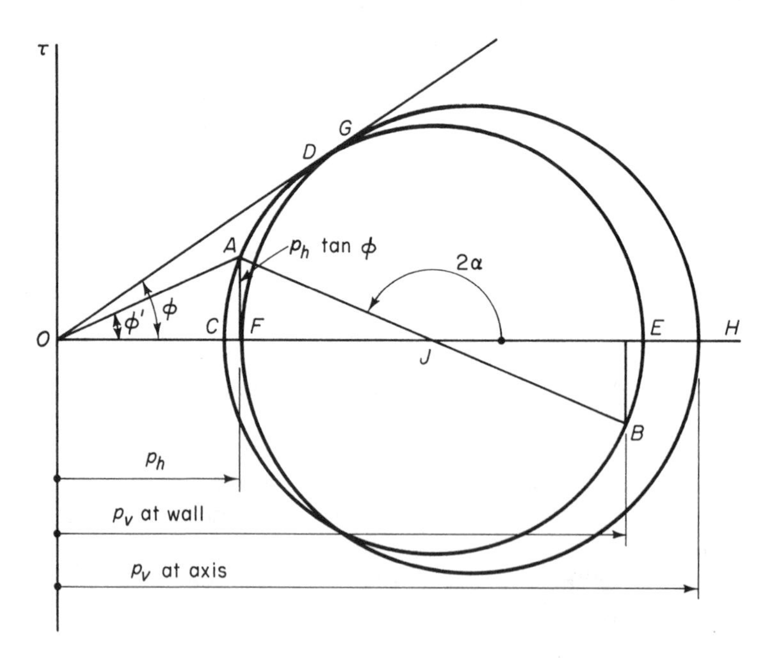

Figure 5-4 Stresses in bin contents, $p_v > p_h$.

elements between the wall and the axis are given by Mohr circles lying between the circles *CDE* and *FGH*. The vertical pressure exceeds the horizontal pressure on every such element, and if each is in a plastic state, the solid can be said to be in the active plastic state at this level in the bin.

Assuming now that p_v at the wall is less than p_h, the Mohr circle *CDAEB* in Fig. 5-5 is obtained for an element at the wall. The principal pressures, which are given by *OC* and *OE*, act at the angles α and $90° + \alpha$ with the horizontal, where α is now half the angle *AJE* in Fig. 5-5. If the horizontal pressure is assumed to be constant over the horizontal cross section containing this element, the circle *FGH* gives the state of stress on an element at the axis of the bin, and the states of stress on all intermediate elements are given by circles between *CDE* and *FGH*. The vertical pressure (*OH* in Fig. 5-5) at the axis is now less than that at the wall, as is the case for all other elements at this section, and since the horizontal pressure exceeds the vertical pressure at all points, the solid can be said to be in a passive plastic state at this level.

Based on the two plastic states of stress just described and the assumption that the shear at a horizontal section increases linearly from zero at the axis of the bin to $p_h \tan \phi'$ at the wall, Nanninga[8] derived the following formula for the corresponding pressure coefficients:

$$K = \frac{\cos^2 \phi}{1 + \sin^2 \phi \pm (4/3a^2) \sin \phi [1 - (1 - a^2)^{3/2}]} \tag{5-4}$$

where $a = \tan \phi'/\tan \phi$ and the plus and minus signs of the third term in the denominator give the active and passive pressure coefficients, respectively. Values of K by this formula reduce to those of Eq. (5-2) if $\phi' = 0$.

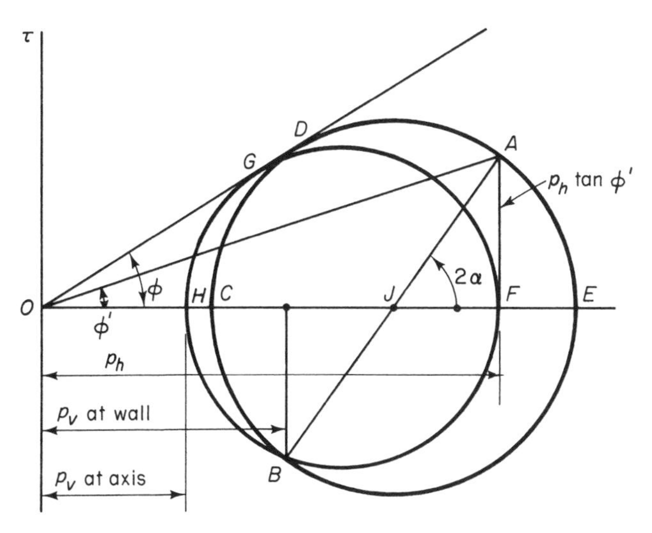

Figure 5-5 Stresses in bin contents, $p_v < p_h$.

5-4. Pressures in Bins

Flow pressures in a mass-flow bin are usually well defined and reproducible because the flow channel is defined and constant. On the other hand, a funnel-flow channel may expand or contract between fillings and even during flow, and some portions of the dead solid may alternately stop and flow because they are in unstable equilibrium. Therefore, pressures are more erratic in funnel-flow bins.

When solid is charged into an empty bin with the gates closed or the feeder at rest, the solid compacts as the initial vertical pressures increase. Therefore, the initial pressure field tends to be active. Janssen's formula (Sec. 5-5) gives a good representation of the corresponding pressures on the wall.

When solid starts flowing out of a bin outlet, it expands vertically within the developing flow channel. The channel usually diverges upward, so that the solid also contracts laterally. Therefore, the pressure field tends to be passive, with the major pressure lines arching across the flow channel. The plane of transition between the developing passive-pressure field and the initial active (Janssen) field is called a *switch*. This switch originates at the outlet and travels upward into the bin (Fig. 5-6).

Nanninga[8] showed that equality of the vertical pressures above and below the switch in a mass-flow cylinder exists only if there is a peak in the distribution of the lateral pressures p_h (Fig. 5-6b), because p_v is a major pressure in the zone above the switch but a minor one in the zone below. The maximum possible value of the ratio of the horizontal pressures immedi-

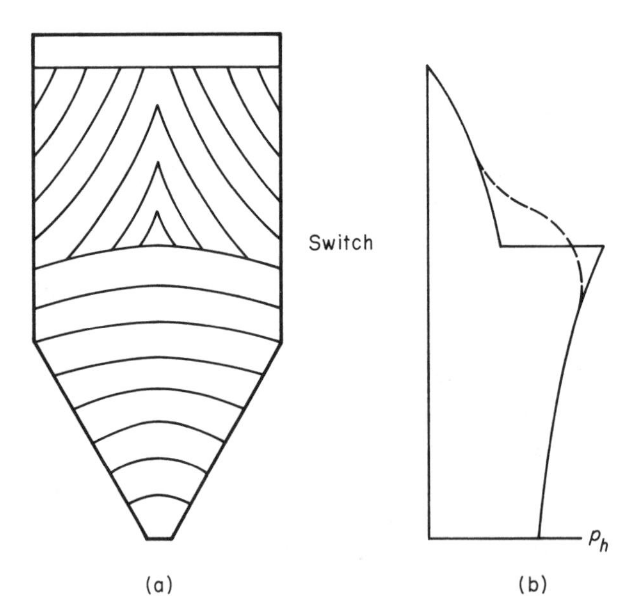

Figure 5-6 Lines of principal pressure during developing flow.

ately above and below the switch is the ratio of the corresponding pressure coefficients K_p and K_a given by Nanninga's Eq. (5-4). Nanninga also shows that the lateral pressure drops off rapidly below the switch and approaches the infinite-depth Janssen pressure asymptotically (Fig. 5-6b). It should be noted, however, that the discontinuity in pressure shown at the switch cannot exist. Equality of horizontal shears immediately above and below the switch requires equality of the corresponding wall friction forces at the switch. Nanninga concludes that the actual pressure distribution must be somewhat as shown by the dashed line in Fig. 5-6b.

These pressure peaks at the boundary between an active-pressure zone and a passive-pressure zone act directly on the walls in a mass-flow bin but tend to be reduced in effect in a funnel-flow bin because of the dead solid in the region of the transition. This effect was reported by Takhtanischev during a series of tests on grain elevators in Baku, Russia.[6] He observed the two types of flow shown in Fig. 5-7 and reported that lateral pressures on the walls were considerably larger with flow like that in Fig. 5-7b.

Walters[9] gives charts of K_p/K_a based on Nanninga's pressure coefficients, which he derives independently. The curves are plotted in terms of the effective angle of friction δ instead of ϕ. The charts show that K_p/K_a can be very large. For example, if $\delta = 50°$ and $\phi' = 25°$, $K_p/K_a = 26.6$ at a switch seven diameters below the top of the solid. This overpressure ratio drops to less than 2 at half a diameter below the switch, which shows the concentrated nature of the peak.

Walker has investigated flow pressures in mass-flow bins (Sec. 5-13). Jenike and Johanson have investigated switch peak pressures in both funnel-flow cylinders and mass-flow cylinders.[14] According to their theory of mass

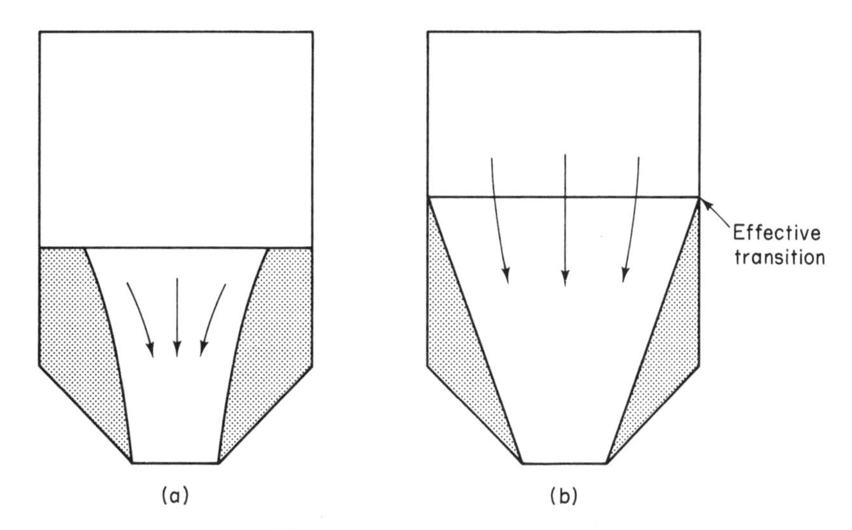

Figure 5-7 Types of funnel flow.

flow (Sec. 5-14), the flow pressure immediately below the switch in the example in the preceding paragraph is only 6.7 times the initial pressure immediately above. This would seem to be a more reasonable value. However, if δ is changed to 25° and the other variables remain the same, the ratio of peak pressure to initial pressure drops to 2.83 according to Nanninga-Walters and 2.58 according to Jenike-Johanson. Thus, pressure peaks appear to be quite sensitive to the effective angle of internal friction.

Flow pressures at any given level in the cylinder above a switch vary widely with time. This appears to be due to changes in conditions at the cylinder wall. In observing continuous flow in a cylinder with transparent walls, one sees patches of thin layers of solid at the walls flowing intermittently. According to Ref. 11, this is because the recoverable strain energy of the flowing mass tends to minimize in accordance with the second law of thermodynamics, and this requires the shape of the flowing vertical mass to deviate slightly from that of the cylinder. Boundary layers form in an effort to develop that shape, but in most cases they are unstable and of momentary duration. The pressure field changes from Janssen to minimum strain energy when a layer forms, but when the layer dissolves, the Janssen field is restored. Therefore, the switch moves up into the cylinder when a boundary layer forms and falls again when it disappears. These movements contribute to the jerky flow often seen in bins.

The development of boundary layers is sensitive to slight imperfections in the shape or finish of the bin because the imperfections provide support for the layers. Test results[10] show that the thickness of the boundary layers in a slightly diverging (downward) cylinder (6 in. in diameter by 16 in. high) is not sufficient to develop a switch, so that the Janssen field is maintained during flow. On the other hand, test results for an 0.5° converging cylinder indicate that thin boundary layers are sufficient to cause pressures approaching the minimum energy values in the lower part of the cylinder.

The results of two tests by Reimbert and Reimbert, conducted one month apart, on a grain bin approximately 13 ft square by 33 ft high are shown in Fig. 5-8. This bin was of horizontal-trough steel construction.[15] The figure shows the ratios of pressures measured during emptying to those measured with the grain (wheat) at rest. These tests illustrate the uncertainties in predicting discharge pressures.

5-5. Janssen's Formula

Figure 5-9 shows the forces acting on a horizontal slice of solid in a bin. Summation of vertical forces gives

$$dp_v A - \gamma A \, dy + qC \, dy = 0 \tag{5-5}$$

where p_v = vertical pressure in bin
q = frictional stress on wall

γ = bulk density of solid
A = cross-sectional area of bin
C = perimeter of bin
y = depth of solid above section

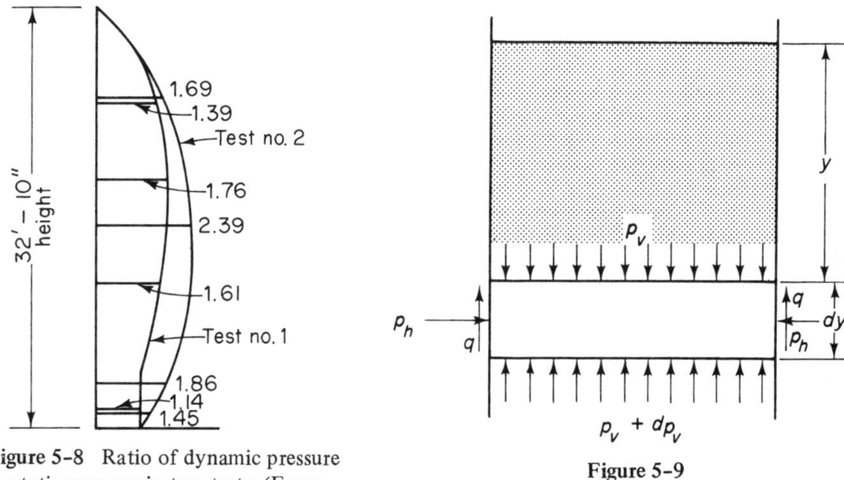

Figure 5-8 Ratio of dynamic pressure to static pressure in two tests. (From Ref. 15.)

Figure 5-9

The relationship between the friction and the pressure p_h on the wall is given by

$$q = \mu' p_h \tag{5-6}$$

where $\mu' = \tan \phi'$ is the coefficient of friction of the material on the wall. Furthermore, assuming the ratio of the horizontal pressure to the vertical pressure to be constant, we have

$$p_h = Kp_v \tag{5-7}$$

where K is the pressure coefficient. Substituting from Eqs. (5-6) and (5-7) into Eq. (5-5) and integrating, we get

$$p_v = \frac{\gamma R}{\mu' K}(1 - e^{-\mu' K y/R}) = \frac{\gamma R}{\mu' K} f\left(\frac{\mu' K y}{R}\right) \tag{5-8a}$$

$$p_h = \frac{\gamma R}{\mu'}(1 - e^{-\mu' K y/R}) = \frac{\gamma R}{\mu'} f\left(\frac{\mu' K y}{R}\right) = Kp_v \tag{5-8b}$$

where $R = A/C$ = hydraulic radius. Values of $f(\mu' Ky/R)$ are given in Table 5-1. Values of K are discussed below.

Pressures during filling. The Rankine active-pressure coefficient, Eq. (5-2a), was proposed by Koenen[16] in 1896 and has been in common use, with no distinction between filling pressures and emptying pressures. Accord-

TABLE 5-1. VALUES OF $f(\mu' Ky/R) = 1 - e^{-\mu' Ky/R}$

$\mu' Ky/R$	0	0.1	0.2	0.3	0.4	0.5	0.6	0.7	0.8	0.9
0	0	0.095	0.181	0.259	0.333	0.393	0.451	0.503	0.551	0.593
1	0.632	0.667	0.699	0.727	0.753	0.777	0.798	0.817	0.835	0.850
2	0.869	0.878	0.889	0.900	0.909	0.918	0.926	0.933	0.939	0.945
3	0.950	0.955	0.959	0.963	0.967	0.970	0.973	0.975	0.978	0.980
4	0.982	0.983	0.985	0.986	0.988	0.989	0.990	0.991	0.992	0.993

ing to Pieper and Wenzel,[17,18] this value gives vertical pressures that are much too large, which results in an underestimate of the vertical compression in the bin wall. Their experiments show that the static pressure coefficient commonly used in soil mechanics

$$K = 1 - \sin \phi \tag{5-9}$$

gives good results for pressures during filling.

According to Jenike, Johanson, and Carson, $K = 0.4$ gives results that compare favorably with filling pressures determined experimentally with many materials.[12]

Based on an analysis of eight sets of tests by various experimenters, Homes[19] obtained good results by using the Janssen formula with $\phi' = 0.67\phi$ and $K = 0.45$ (Table 5-4).

Pressures during emptying. Homes obtained good results for emptying pressures in the tests mentioned by using Janssen's formula with $\phi' = 0.55\phi$ and $K = 0.9$ for the upper segment of the bin of height $H - 1.3D$ and assuming pressures in the remaining segment to vary linearly from the emptying value $1.3D$ above the outlet to the filling value at the outlet.

The Soviet Silo Code CH-65 specifies coefficients by which Janssen pressures computed with K from Eq. (5-2a) are multiplied to determine emptying pressures.[20,21]

ACI 313-77 Standard "Recommended Practice for Design and Construction of Concrete Bins, Silos and Bunkers for Storing Granular Materials" lists the coefficients C_d, given in Table 5-2, by which Janssen pressures computed with K from Eq. (5-2a) or Reimbert pressures (Sec. 5-7) are multiplied to determine emptying pressures.[22] The differences in the values of C_d for the two tends to bring the results into closer agreement, although the Reimbert pressures will still be larger than the Janssen lateral pressures in the upper portions of shallow bins and smaller in the lower portions of deep bins.[23]

German standards. DIN 1055[24] specifies values of K and the angle of wall friction ϕ' to be used in the Janssen formula to determine pressures during filling and emptying (Table 5-3). Values of ϕ' for material with diameters

TABLE 5-2. RECOMMENDED MINIMUM OVERPRESSURE COEFFICIENTS C_d [a]

H/D or H/B	Janssen					Reimbert				
	$\leqslant 2$	$=3$	$\geqslant 4$	$=4$[b]	$\geqslant 5$[b]	$\leqslant 2$	$=3$	$\geqslant 4$	4[b]	$\geqslant 5$[b]
Zone 1[c]	1.35	1.45	1.50	1.60	1.65	1.10	1.20	1.25	1.30	1.35
Zone 2[d]	1.45	1.55	1.60	1.70	1.75	1.20	1.30	1.35	1.40	1.50
Zone 3[d]	1.55	1.65	1.75	1.80	1.90	1.45	1.55	1.60	1.70	1.75
Zone 4[d]	1.65	1.75	1.85	1.90	2.00	1.65	1.75	1.85	1.90	2.00
Zone 5[d]	1.65	1.75	1.85	1.90	2.00	1.65	1.75	1.85	1.90	2.00
Hopper[e]										

[a]Adapted from Ref. 22.

[b]Use these columns for powdery, cohesive materials if emptying is pneumatic.

[c]Upper zone: height = $H_1 = D \tan \phi, B \tan \phi, L \tan \phi$.

[d]These zones $(H - H_1)/4$ high.

[e]Use zone 5 pressure uniform throughout depth or reduce linearly to filling pressure at outlet.

1. H = height above hopper, D = diameter, B = short side of rectangle, L = long side of rectangle.
2. Values of C_d are for bottom of each zone.
3. Interpolate to find values of C_d for H/D or H/B between those in table.
4. For $H_1 < H \leqslant 2H_1$, use C_d for zone 2 for entire depth H.
5. Values of C_d may be too small for mass-flow bins and in the region of a flow-correcting insert.

TABLE 5-3. COEFFICIENTS FROM DIN 1055

Condition	Angle of wall friction ϕ': Granular material, av. dia. > 0.2 mm	Angle of wall friction ϕ': Powder material, av. dia. < 0.06 mm	K
Filling	0.75ϕ	ϕ	0.5
Emptying	0.60ϕ	ϕ	1

between the limiting values in Table 5-3 are to be determined by linear interpolation. The emptying pressures p_h and p_v and the corresponding wall friction q must be multiplied by a coefficient c_1 given by

$$c_1 = 1 + 0.2\left(c_2 + \frac{eC}{1.5A}\right) \tag{5-10}$$

where c_2 = 1 for organic materials
c_2 = 0 for inorganic materials
e = eccentricity of outlet, measured from centroid of cross section
C = perimeter of bin
A = cross-sectional area of bin

except for sugar, for which $c_1 = 1$. Also, values of c_1 by this equation must be multiplied by 1.3 for corn, because European experience shows that corn may contain substantial amounts of dust and broken kernels, which probably results from a number of loadings and unloadings in shipment from foreign sources. This would be unlikely for corn shipped from the field to storage.

Lateral pressures in a bin with an eccentric outlet are not distributed uniformly on the perimeter, and in a circular cylinder this produces bending of the wall in addition to hoop tension. According to Ref. 25, the uniformly distributed pressure $c_1 p_h$ is large enough to compensate for this effect. The effects of eccentric flow are discussed further in Sec. 8-4.

Test results. Table 5-4 gives the averages of the ratios of calculated filling pressures p_h to measured values in eight sets of tests in various experiments.[19] Five of the eight sets of tests were conducted on grain silos ranging from 57 to 213 ft high and with aspect ratios H/D from 3.5 to 7. The other three were on small-scale laboratory models, two with sand and one with grain. Calculated emptying pressures were also compared with test values, but only two comparisons were with standards, one with DIN 1055 before the 1977 supplementary provisions were adopted [Eq. (5-10)] and the other with the 1960 Soviet standard, also since revised. Therefore, the results are not given here.

5-6. Noncircular Cross Sections

According to the assumptions on which the Janssen formula is based, bin pressures are independent of the cross-sectional shape of the bin except as it determines the hydraulic radius. However, tests by Reimbert and Reimbert[15] showed that the lateral pressure at a given depth on the narrow side of a rectangular bin with sides of length L and width B is the same as the pressure at the same depth in a square bin of side B. Therefore, the pressure on the nar-

TABLE 5-4. RATIO OF CALCULATED LATERAL FILLING PRESSURES TO MEASURED VALUES[a]

	Depth below top level				
Method	$0.2H$	$0.4H$	$0.6H$	$0.8H$	H
Janssen[b]	0.80	0.82	0.84	0.87	0.89
DIN 1055	1.02	0.97	0.96	0.93	0.90
Reimbert	0.99	0.89	0.86	0.85	0.88
Homes	1.01	1.01	1.02	1.02	1.01

[a]From Ref. 19.

[b]With K from Eq. (5-2a).

row side is determined by a hydraulic radius $B/4$ rather than the hydraulic radius of the rectangle. The average pressure on the long side is correspondingly smaller and can be shown to be equal to the pressure computed for a square bin of side $B' = B(2 - B/L)$.

Lateral pressures for a bin of regular polygonal cross section can be computed for the inscribed circular cylinder since the hydraulic radius of the polygon is $D/4$, where D is the diameter of the inscribed circle.

5-7. Reimbert Formulas

The following filling-pressure formulas, which are based on an analysis of the results of a number of experiments on models of deep bins, were developed by Reimbert and Reimbert:

$$p_v = \gamma \left[y \left(1 + \frac{y}{y_0} \right)^{-1} + \frac{h_c}{3} \right] \tag{5-11a}$$

$$p_h = \frac{\gamma R}{\mu'} \left[1 - \left(1 + \frac{y}{y_0} \right)^{-2} \right] \tag{5-11b}$$

where h_c = height of cone of surcharge
y = distance below base of cone of discharge
$y_0 = (R/\mu' K) - (h_c/3)$ computed for the value of K by Eq. (5–2a)

For bins of regular polygonal cross section the value of y_0 is to be based on $R = C/4\pi$, where C is the perimeter of the polygon, but the actual hydraulic radius is to be used in Eq. (5–11b).

The Reimberts note that their formulas are intended only for filling pressures. They discuss the overpressures during discharge but give no overpressure coefficients by which the charging pressures may be multiplied except to quote various authorities who report values ranging from 2 to 4.

5-8. Pressures from Dust-Like Materials

During rapid filling of a bin with dust-like material, such as wheat flour, cement, and talc, the material mixes with air and forms a mixture which is approximately hydrostatic in character. The maximum filling pressures from such a mix can be calculated by

$$p_h = \gamma v t_s \tag{5-12}$$

$$p_v = 1.2 \gamma v t_s \tag{5-13}$$

where v = filling velocity (vertical rise of material surface per hour)
t_s = setting time of material

These pressures increase linearly from zero at the surface to the maximum

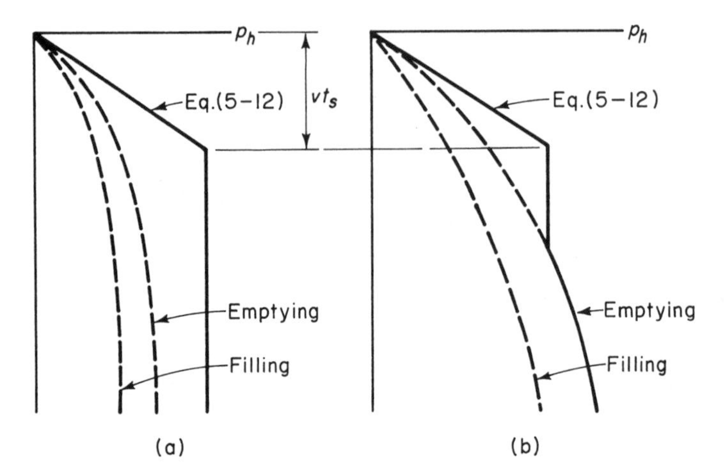

Figure 5–10 Wall pressures from dust-like materials: (a) bin with small hydraulic radius, (b) bin with large hydraulic radius.

value, which occurs at a depth vt_s below the surface. The pressure remains constant at the maximum value below the depth vt_s. The following values of t_s are given in Ref. 26: wheat flour, 0.07 hr; cement, 0.11 hr; and talc, 0.18 hr. Values for other materials must be estimated or determined by test.

The bin must be designed for conventional filling pressures, which develop after the material has settled, and for emptying pressures as well as the rapid-filling pressure. The conventional pressures are usually less than the rapid-filling pressures in bins with a small hydraulic radius but may exceed them in the lower part of a bin with a large hydraulic radius (Fig. 5–10).

Pneumatic emptying. Pneumatic devices are sometimes used to facilitate emptying to prevent a buildup of dust-like material and a subsequent closing of the outlet. If the air pressure p_a exceeds the filling and emptying pressures, it must be used in designing the lower portion of the bin. This pressure must be accounted for over a height Δh above the level at which the pressure is applied given by

$$\Delta h = \frac{1.3p_a}{\gamma} \tag{5–14}$$

The pressure variation can be assumed to be linear (Fig. 5–11).

Homogenizing bins are bins in which dust-like materials are thoroughly mixed with air. The hydrostatic pressures generated during mixing can be computed by

$$p = q = 0.6\gamma y \tag{5–15}$$

The bin design must be checked for these pressures as well as for the conventional filling and emptying pressures.

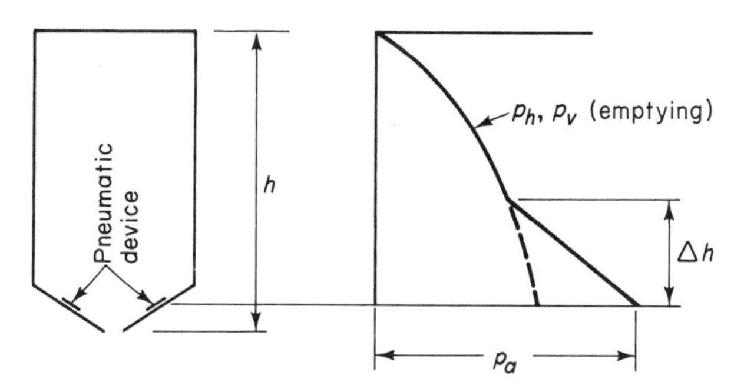

Figure 5-11 Pressure from pneumatic equipment during emptying of dust-like materials.

5-9. Shallow Bins

Coulomb's formula for earth pressure on retaining walls is considered by some to be appropriate for shallow bins. The formula is based on the equilibrium of a wedge of cohesionless material that mobilizes friction on the wall by sliding along it. A second face of the wedge is a failure surface in the soil itself, while the third is at the free surface of the material. For a level fill against a vertical surface the formula is

$$p_h = \gamma y \left[\frac{\cos \phi}{1 + \sqrt{\sin(\phi + \phi') \sin \phi / \cos \phi'}} \right]^2 \tag{5-16}$$

If ϕ is assumed equal to ϕ', Eq. (5-16) gives

$$p_h = \gamma y \left(\frac{\cos \phi}{1 + \sqrt{2} \sin \phi} \right)^2 \tag{5-17}$$

If wall friction is neglected ($\phi' = 0$), Eq. (5-16) reduces to the Rankine active-pressure value; i.e., the pressure coefficient becomes $K = (1 - \sin \phi)/(1 + \sin \phi)$.

Equilibrium of the wedge of material which is the basis for the Coulomb theory is not a correct representation of the equilibrium of material in a circular cylinder. However, it may be appropriate for a long, narrow rectangular bunker, and Reimbert and Reimbert suggest that, for these cases, the pressure be computed by a retaining-wall theory and a bin theory and that the larger of the two be used for design.[15]

The Rankine active pressure is suitable for shallow bins. Since it neglects friction on the wall, it always exceeds both the Coulomb and the Janssen pressures. Furthermore, since flow pressures in shallow bins are not much larger than filling pressures, the Rankine theory can be considered to provide some margin for the dynamic effects of flow in shallow bins. Therefore, there would seem to be no good reason to use the Coulomb theory (except, possibly, for the rectangular bunker, as already noted) in the form of either Eq. (5-16) or (5-17).

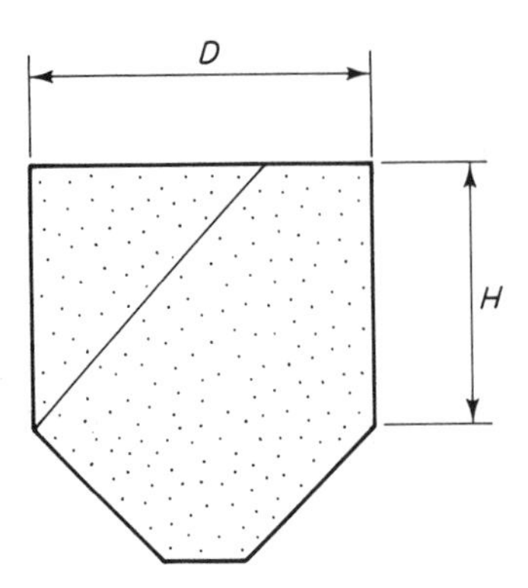

Figure 5-12

According to Jenike, Janssen pressures are adequate for the design of bins for which the effective transition is not lower than one bin diameter below the top. This is generally the case in funnel-flow bins whose total height does not exceed two diameters. Therefore, shallow bins can be designed safely for Janssen or Reimbert pressures, or, somewhat more conservatively, Rankine active pressures.

Various formulas to define the limiting height of a shallow bin have been proposed. According to one of the assumptions of Coulomb's theory, the failure surface in the material must intersect the top of the fill (Fig. 5-12). This condition is satisfied if

$$\frac{H}{D} \leqslant \tan\phi + \sqrt{\frac{\tan\phi(1+\tan^2\phi)}{\tan\phi + \tan\phi'}} \tag{5-18}$$

This equation gives H/D limits of 1.12 and 2.30 for $\phi = \phi' = 20°$ and $\phi = \phi' = 50°$, respectively. Other limiting values that have been proposed are

$$H \leqslant 1.5D \tag{5-19a}$$

$$H \leqslant 1.5\sqrt{A} \tag{5-19b}$$

where A is the cross-sectional area of the bin.

5-10. Vertical Compression in Cylinder Wall

The friction between the solid and the wall of a bin produces a vertical compressive force P in the wall. At any horizontal section through the bin this force is equal to the weight of solid above the section minus the resultant vertical pressure at the section. Thus, $P = \gamma Ay - p_v A$. Dividing P by the circumference gives the force F_c per unit of circumference:

$$F_c = R(\gamma y - p_v) \tag{5-20}$$

where R is the hydraulic radius.

Vertical compression can also be evaluated by integrating the wall friction above the section:

$$F_c = \int_0^y \mu' p_h \, dy \tag{5-21}$$

In the investigation of vertical buckling, DIN 1055 requires the wall friction to be multiplied by the coefficient c_l of Eq. (5-10). Therefore, to compute F_c by Eq. (5-20), p_v rather than $c_1 p_v$ should be used and the resulting value of F_c multiplied by c_1.

Vertical compression by ACI 313-77 is obtained by multiplying the static value of F_c by C_d from Table 5-2. Since the Janssen formula gives larger values of p_v than the Reimbert formula, F_c will be smaller for the Janssen values of p_v. For this reason, ACI 313 requires that p_v in Eq. (5-20) be replaced by $0.8p_v$ for use with the Janssen formula, the implied assumption being that the Reimbert values of p_v are more realistic than the Janssen values. However, Fig. 5-14 suggests that the opposite may be true, so that the ACI results may be too large. A comparison of vertical compression by DIN and ACI is made in Example 5-2, and the resulting values in Table 5-7 suggest that the ACI results may indeed be excessive.

Example 5-1

Determine the filling pressures for the grain silo of Fig. 5-13. This silo was used in one of the eight sets of tests for which the average pressures are compared with predicted values in Table 5-6. The investigation is reported in Ref. 28. The bulk density of the wheat used in the tests ranged from 0.80 to 0.810 ton/m^3, which gives an average of 50.3 pcf. Values of ϕ and ϕ' were not given, but from the reported pressures calculated by the Janssen formulas Homes[19] determined $\phi = \phi' = 22.5°$.

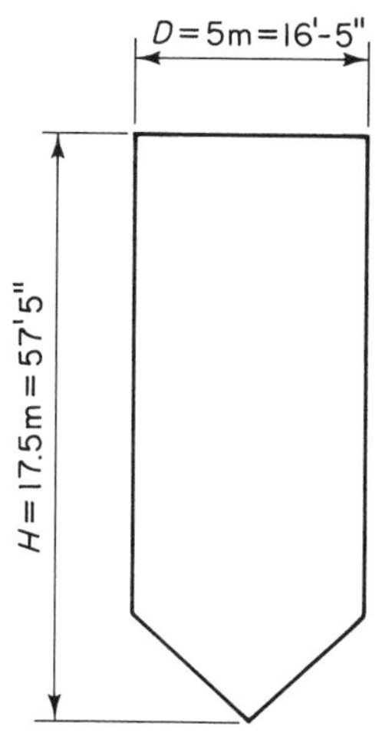

Figure 5-13 Bin for Example 5-1.

TABLE 5-5. JANSSEN PRESSURES (psf) (EXAMPLE 5-1)

y/H	y	$0.045y$	$f(0.045y)$	p_v	p_h
0	0	0	0	0	0
0.2	11.48	0.517	0.403	448	202
0.4	22.97	1.03	0.642	713	321
0.6	34.45	1.55	0.787	874	393
0.8	45.94	2.07	0.874	971	437
1.0	57.42	2.58	0.924	1027	462

(a) Janssen:

$$\gamma = 50 \text{ pcf} \qquad \phi = \phi' = 22.5^\circ \qquad D = 16.42 \text{ ft} \qquad H = 57.42 \text{ ft}$$

$$\mu' = \tan \phi' = 0.414 \qquad K = \frac{1 - \sin \phi}{1 + \sin \phi} = 0.45$$

$$R = \frac{16.42}{4} = 4.10 \qquad \frac{\mu' K}{R} = \frac{0.414 \times 0.45}{4.10} = 0.045$$

From Eqs. (5-8),

$$p_v = \frac{\gamma R}{\mu' K} f\left(\frac{\mu' K y}{R}\right) = \frac{50}{0.045} f(0.045y) = 1110 f(0.045y)$$

$$p_h = K p_v = 0.45 p_v$$

The pressures computed from these equations are given in Table 5-5.

(b) Pieper, Eq. (5-9):

$$K = 1 - \sin \phi = 0.62 \qquad \frac{\mu' K}{R} = \frac{0.414 \times 0.62}{4.10} = 0.063$$

$$p_v = \frac{50}{0.063} f(0.063y) = 794 f(0.063y)$$

$$p_h = 0.62 p_v$$

Since the tabular arrangement of the computation is the same as in part (a), it is omitted here and in parts (c), (d), and (e).

(c) DIN, Table 5-3:

$$K = 0.5 \qquad \mu' = \tan 0.75 \times 22.5^\circ = 0.30 \qquad \frac{\mu' K}{R} = \frac{0.30 \times 0.5}{4.10} = 0.037$$

$$p_v = \frac{50}{0.037} f(0.037y) = 1350 f(0.037y)$$

$$p_h = 0.5 p_v$$

(d) Homes, Sec. 5-5:

$$K = 0.45 \qquad \mu' = \tan \tfrac{2}{3} \times 22.5° = 0.27 \qquad \frac{\mu' K}{R} = \frac{0.27 \times 0.45}{4.10} = 0.030$$

$$p_v = \frac{50}{0.030} f(0.030y) = 1670 f(0.030y)$$

(e) Reimbert, Eqs. (5-11):

$$K = \frac{1 - \sin\phi}{1 + \sin\phi} = 0.45 \qquad \mu' = \tan 22.5° = 0.414$$

Take $h_c = 0$; then

$$y_0 = \frac{R}{\mu' K} = \frac{4.10}{0.414 \times 0.45} = 22.0$$

$$p_v = \gamma\left(\frac{y}{1 + y/y_0}\right) = 50\left(\frac{y}{1 + y/22}\right) = 2870\,\frac{y/H}{1 + 2.61y/H}$$

$$p_h = \frac{\gamma R}{\mu'}\left[1 - \frac{1}{(1 + y/y_0)^2}\right] = 495\left[1 - \frac{1}{(1 + 2.61y/H)^2}\right]$$

Values of p_h and p_v according to the various preceding formulas are shown in Table 5-6. The differences in horizontal pressures are not great, the largest being in the lower part of the bin where Homes' values are about 33% larger than Janssen's at the base and 23% larger at $y/H = 0.6$. Differences in vertical pressures are much larger, the largest being between the Reimbert values and those by Homes.

Filling pressures are compared in Fig. 5-14 with the experimental values reported in Ref. 28. It will be noted that the Janssen formula gives a good prediction of the lateral pressures, although the DIN-Homes values are closer. The vertical pressures according to Homes are in better agreement with the tests than those of the other formulas.

TABLE 5-6. COMPARISON OF FILLING PRESSURES (psf) (EXAMPLE 5-1)

Depth	Janssen		Reimbert		DIN		Pieper		Homes	
y/H	p_h	p_v	p_h	p_v	p_h	p_v	p_h	p_v	p_h	p_v
0	0	0	0	0	0	0	0	0	0	0
0.2	202	448	281	377	233	466	254	410	219	487
0.4	321	713	377	562	386	772	377	608	374	831
0.6	393	873	420	671	486	972	436	703	484	1076
0.8	437	971	443	743	551	1102	465	750	562	1249
1	462	1027	457	795	594	1188	479	773	617	1371

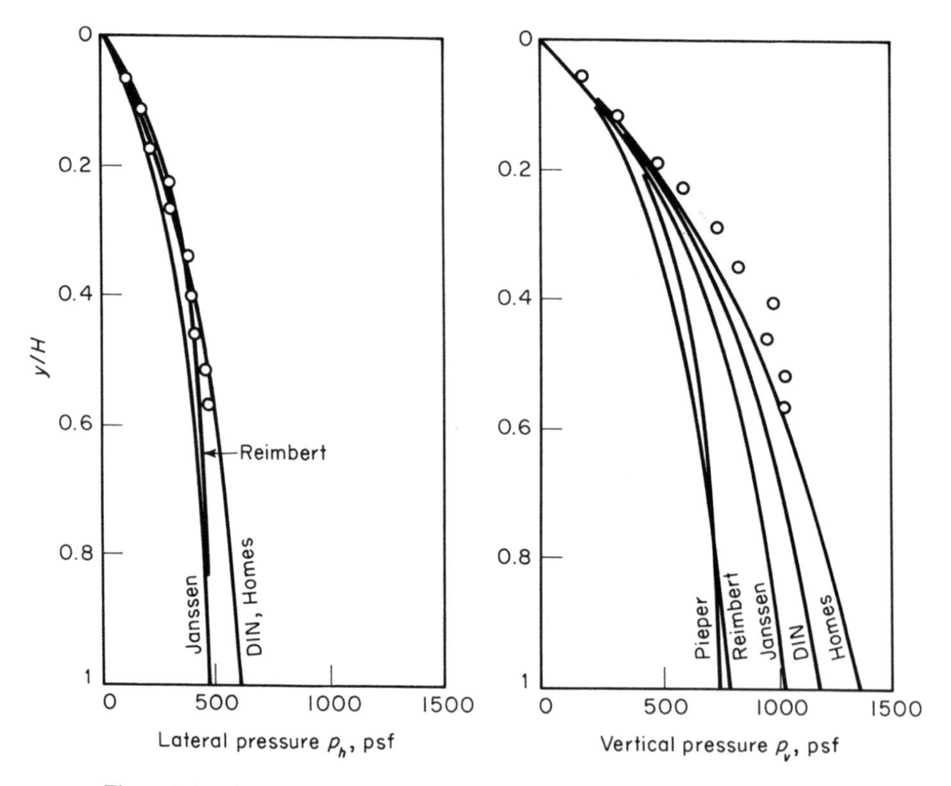

Figure 5-14 Comparison of theoretical filling pressures with experimental values for the bin of Fig. 5-13. (From Ref. 19.)

Example 5-2

Determine the emptying pressures and the corresponding vertical compression in the wall for the bin of Example 5-1.

(a) DIN, Table 5-3:

$$K = 1 \qquad \mu' = \tan 0.6 \times 22.5^\circ = 0.24 \qquad \frac{\mu' K}{R} = \frac{0.24 \times 1}{4.10} = 0.059$$

$$c_1 = 1.2 \qquad \text{[Eq. (5-10)]}$$

$$p_v = 1.2 \frac{50}{0.059} f(0.059y) = 1016 f(0.059y)$$

$$p_h = p_v$$

$$F_c = 1.2 \frac{D}{A} \left(\gamma y - \frac{p_v}{1.2} \right) = 4.92 \left(2870 \frac{y}{H} - \frac{p_v}{1.2} \right)$$

(b) Homes:

$K = 0.9 \qquad \mu' = \tan 0.55 \times 22.5° = 0.22 \qquad \frac{\mu' K}{R} = \frac{0.22 \times 0.9}{4.10} = 0.048$

$$p_v = \frac{50}{0.048} f(0.048y) = 1042 f(0.048y)$$

$$p_h = 0.9 p_v$$

$$F_c = \frac{D}{4}(\gamma y - p_v) = 4.1\left(2870 \frac{y}{H} - p_v\right)$$

(c) ACI 313-77: Either Janssen's formula or Reimberts' formula may be used to compute p_v, except that the Janssen value must be multiplied by 0.8 as noted in Sec. 5-10. The resulting values of F_c are then multiplied by the appropriate values of C_d from Table 5-2.

Values of p_h, p_v, and F_c according to Homes, DIN, and ACI are given in Table 5-7 and pressures are compared in Fig. 5-15 with experimental values reported in Ref. 28. It will be noted that the ACI values of p_v are in better agreement with the test results than the DIN values. Since vertical compression in the wall decreases with increase in p_v, this suggests that F_c should be smaller by ACI than by DIN. However, Table 5-7 shows that this is not the case and that F_c is as much as 50% larger by ACI than by DIN. This is because of the way the factors c_1 and C_d are applied. Thus, multiplying both p_v and F_c by C_d gives values which are not in equilibrium with the weight of the material above the cross section under consideration; equilibrium would require the weight γy in Eq. (5-20) also to be multiplied by C_d. This is also true for the DIN values, but since c_1 for bins with concentric outlets is smaller than C_d, and is in fact unity for inorganic materials, the effect is considerably larger with the ACI values.

TABLE 5-7. COMPARISON OF EMPTYING PRESSURES (psf) AND VERTICAL COMPRESSION IN WALL (plf) (EXAMPLE 5-2)

Depth,	Homes			DIN			ACI 313-77[a]		
y/H	p_h	p_v	F_c	p_h	p_v	F_c	p_h	p_v	F_c
0	0	0	0	0	0	0	0	0	0
0.2	398	442	540	498	498	782	354	475	1,021
0.4	627	697	1850	755	755	2,553	524	781	3,347
0.6	759	843	3610	886	886	4,871	680	1087	6,998
0.8	834	927	5620	949	949	7,405	797	1339	11,482
1.0	617	686	8954	982	982	10,094	823	1431	15,350

[a]Values by Reimbert formulas.

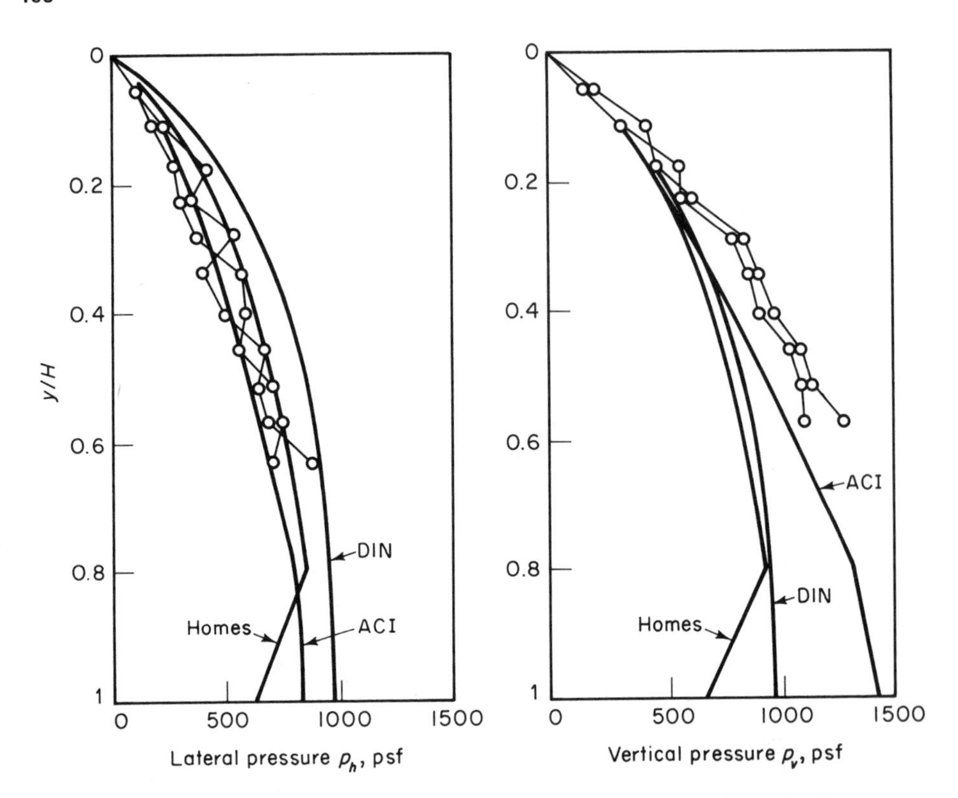

Figure 5-15 Comparison of theoretical flow pressures with experimental values for the bin of Fig. 5-13. (From Ref. 19.)

5-11. Funnel-Flow Bins

Jenike and Johanson developed a procedure for computing the peak pressure at the switch in concentric funnel flow and prepared charts of peak-pressure envelopes.[10,13] However, pressures given by standard specifications are generally adequate for the design of funnel-flow bins with concentric charging and withdrawal.[27] Furthermore, the allowance for eccentricity of outlet in the supplementary provisions of DIN 1055 is claimed to be adequate to provide for the circumferential bending of the cylinder caused by the nonuniform radial pressures.[25]

5-12. Pressures on Hopper Wall

The forces on a triangular element at the wall of a hopper are shown in Fig. 5-16. Projecting these forces on the lines of action of q' and p' gives

$$p' = p_v \sin^2 \theta + p_h \cos^2 \theta + q \sin 2\theta \tag{5-22}$$

$$q' = \tfrac{1}{2}(p_v - p_h) \sin 2\theta + q \cos 2\theta \tag{5-23}$$

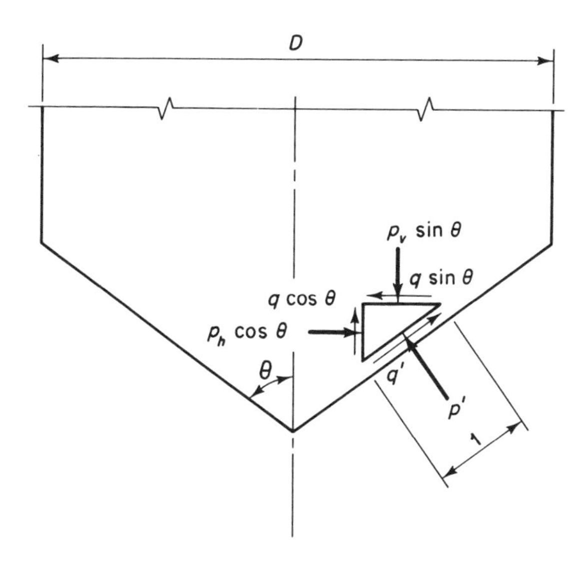

Figure 5-16 Pressures on hopper wall.

If the pressure field is assumed to be a Rankine field, $q = 0$, and these equations reduce to

$$p' = p_v \sin^2 \theta + p_h \cos^2 \theta \qquad (5\text{-}24)$$

$$q' = \tfrac{1}{2}(p_v - p_h) \sin 2\theta \qquad (5\text{-}25)$$

Equations (5-24) and (5-25) are commonly used, with p_v and p_h computed by the Janssen or other formulas. However, it should be noted that formulas based on the equilibrium of forces on a horizontal section of the solid in a cylinder are not correct for a hopper because of the sloping walls. Furthermore, it has been shown experimentally that hopper-wall pressures may be considerably larger than the pressures given by these equations.

The following formulas by Pieper[26] are based on the literature and on the results of tests with quartz sand and wheat on a model bin with wall slopes from 41° to 75° and with various wall roughnesses. In general, pressures on the walls of pyramidal and wedge hoppers can be calculated with sufficient accuracy by these formulas. They can also be used for conical hoppers by computing the pressures on a pyramidal hopper of the same height and perimeter. The pressure is computed in two parts:

1. Pressure from the material in the hopper (Fig. 5-17a):

$$p' = \frac{0.6\gamma KD \cos^2 \theta}{\sqrt{\mu'}} \qquad (5\text{-}26)$$

$$q' = \frac{p'}{2} \qquad (5\text{-}27)$$

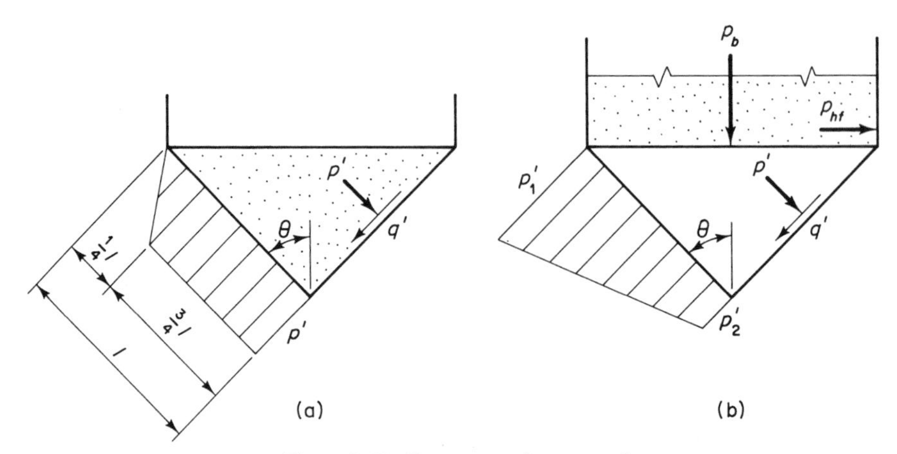

Figure 5-17 Pressures on hopper wall.

2. Pressures from the material above the hopper (Fig. 5-17b):

$$p'_1 = \frac{p_b \sin^2 \theta + p_{hf} \cos^2 \theta}{\sqrt{\mu'}} \tag{5-28a}$$

$$p'_2 = p_b \sin^2 \theta \tag{5-28b}$$

$$q' = \frac{p'}{2} \tag{5-29}$$

in which p_{hf} is the horizontal filling pressure and p_b is the smaller of γH or $c_b p_{vf}$, where p_{vf} is the vertical filling pressure at the depth H and c_b is a coefficient which depends on the material in the bin. For the materials listed in Ref. 26, c_b ranges from 1 to 1.7. Of course, it is on the safe side to use $p_b = \gamma H$.

DIN. The 1977 supplementary provisions[24] prescribe a hopper pressure p' twice the value by Eq. (5-24), computed for the filling values of p_v and p_h at the transition and distributed uniformly from the transition to the vertex. The increase in pressure given by Eq. (5-10) is not required in computing p_v and p_h; that is, $c_1 = 1$ for both organic and inorganic materials and for eccentric as well as concentric outlets.

ACI 313-77. Hopper pressure p' is computed by Eq. (5-24), using the Janssen or Reimbert values of p_h and p_v, and multiplied by C_d (Table 5-2) to obtain emptying pressures. The pressure can be taken uniform with depth at the value for the transition, or it can be assumed to vary linearly from the emptying value at the transition to the filling value at the outlet.

Mass-flow hoppers. Pressures in mass-flow hoppers are discussed in Secs. 5-13 and 5-14.

5-13. Walker Formulas for Mass Flow

The following formulas for mass-flow bins were derived by Walker.[29,30]

Cylinder filling pressures. Walker assumes that the initial pressure field in the cylinder is an active Rankine field but with the pressure coefficient given by the effective angle of friction δ [Eq. (5-3a)] rather than the internal angle of friction ϕ. Therefore, the filling pressures are

$$p_v = \gamma y \tag{5-30a}$$

$$p_h = K p_v \tag{5-30b}$$

where y = depth of solid above section
$K = (1 - \sin \delta)/(1 + \sin \delta)$

Since there is no shear on the walls of the bin according to these equations, they can be expected to yield pressures that are too large except for shallow bins (Sec. 5-9). This conclusion appears to be supported by tests discussed in Sec. 5-15 and is in agreement with conclusions by others that initial wall pressures are well represented by the Janssen formula.

Hopper filling pressures. The stress field in the hopper is also assumed to be one in which the major principal pressure is vertical (Fig. 5-18a). Two cases are considered.

If the hopper-wall friction is too small to support a Rankine active state

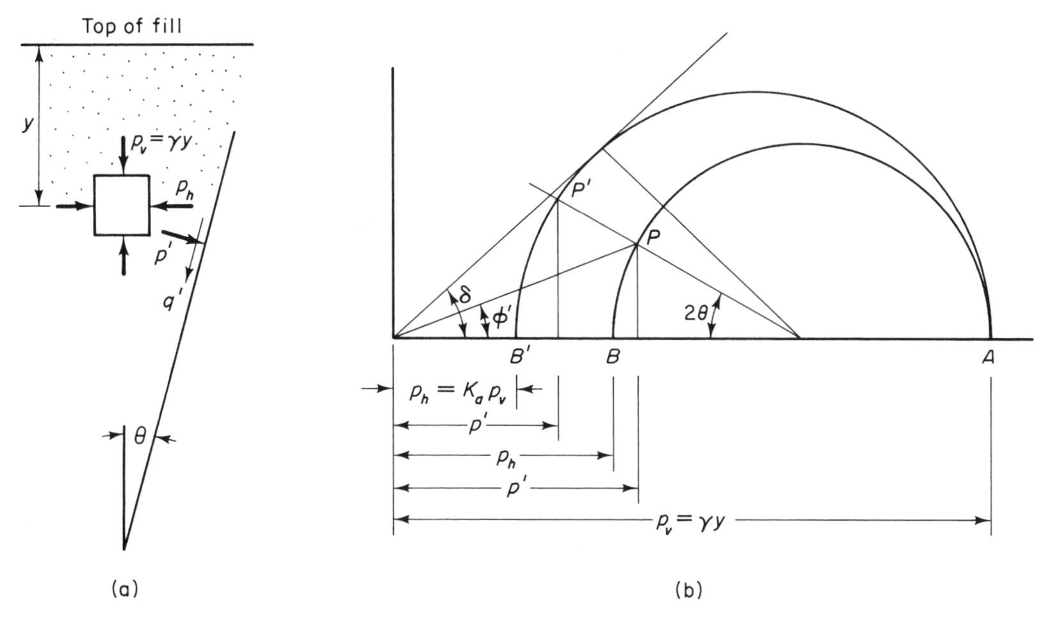

Figure 5-18 Filling pressures in hopper.

in the material, the forces on the hopper wall are given by point P on the Mohr circle APB of Fig. 5-18b. From the geometry of this circle we get

$$p' = \gamma y \frac{\sin 2\theta \cos \phi'}{\sin \phi' + \sin(\phi' + 2\theta)} = \gamma y \frac{\tan \theta}{\tan \theta + \tan \phi'} \tag{5-31a}$$

$$q' = p' \tan \phi' = \mu' p' \tag{5-31b}$$

If the hopper-wall friction is large enough to support a Rankine active state, the forces on the hopper wall are given by point P' of the Mohr circle $AP'B'$ of Fig. 5-18b. For this case,

$$p' = \gamma y \frac{1 - \sin \delta \cos 2\theta}{1 + \sin \delta} \tag{5-32a}$$

$$q' = \gamma y \frac{\sin \delta \sin 2\theta}{1 + \sin \delta} \tag{5-32b}$$

To determine which of Eqs. (5-31) and (5-32) apply, we compute the wall-friction angle $\tan \phi' = q'/p'$ required to develop the active Rankine pressures. Thus, with the values of p' and q' from Eqs. (5-32) we find that Eqs. (5-31) apply whenever the wall-friction angle is such that

$$\tan \phi' \leqslant \frac{\sin \delta \sin 2\theta}{1 - \sin \delta \cos 2\theta} \tag{5-33a}$$

Walker gives this equation in the following simpler form:

$$\sin \delta \geqslant \frac{\sin \phi'}{\sin(\phi' + 2\theta)} \tag{5-33b}$$

Cylinder emptying pressures. Walker assumes that the active Rankine field established during filling is altered during flow by virtue of the wall friction that develops. The major principal pressure shifts from the vertical (Fig. 5-19a), and point P on the Mohr circle tangent to the yield locus determines the wall stresses (Fig. 5-19b). Assuming the vertical pressure p_v to be uniform over a cross section, Walker shows from the equilibrium of a horizontal slice of solid [Eq. (5-5)] and the geometry of the Mohr circle that the pressures p_v and p_h are given by the Janssen formulas [Eqs. (5-8)], with $\mu' K$ given by

$$\mu' K = \frac{\sin \delta \sin \epsilon_1}{1 - \sin \delta \cos \epsilon_1} \tag{5-34}$$

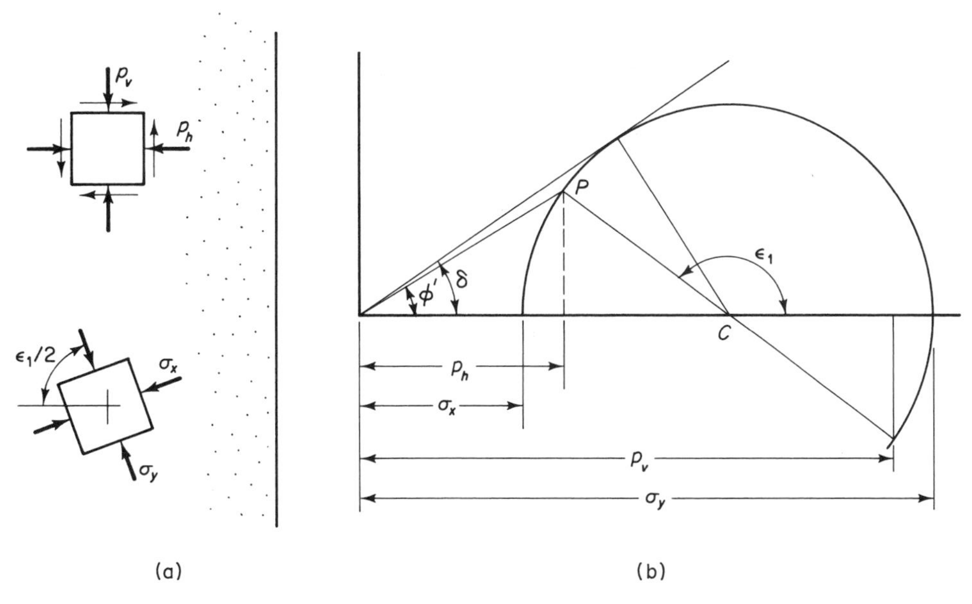

Figure 5-19 Emptying pressures in cylinder.

where

$$\epsilon_1 = \frac{\pi}{2} + \phi' + \cos^{-1} \frac{\sin \phi'}{\sin \delta}$$

To investigate the effect of the variation of p_v over a cross section, Walker follows Nanninga's assumption that the shear varies linearly from the value $\mu' p_h$ at the wall to zero at the axis of the cylinder (Sec. 5-3). His results are identical with those obtained from the Janssen formula with Nanninga's value of K [Eq. (5-4)]. He also shows that, except for rough-walled bins, these pressures differ only slightly from the Janssen pressure based on $K = (1 - \sin \delta)/(1 + \sin \delta)$. For example, if $\delta = 50°$, the Walker-Nanninga pressures are less than 1% more than the Janssen pressures for wall-friction angles up to 40°. However, at $\phi' = 45°$ they are 10% larger, and at $\phi' = 50°$, 20% larger.

Hopper emptying pressures. The material is assumed to be in a yield state and to promote the largest possible hopper-wall reactions because it is virtually wedging itself into the hopper. These two conditions are represented by point P on the Mohr circle tangent to the yield locus, where P is the intersection of the circle with a line OP at the angle ϕ' with the normal-stress axis (Fig. 5-20a). Since the hopper wall pressure p' makes the angle θ with the horizontal pressure p_h (Fig. 5-20b), a line CQ at the angle 2θ with CP determines p_h. The geometry of this figure is similar to that of Fig. 5-19b and

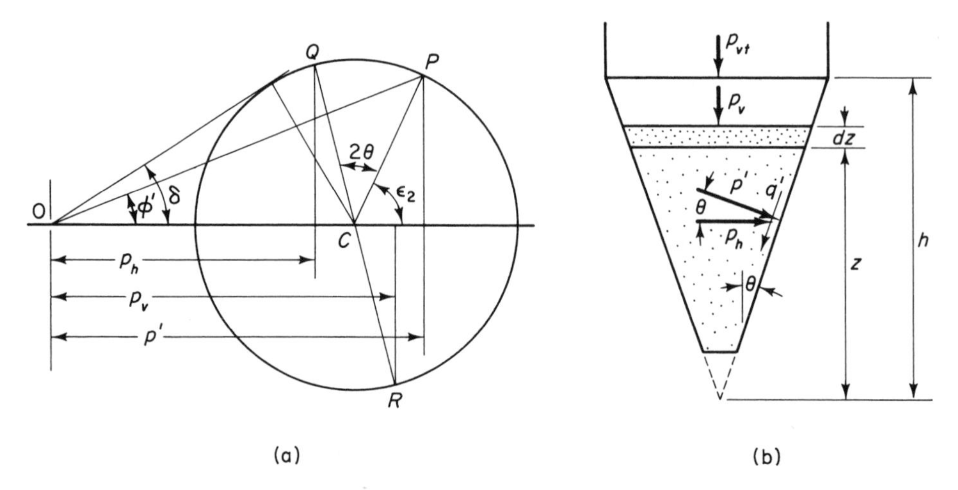

Figure 5-20 Emptying pressures in hopper.

yields the following formula:

$$\frac{p'}{p_v} = \frac{1 + \sin \delta \cos \epsilon_2}{1 - \sin \delta \cos(2\theta + \epsilon_2)} \tag{5-35}$$

where

$$\epsilon_2 = \phi' + \sin^{-1} \frac{\sin \phi'}{\sin \delta}$$

Considering equilibrium of vertical forces on a horizontal slice of material dz thick at the distance z above the apex (Fig. 5-20b) and assuming the vertical pressure to be uniformly distributed, we obtain the following equation:

$$p_v = \frac{\gamma h}{K_w - 1} \frac{z}{h} + \left(p_{vt} - \frac{\gamma h}{K_w - 1}\right)\left(\frac{z}{h}\right)^{K_w} \tag{5-36}$$

where $K_w = \dfrac{1 + m}{\tan \theta} \dfrac{\sin \delta \sin(2\theta + \epsilon_2)}{1 - \sin \delta \cos(2\theta + \epsilon_2)}$

m = 0 for wedge-shaped hoppers
m = 1 for conical or pyramidal hoppers
z = distance above apex
h = distance from apex to transition
p_{vt} = vertical pressure at transition

Equation (5-36) also applies to hoppers with no surcharge by setting $p_{vt} = 0$ and defining h as the depth of material in the hopper.

Example 5-3

Compute the pressures for the bunker of Fig. 5-21. The cylinder is circular and the hopper conical. This bunker was used in tests discussed in Sec. 5-15. The material is iron ore with $\gamma = 112$ pcf, $\delta = 40^\circ$, and $\phi' = 28^\circ$.

Cylinder filling pressure:

$$K = \frac{1 - \sin 40^\circ}{1 + \sin 40^\circ} = 0.217$$

$$p_v = \gamma y = 112y = 112 \times 9.84 = 1102 \text{ psf at transition}$$

$$p_h = Kp_v = 0.217 \times 1102 = 239 \text{ psf at transition}$$

Hopper filling pressure

Use Eq. (5-33b) to determine which of Eqs. (5-31) and (5-32) apply:

$$\frac{\sin \phi'}{\sin(\phi' + 2\theta)} = \frac{\sin 28^\circ}{\sin(28^\circ + 22^\circ)} = 0.613$$

$$\sin \delta = \sin 40^\circ = 0.643 > 0.613 \qquad \text{use Eqs. (5-31)}$$

$$p' = 112y \frac{\tan 11^\circ}{\tan 11^\circ + \tan 28^\circ} = 30.0y$$

$$= 30 \times 9.84 = 295 \text{ psf at transition}$$

$$= 30 \times 16.31 = 489 \text{ psf at outlet}$$

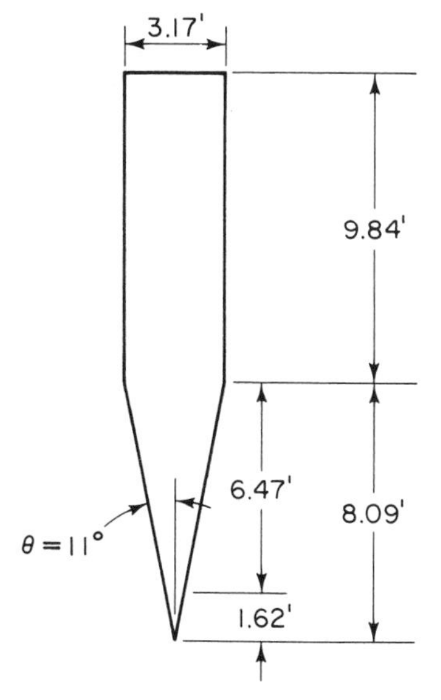

Figure 5-21 Bunker for Example 5-3.

Cylinder flow pressure

From Eq. (5-34),

$$\epsilon_1 = \frac{\pi}{2} + 28^\circ + \cos^{-1}\frac{\sin 28^\circ}{\sin 40^\circ} = 161^\circ$$

$$\mu' K = \frac{\sin 40^\circ \sin 161^\circ}{1 - \sin 40^\circ \cos 161^\circ} = 0.130 \qquad \frac{\mu' K}{R} = \frac{0.130}{3.17/4} = 0.164$$

$$\mu' = \tan 28^\circ = 0.532 \qquad K = \frac{\mu' K}{\mu'} = \frac{0.130}{0.532} = 0.244$$

From Eq. (5-8a),

$$p_v = \frac{\gamma R}{\mu' K} f\left(\frac{\mu' K y}{R}\right) = 683 f(0.164y)$$

Values of p_v are given in Table 5-8 with $f(0.164y)$ from Table 5-1. Values of p_h are given by $p_h = K p_v$ and of F_c by Eq. (5-20).

Hopper flow pressure

From Eq. (5-35),

$$\epsilon_2 = 28^\circ + \sin^{-1}\frac{\sin 28^\circ}{\sin 40^\circ} = 74.9^\circ$$

$$\frac{p'}{p_v} = \frac{1 + \sin 40^\circ \cos 74.9^\circ}{1 - \sin 40^\circ \cos(22^\circ + 74.9^\circ)} = 1.08$$

Then from Eq. (5-36),

$$K_w = \frac{2}{\tan 11^\circ} \frac{\sin 40^\circ \sin(22^\circ + 74.9^\circ)}{1 - \sin 40^\circ \cos(22^\circ + 74.9^\circ)} = 6.10$$

$$p_v = \frac{112 \times 8.09}{6.10 - 1}\frac{z}{h} + \left(546 - \frac{112 \times 8.09}{6.10 - 1}\right)\left(\frac{z}{h}\right)^{6.10}$$

$$= 178\frac{z}{h} + 368\left(\frac{z}{h}\right)^{6.10}$$

Pressures at the transition ($z = 8.09$ ft), the outlet ($z = 1.62$ ft), and three intermediate points are given in Table 5-9.

The Walker pressures computed in this example are compared with experimental results in Fig. 5-24 and discussed in Sec. 5-15.

5-14. Jenike-Johanson Theory for Mass Flow

The following formulas for the pressures in mass-flow bins are derived in Refs. 10 and 12.

TABLE 5-8. CYLINDER FLOW PRESSURES FOR BUNKER OF FIG. 5-21 (WALKER, EXAMPLE 5-3)

y/H	y (ft)	$0.164y$	$f(0.164y)$	p_v (psf)	$p_h = Kp_v$ (psf)	F_c (plf)
0	0	0	0	0	0	0
0.2	1.97	0.323	0.276	189	46	25
0.4	3.94	0.646	0.476	325	79	92
0.6	5.90	0.968	0.620	423	103	188
0.8	7.87	1.29	0.724	494	120	307
1.0	9.84	1.61	0.800	546	133	441

TABLE 5-9. HOPPER FLOW PRESSURES FOR BUNKER OF FIG. 5-21 (WALKER, EXAMPLE 5-3)

z	z/h	p_v (psf)	$p' = 1.08p_v$ (psf)
8.09	1	546	590
6.47	0.8	237	256
4.85	0.6	123	133
3.24	0.4	73	79
1.62	0.2	36	39

Cylinder filling pressure. The Janssen formula gives a good approximation to the initial pressure distribution in the bin. According to Ref. 10, $K = 0.4$ gives results that compare favorably with experimental values for many materials.

Cylinder flow pressure. Figure 5-22b shows a pressure distribution for the situation in which the material below the switch is expanding vertically in the developing flow channel while the material above is still in its initial (Janssen) state. The resulting pressure peak at the switch is evaluated by minimizing the recoverable strain energy in the flow field below the transition. The solution is given in Refs. 10 and 12 in the form of three simultaneous equations. However, they can be reduced to the following formula for p_h:

$$\frac{p_h}{R\gamma/\mu'} = 1 + \frac{(\beta p_{vj}/\gamma - k)[(1+k)e^{\beta(H-y)} - (1-k)e^{-\beta(H-y)}] - 2(1-k^2)}{(1+k)e^{\beta(H-y)} + (1-k)e^{-\beta(H-y)}} \tag{5-37}$$

where R = hydraulic radius of cylinder
H = height of cylinder
y = distance from top of material in cylinder
p_{vj} = Janssen vertical pressure at level y

and where

$$\left.\begin{aligned} k &= \nu\sqrt{\frac{2}{1-\nu}} = K\sqrt{\frac{2}{1+K}} \\ \beta &= \frac{\mu'}{R}\frac{1}{\sqrt{2(1-\nu)}} = \frac{\mu' K}{Rk} \end{aligned}\right\} \quad \text{circular cylinders}$$

$$\left.\begin{aligned} k &= \frac{\nu}{1-\nu} \qquad (= K \text{ approximately}) \\ \beta &= \frac{\mu'}{R} \end{aligned}\right\} \quad \text{rectangular cylinders}$$

ν = Poisson's ratio

K = Janssen pressure coefficient

It will be noted that k and β for a circular cylinder are given in terms of K. This is possible because $\nu = K/(1 + K)$ for axially symmetric flow. This formula does not hold for rectangular cylinders (plane flow), and since there are no published values of ν for particulate solids, it becomes necessary to assume the value of k unless ν is determined by tests. If ν is not determined by tests, it is convenient to take it the same as for axially symmetric flow in which case $k = K$.

The envelope of pressure peaks is determined by computing p_h at various levels. However, Eq. (5-37) does not give $p_h = 0$ at $y = 0$, as it should. According to Jenike and Johanson, this is because the switch is arrested at $y = 0.5D$ to D below the top of the material in the bin so that the formula does not apply in this region. Jenike[27] suggests that from the top of the bin to a section one diameter below the top the pressure normal to the wall be

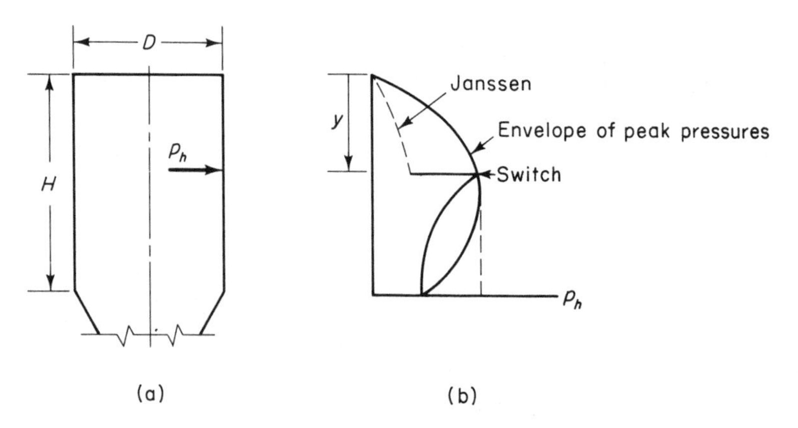

Figure 5-22 Mass-flow pressures.

taken at 1.5 times the Janssen pressure with K the larger of 0.4 or $(1 - \sin \delta)/(1 + \sin \delta)$.

Another modification of the peak-pressure envelope is required because p_h according to Eq. (5-37) reduces to the Janssen lateral pressure at the bottom of the cylinder because the material in the hopper was not taken into account in deriving the equation. In effect, then, the solution is based on the assumption that flow begins to develop at the bottom of the cylinder instead of at the hopper outlet. A few cases that were investigated by minimizing the strain energy in the entire bin showed that pressures in the upper part of the cylinder were about the same as by Eq. (5-37) but did not decrease in the lower part (Ref. 12). Therefore, the peak-pressure envelope is taken as a straight (vertical) line below the point at which p_h is maximum (dotted line in Fig. 5-22b).

Charts of peak-pressure envelopes according to Eq. (5-37) are given in Refs. 10 and 12 for circular cylinders with $H/D = 1$, 2, 3, 4, and 5 and for rectangular cylinders with $H/D = 1$ and 2. The charts are based on $K = 0.4$. Values of μ' range from 0.25 to 0.75 ($0.10 \leqslant \mu' K \leqslant 0.30$). Although p_h is a function of μ', K, and y independently in Eq. (5-37), it is very nearly a function of $\mu' K$ and y so that the charts can be used for other values of K.

Jenike[27] recommends that the pressures given by Eq. (5-37) be reduced by 15% because of the rounding of the pressure peaks as shown in Fig. 5-6b. He also suggests that K be taken the same as in the upper portion of the cylinder, that is, the larger of 0.4 or $(1 - \sin \delta)/(1 + \sin \delta)$.

Equation (5-37) can be written to exhibit $p_h/\gamma H$ as a function of K, $\mu' KH/R$, and y/H. Values in Tables 5-10 and 5-11 have been computed using the equation in this form but with a uniform pressure in the region below the maximum as shown in Fig. 5-22b and with the 15% reduction already noted. Although the tables are based on $K = 0.4$, the error in using them for other values is small. Because the parameter $\mu' KH/R$ is involved, they cover a wider range of cylinder proportions than the charts.

Hopper pressures. Jenike and Johanson developed a procedure for computing pressures in mass-flow hoppers.[10,12] Later, Jenike suggested the following formula[27]:

$$p' = K\gamma \left[\frac{z}{n - 1} + \left(h_c - \frac{h}{n - 1} \right) \left(\frac{z}{h} \right)^n \right] \tag{5-38}$$

where $n = (1 + m)[(1 + \tan \phi' / \tan \theta) K - 1]$
z = distance above apex
h = distance from apex to transition
h_c = height of solid above transition
m = 0 for plane flow
m = 1 for axisymmetric flow

TABLE 5-10. AXIALLY SYMMETRIC MASS FLOW (CIRCULAR CYLINDERS). VALUES OF $p_h/\gamma H$ BY EQ. (5-37), REDUCED 15%.

$\mu' KH/R$ \ y/H	0.1	0.2	0.3	0.4	0.5	0.6	0.7	0.8	0.9	1.0
0.2	0.125	0.157	0.187	0.213	0.236	0.257	0.274	0.288	0.300	0.309
0.4	0.175	0.210	0.240	0.264	0.281	0.293	0.299	0.299	0.299	0.299
0.6	0.195	0.235	0.268	0.293	0.309	0.317	0.317	0.317	0.317	0.317
0.8	0.198	0.243	0.280	0.306	0.324	0.330	0.330	0.330	0.330	0.330
1.0	0.193	0.241	0.281	0.309	0.328	0.334	0.334	0.334	0.334	0.334
1.2	0.184	0.235	0.275	0.306	0.326	0.333	0.333	0.333	0.333	0.333
1.4	0.175	0.226	0.268	0.299	0.320	0.328	0.328	0.328	0.328	0.328
1.6	0.166	0.218	0.258	0.290	0.311	0.320	0.320	0.320	0.320	0.320
1.8	0.157	0.208	0.249	0.280	0.301	0.311	0.311	0.311	0.311	0.311
2.0	0.149	0.200	0.240	0.269	0.290	0.301	0.301	0.301	0.301	0.301
3.0	0.120	0.165	0.198	0.222	0.238	0.247	0.248	0.248	0.248	0.248
4.0	0.103	0.142	0.168	0.185	0.196	0.203	0.204	0.204	0.204	0.204
6.0	0.083	0.112	0.128	0.138	0.142	0.145	0.145	0.145	0.145	0.145
8.0	0.071	0.094	0.103	0.107	0.110	0.111	0.111	0.111	0.111	0.111
10.0	0.063	0.079	0.085	0.088	0.088	0.088	0.088	0.088	0.088	0.088
12.0	0.056	0.069	0.072	0.074	0.074	0.074	0.074	0.074	0.074	0.074
14.0	0.051	0.060	0.063	0.063	0.064	0.064	0.064	0.064	0.064	0.064
16.0	0.047	0.054	0.055	0.055	0.055	0.055	0.055	0.055	0.055	0.055

Note: R = hydraulic radius.

TABLE 5-11. PLANE MASS FLOW (NARROW RECTANGULAR CYLINDERS). VALUES OF $p_h/\gamma H$ BY EQ. (5-37), REDUCED 15%.

$\mu' KH/R$ \ y/H	0.1	0.2	0.3	0.4	0.5	0.6	0.7	0.8	0.9	1.0
0.2	0.173	0.206	0.234	0.258	0.277	0.292	0.303	0.309	0.310	0.310
0.4	0.240	0.280	0.311	0.334	0.349	0.354	0.354	0.354	0.354	0.354
0.6	0.259	0.308	0.347	0.374	0.389	0.392	0.393	0.393	0.393	0.393
0.8	0.255	0.311	0.356	0.389	0.408	0.412	0.412	0.412	0.412	0.412
1.0	0.242	0.303	0.352	0.389	0.412	0.418	0.418	0.418	0.418	0.418
1.2	0.228	0.290	0.342	0.381	0.406	0.417	0.417	0.417	0.417	0.417
1.4	0.213	0.276	0.328	0.369	0.396	0.408	0.408	0.408	0.408	0.408
1.6	0.200	0.263	0.315	0.354	0.383	0.396	0.397	0.397	0.397	0.397
1.8	0.187	0.250	0.300	0.340	0.368	0.383	0.384	0.384	0.384	0.384
2.0	0.177	0.238	0.287	0.326	0.353	0.368	0.370	0.370	0.370	0.370
3.0	0.141	0.196	0.235	0.264	0.285	0.298	0.300	0.300	0.300	0.300
4.0	0.121	0.168	0.200	0.220	0.234	0.242	0.245	0.245	0.245	0.245
6.0	0.098	0.133	0.152	0.163	0.168	0.172	0.173	0.173	0.173	0.173
8.0	0.084	0.111	0.122	0.128	0.130	0.131	0.131	0.131	0.131	0.131
10.0	0.074	0.094	0.101	0.104	0.105	0.105	0.105	0.105	0.105	0.105
12.0	0.066	0.081	0.086	0.088	0.088	0.088	0.088	0.088	0.088	0.088
14.0	0.060	0.071	0.075	0.075	0.076	0.076	0.076	0.076	0.076	0.076
16.0	0.055	0.064	0.065	0.065	0.065	0.065	0.065	0.065	0.065	0.065

Note: R = hydraulic radius.

This is similar in form to Walker's Eqs. (5–35) and (5–36), with K equivalent to the ratio p'/p_v in Eq. (5–35), n corresponding to K_w in Eq. (5–36), and p_{vt} in Eq. (5–36) replaced by an equivalent head of solid γh_c. The equivalent head is computed by Janssen's formula.

According to Jenike, K varies in an unpredictable way between values for flow and values for the solid at rest. He suggests that it be assumed to vary from the maximum flow value K_{max} at the transition to the static value $K = (1 + \tan\phi'/\tan\theta)^{-1}$ lower in the hopper and that it be chosen at each level so as to maximize p'. K_{max} is read from charts as a function of δ, θ, and ϕ'. With the static value of K, $n = 0$, and Eq. (5–38) reduces to

$$p' = \gamma(h + h_c - z)(1 + \tan\phi'/\tan\theta)^{-1} \tag{5-39}$$

This is the same as Walker's Eq. (5–31a). This formula applies below the level at which the maximum flow pressure equals the filling pressure.

Vertical compression in wall. Vertical pressures in the material are smallest when the pressure field is passive, i.e., during flow, and when the switch is at the top. This condition gives maximum values of the vertical compression in the wall, and although the switch is arrested before it reaches the top, it is on the safe side to assume it there.

Vertical compression is computed by Eq. (5–20). The pressure coefficient is not constant in mass flow, however, so that p_v cannot be determined from Eq. (5–37). Instead, with the switch at the top, it is given by

$$\frac{p_v}{\gamma/\beta} = k + \frac{(1 - k^2)(e^{\beta y} - e^{-\beta y}) - k[(1 + k)e^{\beta(H-y)} + (1 - k)e^{-\beta(H-y)}]}{(1 + k)e^{\beta H} + (1 - k)e^{-\beta H}} \tag{5-40}$$

The notation in this equation is the same as for Eq. (5–37).

Charts for determining values of F_c computed by Eqs. (5–20) and (5–40) are given in Ref. 10 for circular cylinders with $H/D = 1$, 2, 3, 4, and 5 and for rectangular cylinders with $H/D = 1$ and 2. However, Eq. (5–40) can be written in terms of the parameters K, $\mu' KH/R$, and y/H, as for Eq. (5–37), from which Table 5–12 for circular cylinders and Table 5–13 for narrow rectangular cylinders were prepared.

Example 5–4

Compute the pressures in the bunker of Fig. 5–21. The cylinder is circular and the hopper conical. This bunker was used in tests discussed in Sec. 5–15. The material is iron ore with $\gamma = 112$ pcf, $\delta = 40°$, and $\phi' = 28°$.

Cylinder filling pressure

Using $K = 0.4$ as recommended by Jenike, we obtain

$$\mu' = \tan 28° = 0.532 \qquad \mu' K = 0.532 \times 0.4 = 0.213$$

$$R = \frac{D}{4} = \frac{3.17}{4} = 0.793 \qquad \frac{\mu' K}{R} = \frac{0.213}{0.793} = 0.269 \qquad \frac{\gamma R}{\mu' K} = \frac{112 \times 0.793}{0.213} = 417$$

TABLE 5-12. AXIALLY SYMMETRIC MASS FLOW (CIRCULAR CYLINDERS). VALUES OF $F_c/\gamma HR$ BY EQS. (5-20) AND (5-40), REDUCED 15%.

$\mu' KH/R$ \ y/H	0.1	0.2	0.3	0.4	0.5	0.6	0.7	0.8	0.9	1.0
0.2	0.005	0.012	0.020	0.030	0.041	0.053	0.065	0.079	0.093	0.107
0.4	0.015	0.033	0.054	0.076	0.099	0.124	0.150	0.176	0.202	0.227
0.6	0.025	0.053	0.085	0.119	0.155	0.192	0.229	0.264	0.299	0.332
0.8	0.032	0.070	0.111	0.155	0.201	0.247	0.293	0.337	0.378	0.416
1.0	0.037	0.082	0.131	0.183	0.236	0.290	0.343	0.394	0.439	0.480
1.2	0.042	0.092	0.146	0.204	0.264	0.324	0.383	0.438	0.488	0.530
1.4	0.045	0.099	0.158	0.221	0.286	0.350	0.413	0.473	0.526	0.570
1.6	0.048	0.104	0.167	0.234	0.303	0.371	0.438	0.500	0.556	0.602
1.8	0.049	0.109	0.174	0.244	0.316	0.388	0.457	0.523	0.581	0.627
2.0	0.050	0.112	0.180	0.252	0.327	0.401	0.473	0.541	0.601	0.649
3.0	0.054	0.124	0.199	0.279	0.360	0.441	0.520	0.596	0.663	0.715
4.0	0.058	0.130	0.210	0.292	0.377	0.460	0.542	0.621	0.694	0.748
6.0	0.062	0.140	0.224	0.308	0.393	0.478	0.562	0.644	0.723	0.782
8.0	0.065	0.146	0.231	0.315	0.400	0.485	0.570	0.655	0.736	0.799
10.0	0.068	0.150	0.235	0.320	0.405	0.490	0.575	0.660	0.743	0.809
12.0	0.070	0.154	0.239	0.324	0.409	0.494	0.579	0.664	0.747	0.816
14.0	0.071	0.156	0.241	0.326	0.411	0.496	0.581	0.666	0.751	0.821
16.0	0.073	0.158	0.243	0.328	0.413	0.498	0.583	0.668	0.752	0.825

Note: R = hydraulic radius.

TABLE 5-13. PLANE MASS FLOW (NARROW RECTANGULAR CYLINDER). VALUES OF $F_c/\gamma HR$ BY EQS. (5-20) AND (5-40) REDUCED 15%.

$\mu' KH/R$ \ y/H	0.1	0.2	0.3	0.4	0.5	0.6	0.7	0.8	0.9	1.0
0.2	0.008	0.017	0.027	0.039	0.051	0.065	0.078	0.093	0.107	0.122
0.4	0.020	0.044	0.071	0.098	0.126	0.155	0.183	0.211	0.237	0.262
0.6	0.032	0.069	0.108	0.149	0.191	0.233	0.273	0.312	0.347	0.377
0.8	0.041	0.087	0.136	0.188	0.241	0.292	0.342	0.388	0.429	0.464
1.0	0.047	0.099	0.156	0.216	0.275	0.335	0.392	0.444	0.490	0.528
1.2	0.050	0.108	0.171	0.235	0.302	0.366	0.428	0.486	0.536	0.575
1.4	0.054	0.115	0.181	0.250	0.320	0.389	0.456	0.518	0.570	0.611
1.6	0.055	0.119	0.189	0.261	0.335	0.407	0.477	0.541	0.598	0.640
1.8	0.056	0.122	0.195	0.269	0.346	0.421	0.493	0.560	0.619	0.662
2.0	0.058	0.125	0.199	0.276	0.354	0.432	0.507	0.575	0.636	0.681
3.0	0.061	0.134	0.214	0.296	0.379	0.462	0.542	0.620	0.688	0.737
4.0	0.064	0.140	0.223	0.307	0.391	0.475	0.558	0.639	0.712	0.765
6.0	0.068	0.149	0.233	0.317	0.402	0.487	0.572	0.655	0.734	0.793
8.0	0.071	0.153	0.238	0.323	0.408	0.493	0.578	0.662	0.745	0.808
10.0	0.072	0.156	0.241	0.326	0.411	0.496	0.581	0.666	0.750	0.816
12.0	0.074	0.159	0.244	0.329	0.414	0.499	0.584	0.669	0.753	0.822
14.0	0.076	0.161	0.246	0.331	0.416	0.501	0.586	0.671	0.755	0.825
16.0	0.077	0.162	0.247	0.332	0.417	0.502	0.587	0.672	0.757	0.829

Note: R = hydraulic radius.

TABLE 5-14. CYLINDER FILLING PRESSURES FOR BUNKER OF FIG. 5-21 (JENIKE-JOHANSON)

y/H	y (ft)	$0.269y$	$f(0.269y)$	p_v (psf)	$p_h = Kp_v$ (psf)
0	0	0	0	0	0
0.2	1.97	0.530	0.411	171	68
0.4	3.94	1.06	0.653	272	109
0.6	5.90	1.59	0.796	332	133
0.8	7.87	2.12	0.880	367	147
1.0	9.84	2.65	0.930	388	155

From Eq. (5-8a),

$$p_v = \frac{\gamma R}{\mu' K} f\left(\frac{\mu' K y}{R}\right) = 417 f(0.269y)$$

Pressures are given in Table 5-14. Values of $f(0.269y)$ are from Table 5-1.

Cylinder flow pressure

Use Tables 5-10 and 5-12 with

$$\frac{\mu' KH}{R} = \frac{0.213 \times 9.84}{3.17/4} = 2.64 \qquad \gamma H = 112 \times 9.84 = 1102$$

$$\gamma HR = 1102 \times 3.17/4 = 873$$

The resulting pressures are given in Table 5-15 and compared with experimental results in Fig. 5-24.

Hopper pressures

Since hopper pressures involve a value of K from charts in Ref. 27, calculations are not given here. However, the results are compared with experimental values in Fig. 5-24.

TABLE 5-15. CYLINDER FLOW PRESSURES FOR BUNKER OF FIG. 5-21 (JENIKE-JOHANSON)

y/H	$\frac{p_h}{\gamma H}$	p_h (psf)	$\frac{F_c}{\gamma HR}$	F_c (plf)
0	0	0	0	0
0.2	0.178	196	0.120	105
0.4	0.239	263	0.269	235
0.6	0.266	293	0.427	373
0.8	0.267	294	0.576	503
1.0	0.267	294	0.691	603

5-15. Mass-Flow Bin Tests

Walker and Blanchard conducted tests for the Central Electricity Generating Board of England on experimental stainless-steel coal hoppers of up to 5 tons of capacity.[30] The set included conical and pyramidal hoppers with half-apex angles of 15°, 30°, and 45°. The conical hoppers had 6-in.-diameter outlets and 72-in-diameter inlets; the pyramidal hoppers had 6-in. square outlets and 72-in. square inlets. The 30° and 45° hoppers were attached to vertical cylinders to carry a surcharge. The coal was sieved through a $\frac{1}{8}$-in. mesh and had an average bulk density of 50.7 pcf. It was kept at a free-moisture content of about 3%, but a few tests were run with 11–17% moisture. The effective internal angle of friction was 41°, and the angle of friction on stainless steel was 16°.

The pressure probes were 3-in.-diameter stainless-steel diaphragms 0.006 in. thick mounted on cast blocks which left the central $1\frac{1}{2}$ in. of diameter unsupported. A displacement transducer, attached to the same mounting plate as the cast block, made contact with the center of the diaphragm. The deflection of the diaphragm was about 0.001 in. at the maximum pressure. The probes, which were calibrated by air pressure, were mounted flush with the inside of the hopper wall and as many as 10 were used to monitor each run. Wall-pressure probe units were embedded in the coal in some runs to measure internal vertical pressures. Most of the tests were monitored by a 100-channel data logger which scanned the channels at the rate of 2/sec. For a few runs outputs of two or three probes were continuously recorded in analogue form.

Several different loading techniques were used. The normal one consisted in allowing the coal to fall freely off the loading belt in a steep trajectory that hit the far wall of the hopper about one-third the way up from the outlet. Analogue traces for two probes, one 78 in. and the other 28 in. above the apex, showed a smooth buildup of pressure during the normal loading and subsequent emptying of a 15° pyramidal hopper. When discharge began, the pressure on the higher probe increased and oscillated about a mean value which remained fairly constant for about two-thirds of the discharge and then dropped gradually to zero as the coal surface level passed the diaphragm. Pressure varied similarly at the lower probe except that it dropped when flow began. There were deviations from the mean of as much as ±40% at both probes. It is not known how localized these peak pressures are.

Figure 5–23 shows the range of wall pressures measured in five runs on a 15° conical hopper. Figure 5–23b shows the initial pressures after the hopper had been filled to a height of 133 in. above the apex. Figures 5–23c and d show flow pressures when the level of the coal was 126 and 94 in. above the apex. Also shown are comparisons with the predicted pressures by Walker's theory.

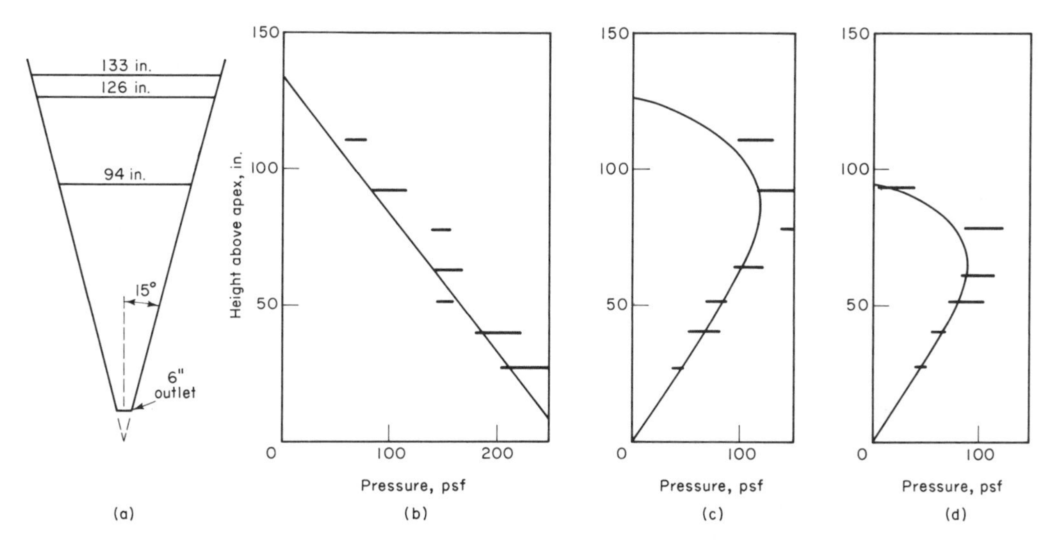

Figure 5-23 Comparison of Walker mass-flow theory with tests: (a) conical hopper: fill height 133 in., first flow pressures 126 in., second flow pressures 94 in.; (b) range of average wall pressures during five fillings; (c) and (d) range of average wall pressures during five emptyings. (From Ref. 30.)

Test results for the 15° pyramidal hopper were in good agreement with predicted values except for the maximum first-flow pressures (i.e., pressures in Fig. 5-23c 93 in. above the apex), which were about double the theoretical values. First-flow pressures elsewhere were about the same for both the conical and the pyramidal hopper. Initial pressures (Fig. 5-23b) and second-flow pressures (Fig. 5-23d) were practically the same throughout for both.

Agreement between the theoretical curves and the experimental averages was only moderately good for the 30° hoppers. Initial and first-flow pressures were generally greater than the predicted values throughout the pyramidal hopper and generally less for the conical hopper. There was a wide scatter of measured pressures in the pyramidal hopper. Scatter was smaller in the conical hopper but still larger than for the 15° hoppers. The observed flow pattern was somewhere between mass flow and pipe flow, with the pyramidal hopper especially inclined toward the latter. Coal at the walls was disturbed intermittently by a series of surface collapses while it had sheared continuously in the 15° hoppers.

Measured and theoretical vertical pressures for the 15° and 30° hoppers are given in Table 5-16. Agreement is generally very good for the initial pressures and moderately good for the smaller flow pressures.

According to Ref. 30, flow pressures do not depend on flow actually taking place but only on its having taken place. In other words, once they are established, flow pressures remain even though discharge is interrupted (overnight in one case). This suggests that if a hopper is to be left filled for

TABLE 5-16. VERTICAL PRESSURES IN HOPPERS[a]

Apex half angle	Height above apex (in.)	Initial pressure (psf)		Flow Pressure (psf)	
		Theoretical	Measured	Theoretical	Measured
15°	93	173	230	58	60
	78	230	245	58	45
	63	300	300	43	30
	51	345	331	29	15
30°	68	130	190	101	60
	47	216	220	86	130
	35	274	320	58	70

[a]From Ref. 30.

some time with a material which strengthens on standing, some material should be drawn off after filling so as to develop a flow stress field with its lower pressures near the outlet.

Tests on the bunker of Examples 5-3 and 5-4 (Fig. 5-21) were conducted by Clague and Wright for the British Steel Corporation.[31] The bunkers were of welded mild steel construction, which gave inside surfaces that were rusted but reasonably smooth. The material was Cerro Bolivar iron ore which had a bulk density of 112 pcf and the size range shown in Table 5-17. The effective internal angle of friction was 41°, and the angle of friction on the rusted steel surface 28°. The hoppers were designed for mass flow according to Jenike's method. Pressures were measured with the probes used by Walker and Blanchard in the tests discussed earlier in this section. The measured pressures were checked by comparing a summation of pressure times elemental bunker wall area for the whole of the bunker wall with the weight of the ore in the bunker, and the agreement was generally good.

The following variables were considered:

1. Hopper shape (conical and chisel).
2. Surcharge on hopper.
3. Material state (static after filling, and flow).
4. Duration of storage before discharge.
5. Use of an impact breaker to reduce pressure buildup near the outlet. The breaker was a flat steel plate about two-thirds of the way up the hopper.
6. Size of outlet.

Twenty tests with various combinations of these variables were run. Most were run six times. Reported flow pressures are mean values of fluctuating pressures that varied from a minimum of ±5% to a maximum of ±30% from the mean.

TABLE 5-17. SIZE RANGE OF CERRO BOLIVAR IRON ORE

Size (in.)	$\frac{1}{8}$	$\frac{1}{8}-\frac{1}{4}$	$\frac{1}{4}-\frac{3}{8}$	$\frac{3}{8}-\frac{3}{4}$
Percent (wt)	69	16	8	7

Correlation between the measured average pressures and the Walker values was very good for bins consisting of only the hopper but variable for tall bins. Pressures in tall bins tended to be about half the Walker values if the ore was in a static state. Dynamic pressures at the transition sometimes exceeded the Walker values and generally were about five times the Janssen value. Pressures in the chisel hopper tended to be markedly lower than those in the conical hopper and were generally less than predicted by Walker's formula. This discrepancy may be because Walker assumes the wedge-hopper to be infinitely long. The experiments showed that static filling pressures near the hopper outlet were reduced by the impact breaker.

Experimental pressures for the bunker of Fig. 5-21 are compared in Fig. 5-24 with pressures according to Walker, Jenike-Johanson, DIN, and ACI. The Walker pressures are computed in Example 5-3 and the Jenike-

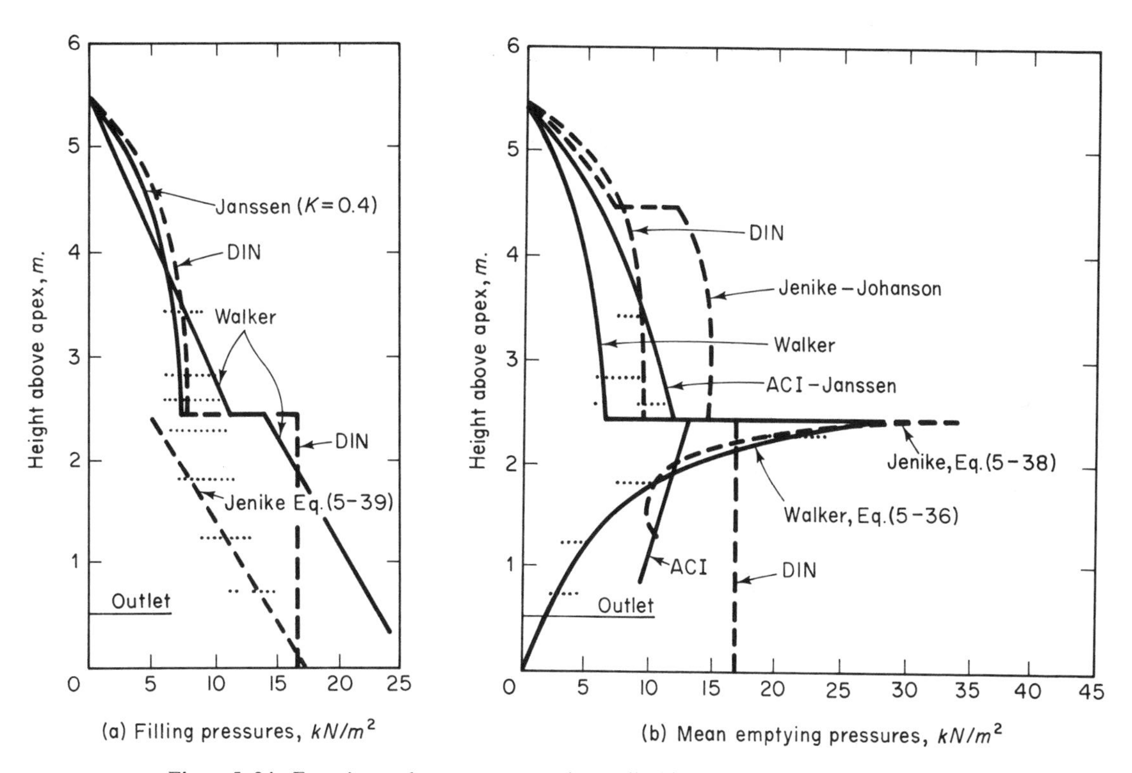

Figure 5-24 Experimental pressures normal to wall of bunker of Fig. 5-21. (From Ref. 31.)

Johanson cylinder pressures in Example 5–4. Jenike's Eq. (5–39) is used for the hopper filling pressures and Eq. (5–38) for the hopper emptying pressures.

The Janssen values, with $K = 0.4$ as recommended by Jenike and Johanson, and the DIN values are in good agreement with the average filling pressures in the cylinder. Cylinder emptying pressures tend to be underestimated by Walker's formula, while the DIN and ACI values are a good upper bound. The Jenike-Johanson emptying-pressure curve in the cylinder is about double the average test values. However, it is the locus of switch-pressure peaks, while the test values are averages that were exceeded by as much as 30%.

Jenike's Eq. (5–39) gives a good prediction of average filling pressures in the hopper, while Walker's formula overestimates them by a considerable margin. Emptying pressures by both Walker's formula and Jenike's formula are in good agreement with the test results except that the Jenike value at the transition may be too large. Jenike's Eq. (5–38) is plotted only to the level at which the emptying pressure equals the filling pressure by Eq. (5–39), roughly at midheight of the hopper. Thus, Eq. (5–39) for the lower part of the hopper and Eq. (5–36) or (5–38) for the upper part give a good envelope of the maximum pressures. Neither the DIN nor the ACI hopper pressure distributions, computed as explained in Sec. 5–12, is representative of the test results.

Jenike and Johanson conducted tests on model bins.[10,12] Three different hoppers were used: a conical hopper with an apex half angle of 12°, one with a half angle of 20°, and a transition hopper. The hoppers were of stainless steel and were attached to a plexiglass cylinder about 12 in. in diameter by 36 in. high (Fig. 5–25a). Tests were run with Ottawa sand and corn grits.

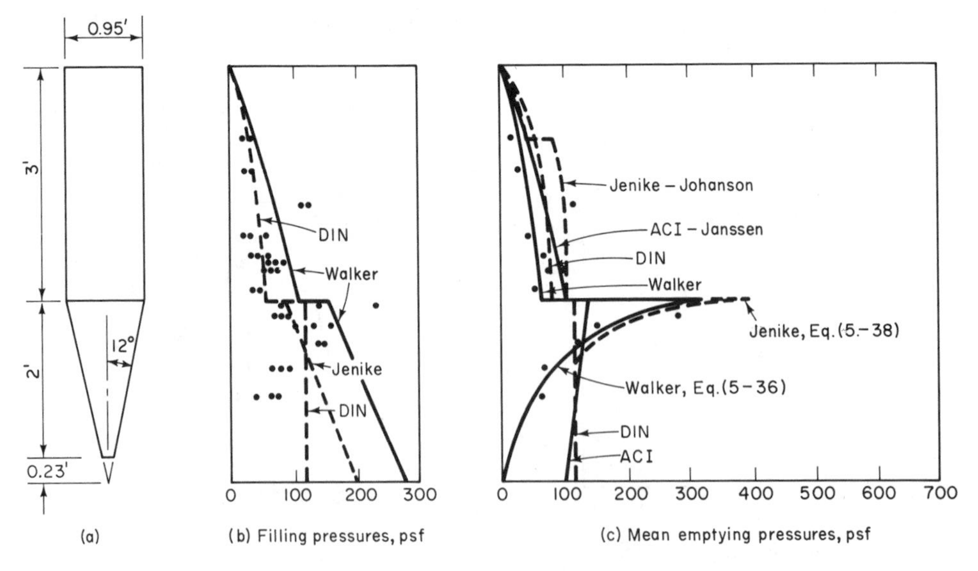

Figure 5–25 Comparison of theoretical pressures with tests. (From Ref. 10.)

Pressure transducers with an effective sensing diameter of $\frac{1}{8}$ in. were mounted flush with the walls. Transducers and recorders were available for only four simultaneous readings so that tests had to be repeated to measure all the test points reported. Therefore, there were no simultaneous measurements throughout the bin as in the tests described earlier in this section. Usually all four transducers were in the cylinder or in the hopper.

Measured pressures were checked as in the Clague and Wright experiments by comparing a summation of pressure times elemental wall area with the weight of the contained solid. According to this calculation, wall pressures for the Ottawa sand and the corn grits were only 70 and 30%, respectively, of the transducer values. Therefore, the measured pressures were adjusted by these amounts. Jenike and Johanson attribute the difference to the fact that the particles were large compared to the pressure-sensing diaphragm. The particle size was 0.028 ± 0.005 in. for the Ottawa sand and 0.053 ± 0.030 in. for the corn grits. The corresponding ratios of diaphragm diameter to average particle size are 4.5/1 and 2.4/1 and to maximum particle size 3.8/1 and 1.5/1. (The ratio of diaphragm diameter to maximum particle size was 24/1 in the Walker and Blanchard tests and 4/1 in the Clague and Wright tests. In the latter, however, the ratio was 8/1 or more for 93% of the particles.)

Three tests were conducted in quick succession for each case to determine the repeatability of the results. Flow-pressure readings oscillated rapidly, as in the Walker and Blanchard tests. Test results for the Ottawa sand, for which $\gamma = 106$ pcf, $\delta = 30°$, and $\phi' = 17°$ for the plexiglass cylinder and 12° for the stainless-steel hopper, are shown in Fig. 5-25. The flow pressures shown are averages of the lowest and the highest pressure measured in the three runs rather than the average of the mean pressures of the three runs. (The individual means were not reported.)

It will be noted that cylinder filling pressures are in good agreement with the DIN values except at one point. The pressure at this point is suspect, however, since the average flow pressure at this level is also considerably larger than the averages elsewhere in the cylinder. The Janssen formula, with $K = 0.4$, gives results so close to the DIN values that it could not be shown on the figure. As in Fig. 5-24b, the Jenike-Johanson emptying pressures for the cylinder are about double the Walker values with DIN and ACI midway between the two.

There is a wide scatter of hopper filling pressures near the transition. However, filling pressures in the lower part of the hopper are more significant since it is here that they tend to exceed emptying pressures. Therefore, either Jenike or DIN gives a better prediction than Walker, which is also the case in Fig. 5-24a. Mean emptying pressures in the hopper are in good agreement with both the Walker and the Jenike curves, although the latter gives values at the transition that appear to be too large, as is the case in Fig. 5-24b. The Jenike curve is shown only to the level at which the emptying pressure equals the Jenike filling pressure. Neither the DIN nor the ACI hopper emptying pressure is representative of the test results.

Variations of the fluctuating flow pressures from the means of the individual runs cannot be determined from the data in Ref. 10 because only the single lowest and single highest pressures at each level are reported. Departure from the means of these values was generally on the order of ±30% but ranged from ±15 to upward of ±50% and at one level was ±69%. Occasional still larger extremes were ignored in reporting the results because it was felt that they may have been caused by a single particle moving across the diaphragm.

It is clear from the tests discussed in this section that the diameter of the pressure-probe diaphragm relative to particle size is important. The pressure of the solid on the wall can be conceived as consisting of forces concentrated at points of contact of its particles, and the larger the particle, the larger the spacing of these concentrated forces. The effect of the particle forces in producing stress in the walls of the bin is related to their spacing, and if the diameter of the sensor is small enough to register the pressure of a single particle rather than an average of a reasonable number of particles, the recorded pressure is certain to exaggerate the effective pressure. This is shown by the fact that pressures in the Jenike-Johanson experiments, which were measured with a $\frac{1}{8}$-in. diaphragm, had to be reduced by 30% for sand and 70% for the coarser corn grits to make them comparable to pressures that would have had to exist to equilibrate the weight of the solid. The tests by Walker and Blanchard and those by Clague and Wright, in which a $1\frac{1}{2}$-in. diaphragm was used, required no adjustment.

It is difficult to assess the significance of the fluctuating flow pressures in terms of the stresses they produce in the walls of a bin. If the pressure peaks are highly concentrated, as the evidence seems to suggest, they have the same effect as a smaller pressure distributed uniformly over an effective area of wall. Furthermore, it is highly unlikely that they are uniformly distributed circumferentially. Therefore, the range of fluctuation of the effective stresses in the wall may be smaller than the pressure fluctuations observed in the experiments.

Vertical pressure in the solid is important because it determines the vertical compression in the wall. Vertical pressures were measured in the Walker-Blanchard tests but not in the Clague-Wright and Jenike-Johanson tests. Walker's formula is in reasonably good agreement with the Walker-Blanchard tests on hoppers with no surcharge (Table 5–16). The Jenike-Johanson formulation for vertical pressure is based on the conservative assumption that the switch is at the top of the bin (Sec. 5–14) and gives smaller vertical pressures in the solid, which results in much larger vertical compression in the wall (Tables 5–8 and 5–15).

In choosing a basis for calculating bin pressures, it should be noted that a design based on mean flow pressures and one based on peak flow pressures will have essentially the same reliability (probability of failure) if the ratio of peak pressures to mean pressures is reasonably constant and proportional factors of safety are used. In the tests described in this section the Jenike-Johanson pressure fluctuations were generally of the order of 30 to 40% of the mean, while Clague and Wright reported deviations of ±30% and Walker and Blanchard ±40%.

5-16. Wind Loads

The local pressure at a point on an immovable surface in the path of a wind stream normal to the surface is given by

$$q = \tfrac{1}{2}\rho v^2 \tag{5-41a}$$

where q = pressure
ρ = mass density of air
v = velocity of air

This pressure, which is called the *velocity* pressure, *dynamic* pressure, or *stagnation* pressure, varies from point to point on the surface. The resultant average pressure p normal to the projected area of an object in the path of the wind is

$$p = \tfrac{1}{2}C_p \rho v^2 \tag{5-41b}$$

where C_p is a *pressure coefficient* or *shape factor.* Air at sea level weighs 0.0756 pcf at 15°C (59°F). Substituting the mass density 0.0765/32.2 into Eq. (5-41b) and converting v from units of feet per second to velocity V in miles per hour gives pressure p, pounds per square foot, as

$$p = 0.00256 C_p V^2 \tag{5-42}$$

Weather-station wind velocities are averages of the fluctuating velocities encountered during a prescribed period of time. Weather stations in the United States report the average velocity during the time it takes a column of air 1 mi long to pass the anemometer; i.e., winds of 30 mph are averaged over a 2-min period, 60 mph winds over 1 min, etc.* In the United States the highest velocity in 1 day is called the *fastest mile.* The *annual extreme mile* is the largest of the daily maximums.

The period of time during which a specified wind velocity has a probability of being exceeded at least once is called a *mean recurrence interval* or *return period.* Return period can also be defined as the expected average number of years between recurrences of velocities greater than the specified value. Charts of annual extreme fastest-mile, open-country velocities in the United States with return periods of 2, 10, 25, 50, and 100 years have been

*Most weather stations in Canada average velocities over a 1-hr period, while those in Great Britain average for a 1-min period.

published.[32] The 50-year map is shown in Fig. 5–26. This return period is recommended for all buildings and structures except those highly sensitive to wind and those which may involve unusually high loss of life or property, for which a 100-year period is recommended, and those with a low degree of hazard to life or property, which may be designed for a 25-year period.

Wind turbulence results in pressures greater than are predicted by using the fastest-mile velocity in Eq. (5–42). The increased pressure is obtained by multiplying the basic velocity pressure by a gust factor C_g.

Wind velocity increases with height because the flow of air near the ground is slowed by surface roughness. A number of roughness categories, with formulas for computing the vertical distribution of velocity, are given in Refs. 33 and 34. Three categories that apply in the design of bins are given in Table 5–18; the corresponding velocity pressures are given in Table 5–19. These pressures may be applied in a stepwise distribution, with the pressure at midheight of each zone distributed uniformly over the height.

Pressure coefficients C_p for bins are given in Table 5–20 and for supporting members and appurtenances (columns, column bracing, railings, etc.) in Table 5–21.

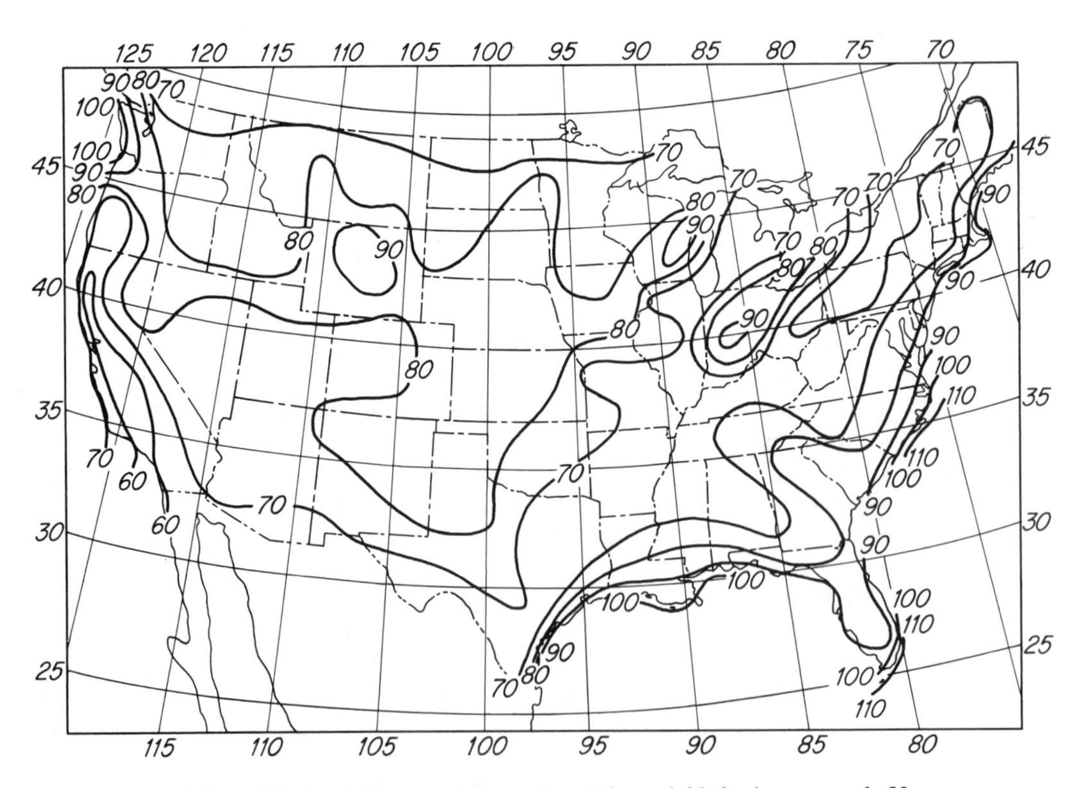

Figure 5–26 Annual extreme fastest-mile wind speed 30 ft above ground, 50-year mean recurrence interval. (From Ref. 32.)

TABLE 5-18. EXPOSURE CATEGORIES

Exposure	Description
1	Urban and suburban areas; well-wooded areas
2	Flat open country with scattered low obstructions; grasslands
3	Flat coastal areas adjoining large bodies of open water; low unsheltered islands

TABLE 5-19. VELOCITY PRESSURES q (psf): V = 100 mph[a]

Height above ground (ft)	Exposure 1	Exposure 2	Exposure 3
0–15	16	27	35
20	17	29	37
40	21	33	41
60	24	37	44
100	29	41	48
140	32	44	50

[a]For other velocities V_1, multiply values in table by $(V_1/100)^2$.

TABLE 5-20. PRESSURE COEFFICIENTS C_p FOR BINS

Shape	Type of surface	h/d	
		1	7
Square (wind normal to face)	All	1.3	1.4
Square (wind along diagonal)	All	1.0	1.1
Hexagonal, octagonal	All	1.0	1.2
Round	Moderately smooth	0.5	0.6
	Rough ($d'/d \approx 0.02$)	0.7	0.8

Note: h = height (ft), d = diameter or least horizontal dimension (ft), d = depth (ft) of protruding elements such as stiffeners

TABLE 5-21. PRESSURE COEFFICIENTS C_p FOR INDIVIDUAL MEMBERS

Shape	C_p
Flat-sided	2
Cylindrical, $d\sqrt{q} < 2.5$	1.2
Cylindrical, $d\sqrt{q} > 2.5$	0.8

Note: d = diameter (ft), q = velocity pressure (psf)

It is important to note that wind pressures specified by building codes usually include an allowance for both pressure coefficient and gust factor, and it would be overly conservative to use them with the coefficients in Tables 5-20 and 5-21. For example, the wind-pressure map of the Uniform Building Code[35] gives pressures which correspond closely to values using the 50-year wind velocity from Fig. 5-26 with the gust factor $C_g = 1.3$ and pressure coefficient $C_p = 1.3$. The latter is (approximately) the sum of a pressure coefficient 0.8 for the positive pressure on the windward wall of a rectangular building and a pressure coefficient 0.5 for the negative (suction) pressure on the leeward wall. The Uniform Code gives *multiplying factors* for chimneys and tanks that are factors by which the pressure coefficient for rectangular buildings must be multiplied to give the pressure coefficient for a tank; i.e., they are (approximately) the coefficients of Table 5-20 divided by 1.3.

Open-top bins. To determine the wind loads on circular cylindrical bins during construction and in service, a number of wind-tunnel tests were performed by Esslinger et al.[36] on cylinders of the dimensions shown in Fig. 5-27a. Pressures were determined for three conditions: open top, deck with inlet open, and deck with the inlet closed.

The pressure coefficient C_p for a bin with a closed deck is shown in Fig. 5-28a. Figures 5-28b and c show the differential-pressure coefficients ΔC_p for a bin with the deck inlet open and an open bin, respectively. The differential pressure is defined as $p_i - p_o$, where p_i is the pressure on the inner surface and p_o the pressure on the outer surface. It will be noted that the maximum resultant pressure on the windward surface of an open bin is about double the pressure on a bin with a deck; the difference is due to the internal suction in an open bin, which is numerically additive to the pressure on the outer surface. Furthermore, the pressure on an open bin is compressive on the windward as well as on the leeward surface, while the pressures on a bin with a deck are compressive on the windward surface and suction on the leeward surface. This suggests that an open bin is likely to be more

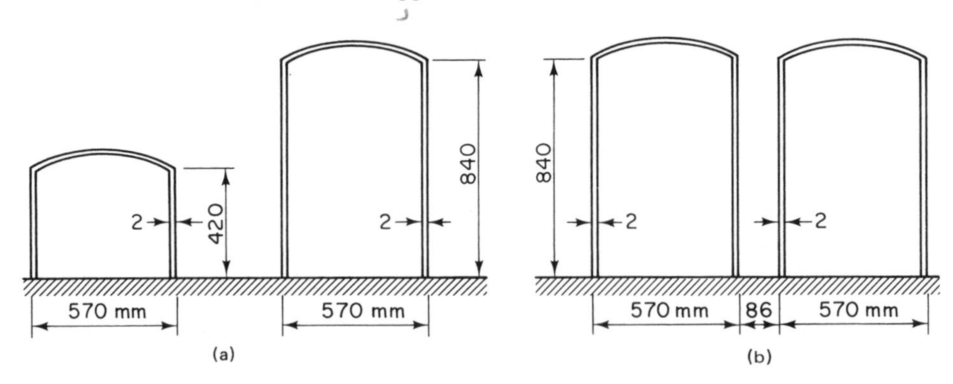

Figure 5-27 Cylinders for wind-tunnel tests. (From Ref. 36.)

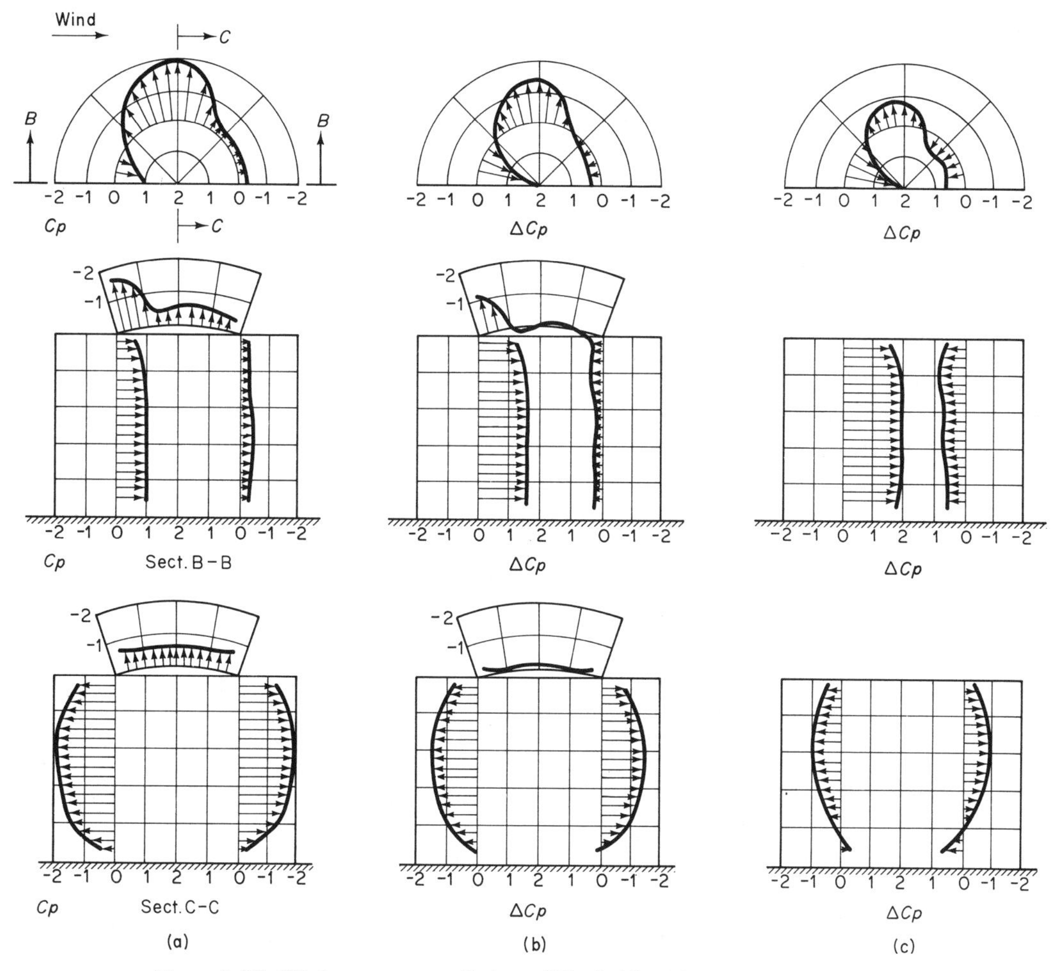

Figure 5-28 Wind pressure on cylinders of Fig. 5-27a: (a) deck inlet closed, (b) differential-pressure coefficient with deck inlet open, (c) differential-pressure coefficient for open bin. (From Ref. 36.)

susceptible to buckling than one with a closed deck. This is a consideration in erection.

Figure 5-28b shows that the pressures on an empty bin with an open inlet on the deck are approximately intermediate between the pressures on an open bin and one with the deck inlet closed.

Wind pressures on closely spaced bins. Esslinger et al. also made wind-tunnel tests on the two cylinders shown in Fig. 5-27b. Results are shown in Fig. 5-29. Maximum pressures on the windward side are about 30% larger in Fig. 5-29a and 50% larger in Fig. 5-29b than the corresponding pressures for

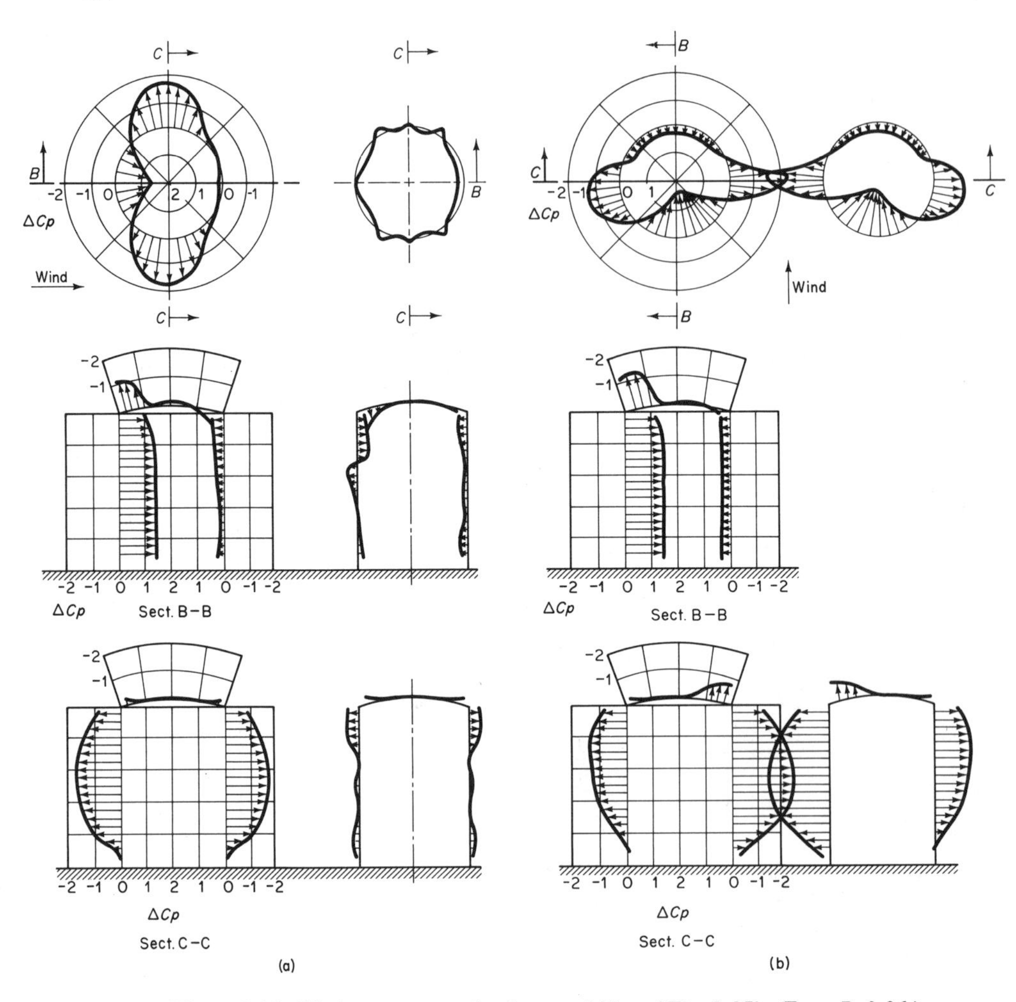

Figure 5-29 Wind pressures on closely spaced bins of Fig. 5-27b. (From Ref. 36.)

the single bin of Fig. 5-28, and the differential-pressure coefficient ΔC_p for wind in the direction shown in Fig. 5-29b is about -2.5 compared to about -2 for Fig. 5-28a.

Example 5-5.

Compute the wind pressures and base moments for the bin of Fig. 5-30 for exposure C in an 80-mph wind (this is the 50-year return wind for most of the continental United States; see Fig. 5-26).

For a stepwise variation in pressure as described in Sec. 5-16, use the wind pressure at 0-15 ft for the hopper and columns, at 30 ft for the lower 20 ft of bin, and at 50 ft for the upper 20 ft. Then with corresponding pressures from Table 5-19, multiplied by $(80/100)^2$, the design pressures are 17, 20, and 22 psf at the

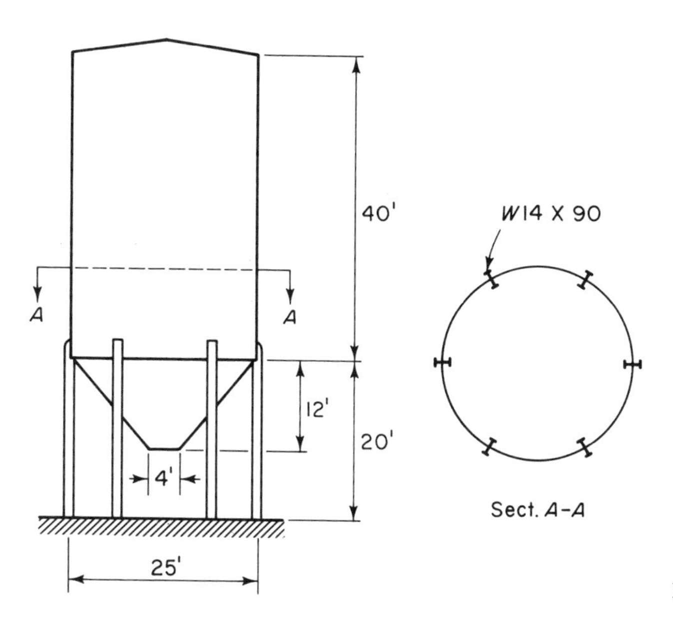

Figure 5-30 Example 5-5.

TABLE 5-22. WIND PRESSURES FOR BIN OF FIG. 5-30

Area	Pressure (kips)	Arm (ft)	Moment (ft-kips)
Upper 20 ft of bin	22 × 25 × 20 × 0.5 = 5.5	50	275
Lower 20 ft of bin	20 × 25 × 20 × 0.5 = 5.0	30	150
Hopper	17 × 14.5 × 12 × 0.5 = 1.5	15	23
Six columns	17 × 1.17 × 20 × 6 × 2 = 4.8	10	48
	16.8		496

three levels. The pressure coefficients are from Tables 5-20 and 5-21. The resultant wind pressures and the moments at the base of the structure are given in Table 5-22.

Some designers simplify wind pressure by assuming that it does not vary with height. A uniform pressure of 30 psf is often used. Table 5-19 shows that this may be unconservative for tall bins which must withstand winds of high velocity.

5-17. Snow Loads

Snow load on a roof is usually less than that on the ground. A snow load of about 80% of the ground snow load is recommended in Ref. 33 for roofs with slopes of 30° or less on unheated structures in windy areas with little shelter, provided the roof surface is one that will allow snow to slide off. However, it is not to be taken less than the ground load if the ground load is less than 20 psf, or less than 20 psf for ground loads more than 20 psf. For

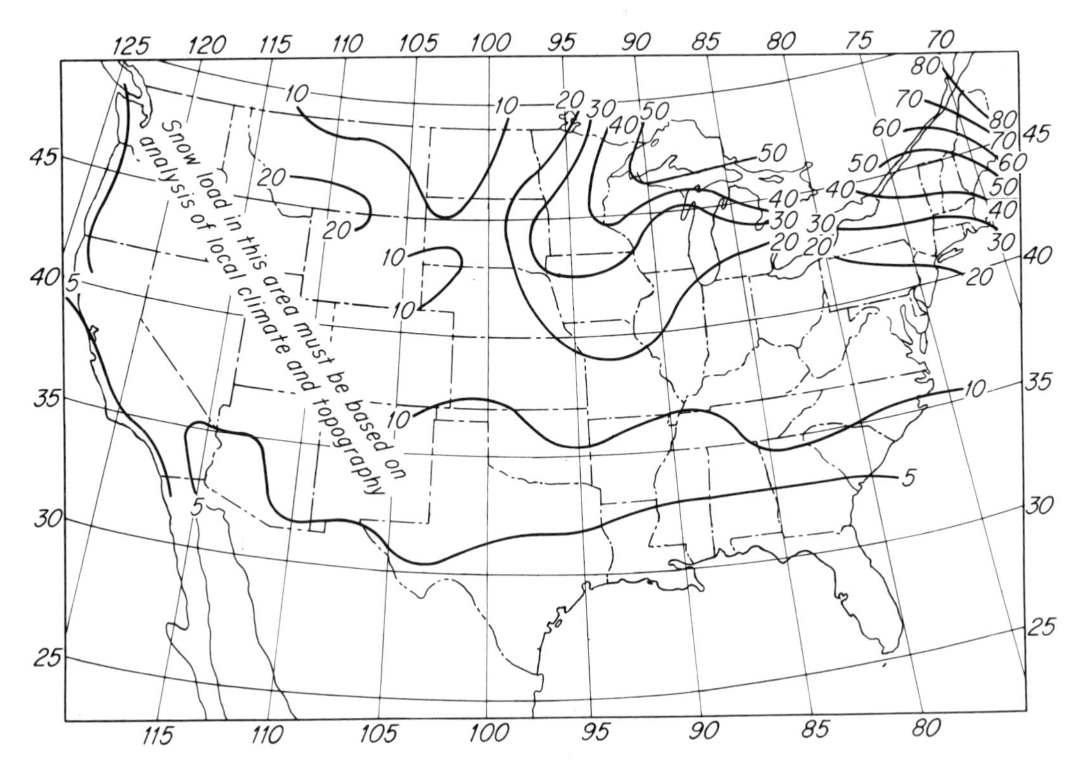

Figure 5-31 Snow load on the ground, 50-year mean recurrence interval. (National Weather Service, NOAA, U.S. Dept. of Commerce, Map 12158, Jan. 1969.)

slopes exceeding 30° the snow load is determined by interpolating linearly between the value for a 30° slope and zero for a 70° slope.

Figure 5-31 gives ground snow loads having a 50-year return period. These loads are based on the annual maximum water equivalent of snow on the ground. The differences between these and ground snow for a 25-year interval are generally small. There are extreme local variations in depth of snow in the region for which no loads are shown. Snow loads in this region must be based on local data.

In general, bin roofs should be designed for some live load, even if no snow or wind load need be considered, in order to protect the roof against collapse because of insufficient intake of air during discharge of the bin. This condition can result from inadequate provision for the movement of air or from blockage of the means provided.

5-18. Earthquake Loads

The response of a structure to an earthquake depends on its natural periods of vibration, its damping characteristics, the mechanical properties of the structural material, and the nature of the foundation material. According to the Uniform Building Code[35] (UBC), tanks supported directly on the ground and elevated tanks supported on four or more cross-braced columns and not supported by a building must be designed for a lateral seismic force V not less than

$$V = ZIKCSW \tag{5-43a}$$

where Z = seismic-zone coefficient = 0, $\frac{3}{16}$, $\frac{3}{8}$, $\frac{3}{4}$, and 1 for zones 0, 1, 2, 3, and 4, respectively (Fig. 5-32)
I = structure importance factor
K = 2 for the ground-supported tank
K = 2.5 for the column-supported tank
C = coefficient from Eq. (5-44)
S = site-structure resonance coefficient
W = total weight including contents

KC is to be taken not less than 0.12 but need not exceed 0.25. Therefore, in lieu of an evaluation of C, Eq. (5-43a) can be written as

$$V = 0.25ZISW \tag{5-43b}$$

For elevated tanks not supported in the manner described above and for tanks supported by buildings, the lateral force is determined by

$$V = 0.3ZIW \tag{5-43c}$$

The lateral force should be distributed vertically in proportion to the vertical distribution of W.

The importance factor I is 1.5 for essential facilities, which are defined as those which must be safe and usable for emergency purposes after an earthquake, and 1 for others (an intermediate value applicable only to buildings is also specified).

Formulas for S as a function of the ratio of the fundamental period of the structure to the characteristic site period are given in the UBC. In lieu of formulas, ANSI A58.1 specifies resonance coefficients for three soil profiles S_1, S_2, and S_3 as follows[33]:

1. $S = 1$ for soil profile S_1, which is a profile with rock of any characteristic, either shale-like or crystalline, or one with stiff soil conditions where the soil depth is less than 200 ft and the soil types overlying rock are stable deposits of sands, gravels, or stiff clays.
2. $S = 1.2$ for soil profile S_2, which is a profile with deep cohesionless or stiff clay conditions, including sites where the soil depth exceeds 200 ft and the soil types overlying rock are stable deposits of sands, gravels, or stiff clays.
3. $S = 1.5$ for soil profile S_3, which is a profile with soft to medium-stiff clays and sands characterized by 30 ft or more of soft to medium-stiff clays without intervening layers of sand or other cohesionless soils.

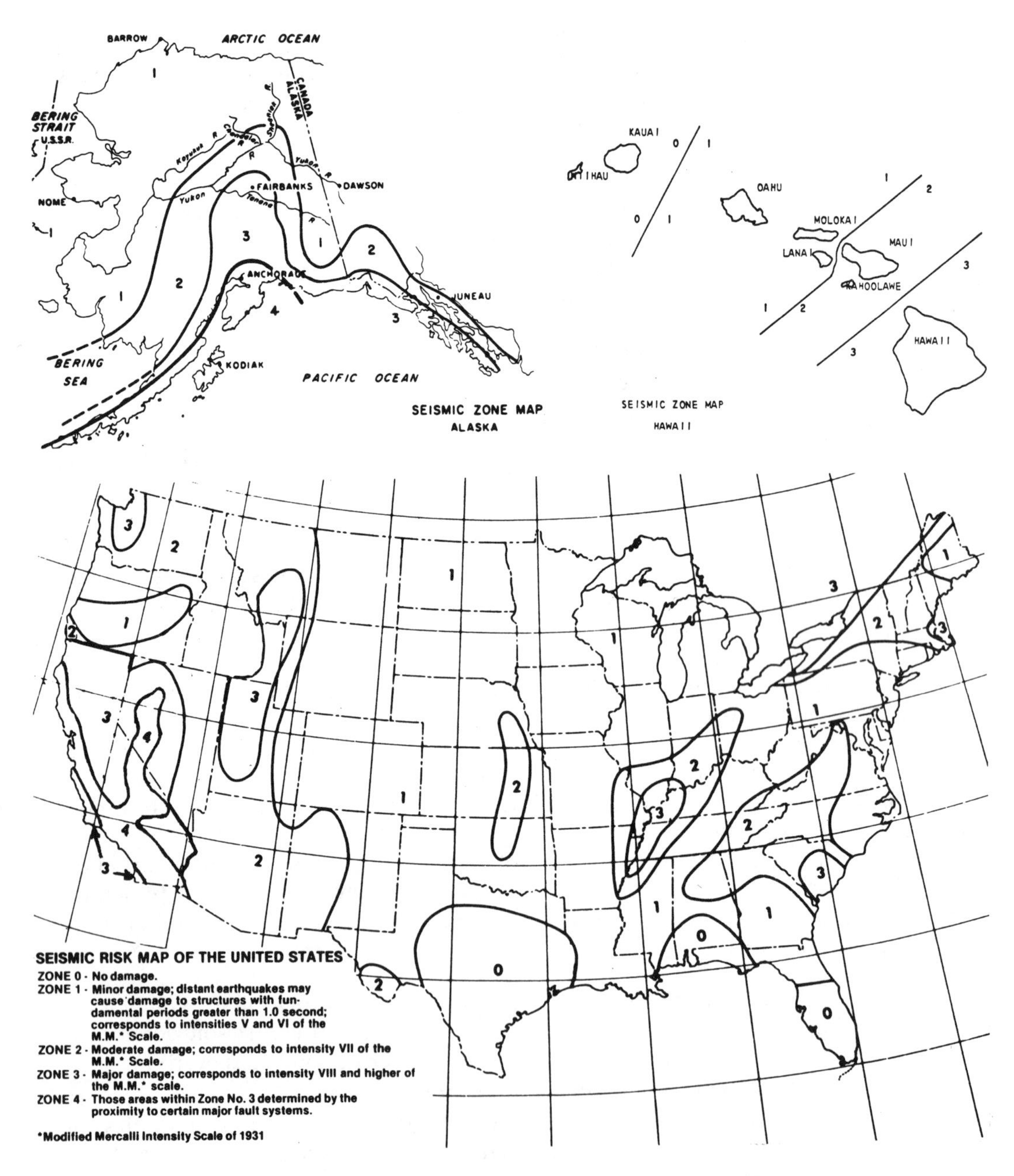

Figure 5-32 Seismic risk zones. (From Ref. 35.)

If the soil properties are not known in sufficient detail to determine the profile or if the known profile does not fit any of the three types described, $S = 1.2$.

The UBC coefficient C is given by

$$C = \frac{1}{15\sqrt{T}} \tag{5-44}$$

but need not exceed 0.12 and the product CS need not exceed 0.14. T is the fundamental period of vibration, in seconds, in the direction considered.

ACI 313-77 does not require an evaluation of the soil profile or specify an importance factor and permits design to be based on smaller seismic forces than are given by the preceding formulas, namely, for bins with stored material resting directly on the ground,

$$V = 0.1ZW \tag{5-45a}$$

and for bins with above-ground bottoms,

$$V = 0.2ZW \tag{5-45b}$$

Energy absorbed through intergranular movement and friction tends to reduce the dynamic effect of the stored material in bins,[37] and ACI 313-77 allows W to be computed for the weight of the bin plus 80% of the contents instead of the full contents as in the UBC provision for tanks.

REFERENCES

1. I. Roberts, Pressures of Stored Grain, *Engineering,* Vol. 34, Oct. 27, 1882, p. 399.
2. I. Roberts, Determination of the Vertical and Lateral Pressures of Granular Substances, *Proc. R. Soc. London,* Vol. 36, 1884, pp. 225–240.
3. H. A. Janssen, Versuche über Getreidedruck in Silozellen, *Z. Ver. Dtsch. Ing.,* Vol. 39, Aug. 31, 1895, pp. 1045–1049.
4. W. Airy, The Pressure of Grain, Minutes of *Proc. Inst. Civ. Eng. London,* Vol. 131, 1897, pp. 347–358.
5. M. S. Ketchum, *The Design of Walls, Bins, and Grain Elevators,* McGraw-Hill, New York, 1909.
6. A. M. Turitzin, Dynamic Pressure of Granular Materials in Deep Bins, *J. Struct. Div. ASCE,* April 1963.
7. A. W. Jenike and J. R. Johanson, Bin Loads, *J. Struct. Div. ASCE,* April 1968.
8. N. Nanninga, Does the Conventional Method of Calculating Bin Pressures Give Correct Results? (in Dutch), *Ingenieur* (The Hague), Vol. 44, Nov. 1956, pp. 190–194.
9. J. K. Walters, A Theoretical Analysis of Stresses in Silos with Vertical Walls, *Chem. Eng. Sci.,* Vol. 28, 1973, pp. 13–21.

10. A. W. Jenike and J. R. Johanson, Annual Report, Project 126–Fourth Part, Mass-Flow Bins, and Project 126A–First Part, Funnel-Flow Bins, American Iron and Steel Institute, Washington, D.C., April 1971, unpublished.
11. A. W. Jenike, J. R. Johanson, and J. W. Carson, Bin Loads–Part 2: Concepts, *Trans. ASME, J. Eng. for Ind.*, Feb. 1973.
12. A. W. Jenike, J. R. Johanson, and J. W. Carson, Bin Loads–Part 3: Mass-Flow Bins, *Trans. ASME, J. Eng. for Ind.*, Feb. 1973.
13. A. W. Jenike, J. R. Johanson, and J. W. Carson, Bin Loads–Part 4: Funnel-Flow Bins, *Trans. ASME, J. Eng. for Ind.*, Feb. 1973.
14. A. W. Jenike and J. R. Johanson, Annual Report, Project 126–First Part, Mass-Flow Hoppers, American Iron and Steel Institute, Washington, D.C., April 1967, unpublished.
15. M. Reimbert and A. Reimbert, *Silos–Theory and Practice* (translated from Editions Eyrolles, Paris, 1961), Trans Tech Publications, Clausthal-Zellerfeld, West Germany, 1976.
16. M. Koenen, Berechnung des Seiten- und Bodendrucks in Silos, *Zentralbl. Bauverwaltung,* Vol. 16, 1896, pp. 446–449.
17. K. Pieper and F. Wenzel, Comments on DIN 1055 Design Loads for Buildings, Loads in Silo Bins, *Beton und Stahlbetonbau,* 1962, pp. 6–11.
18. K. Pieper and F. Wenzel, *Druckverhältnisse in Silozellen,* W. Ernst & Sohn, Berlin, 1964.
19. A. G. Homes, Lateral Pressures of Granular Material in Silos, *Trans. ASME, J. Eng. for Ind.,* Feb. 1973.
20. Instructions for Design of Silos for Granular Materials, *Soviet Code CH-302-65,* Gosstroy, USSR, Moscow, 1965.
21. S. S. Safarian, Design Pressure of Granular Material in Silos, *J. Am. Concr. Inst.*, Aug. 1969.
22. *ACI Standard 313-77,* Recommended Practice for Design and Construction of Concrete Bins, Silos and Bunkers for Storing Granular Materials, American Concrete Institute, Detroit, 1977.
23. Discussion of Proposed ACI Standard on Recommended Practice for Design and Construction of Concrete Bins, Silos and Bunkers for Storing Granular Materials, *J. Am. Concr. Inst.,* June 1976, p. 359.
24. Deutsche Normen, *DIN 1055,* Blatt 6, Lasten in Silozellen, Nov. 1964. Also supplementary provisions, May 1977.
25. Peter Martens, Lasten in Silozellen, *Die Mühle + Mischfuttertechnik,* Jan. 1978.
26. K. Pieper, *Proposed Revision of DIN 1055,* Teil 6, Lasten in Silozellen, March 1980.
27. A. W. Jenike, Load Assumptions and Distributions in Silo Design, Norwegian Society of Chartered Engineers Conference on Construction of Concrete Silos, Oslo, Jan. 1977.
28. A. P. Kovtum and P. N. Platonov, The Pressure of Grain on Silo Walls (in Russian), *Mukomol'no Elevat. Promst.,* Vol. 25, No. 12, Dec. 1959, pp. 22–24.
29. D. M. Walker, An Approximate Theory for Pressures and Arching in Hoppers, *Chem. Eng. Sci.,* Vol. 21, 1966, pp. 975–997.

30. D. M. Walker and M. H. Blanchard, Pressures in Experimental Coal Hoppers, *Chem. Eng. Sci.,* Vol. 22, 1967, pp. 1713–1745.
31. K. Clague and H. Wright, Pressures in Bunkers, ASME Paper No. 72-MH-31, American Society of Mechanical Engineers, New York, 1972.
32. H. C. S. Thom, New Distributions of Extreme Winds in the United States, *J. Struct. Div. ASCE,* July 1968.
33. *ANSI A58.1-1982, Minimum Design Loads for Buildings and Other Structures,* American National Standards Institute, Inc., New York, 1982.
34. A. G. Davenport, Rationale for Determining Wind Design Velocities, *J. Struct. Div. ASCE*, May 1960.
35. Uniform Building Code, International Conference of Building Officials, Whittier, Calif., 1982.
36. M. Esslinger, S. R. Ahmed, and H. H. Schroeder, Stationäre Windbelastung Offener und Geschlossener Kreiszylindrischer Silos, *Der Stahlbau,* Dec. 1971.
37. A. R. Chandrasekaran and P. C. Jain, Effective Live Load of Storage Materials under Dynamic Conditions, *Indian Concr. J.,* Sept. 1968.

6

DESIGN OF STRUCTURAL COMPONENTS

6-1. Introduction

The design of beams, columns, and plate elements, both stiffened and unstiffened, for bins and their supporting structure is considered in this chapter. The American Iron and Steel Institute (AISI) Specification for the Design of Cold-Formed Steel Structural Members is the standard in the United States for the design of cold-formed components, while the American Institute of Steel Construction (AISC) Specification for the Design, Fabrication and Erection of Structural Steel for Buildings is the standard for the design of hot-rolled components for building frames as well as many other types of structures. A number of the AISI and AISC formulas are identical or lead to essentially identical results. Many of the formulas have been adopted in other specifications, as, for example, the American Petroleum Institute (API) Standard 650, Welded Steel Tanks for Oil Storage.

6-2. Beams

The shape factor of a structural-member cross section is the ratio of the plastic moment ZF_y to the yield moment SF_y, where Z is the plastic section modulus and S the elastic section modulus. The shape factor is useful in establishing allowable bending stresses, for use in allowable-stress design, that are consistent with respect to the bending strength, which is given by the plastic moment. Shape factors for certain cross sections are as follows:

I:	1.10–1.18
Solid rectangle:	1.5
Thin-walled rectangular tube:	1.12
Solid circular:	1.70
Thin-walled circular tube:	1.27

The AISC allowable bending stress for laterally supported beams of I or box section, loaded in the plane of the minor axis, whose flanges and webs do not buckle locally until the yield moment is reached is $0.6F_y$. If the flanges and webs are thick enough so as not to buckle until the plastic moment is reached, the allowable stress is $0.66F_y$. The ratio of these two allowable stresses is 1.10, which is the lower limit of the range of shape factors for I and box sections. Similarly, the allowable stress for solid rectangular sections and for I-shaped members bent about their minor axes is $0.75F_y$, which is a partial allowance for the shape factor 1.5 of such sections. Sections to which these larger allowable stresses apply are called compact sections. The AISI specification does not cover compact sections because cold-formed members are usually of such proportions as to be unable to develop the plastic moment.

Local buckling. The allowable stresses quoted above govern if the limitations on flange and web slenderness given in Table 6–1 are satisfied. These limits are needed to assure the capacity of noncompact sections to reach yield strain on the extreme fiber and of compact sections to develop the plastic moment. However, provisions are made for more slender elements to be used (Sec. 6–7).

Elements with one unloaded edge free and the other unloaded edge supported are called *unstiffened elements* in the AISI and AISC specifications, while those with both unloaded edges supported are called *stiffened elements.*

Lateral buckling. Beams which are not supported laterally may twist and bend sideways before reaching the yield moment. This requires a reduction in the allowable bending stress. The reduced allowable stress for the Z cross section and for I, channel, box, and other cross sections symmetrical about the axis of bending and which are loaded through the shear center can be determined by treating the beam as a column with an equivalent radius of gyration given by[1]

$$r_{eq}^2 = C_b \frac{\sqrt{I_y}}{S_x} \sqrt{C_w + 0.04J(KL)^2} \qquad (6\text{–}1)$$

where C_b = coefficient based on variation in moment along the beam
I_y = y-axis moment of inertia
S_x = x-axis section modulus
C_w = section warping constant
J = section torsion constant
K = effective length coefficient
L = span of beam

Values of C_w and J for various shapes are given in the AISC and AISI manuals. [2,3] For open sections, $J = \Sigma\ bt^3/3$ of all the elements of the section. Formulas for C_w are given in various books and manuals[1,3,4]; C_w can be

TABLE 6-1. WIDTH-THICKNESS LIMITS FOR BEAM ELEMENTS[a]

	AISI	AISC	
Type of element	$F_b = 0.6F_y$	$F_b = 0.6F_y$	$F_b = 0.66F_y$
Flange Unstiffened	$\frac{63.3^b}{\sqrt{F_y}}$	$\frac{95^c}{\sqrt{F_y}}$	$\frac{65^c}{\sqrt{F_y}}$
Stiffened	$\frac{171^{d,e}}{\sqrt{f}}$	$\frac{253^{f,g}}{\sqrt{F_y}}$	$\frac{190^f}{\sqrt{F_y}}$
Web	200^h	$\frac{760^i}{\sqrt{F_b}}$	$\frac{640^j}{\sqrt{F_y}}$

[a] f (service-load stress) and F_y in ksi.

[b] w/t, where w = flat projection of flange from web, excluding fillets and stiffening lip.

[c] b/t, where b = full nominal dimension of angle legs, flanges of channels and zees and half the nominal flange width of Is and tees.

[d] w/t, where w = flat width of flange between webs, excluding fillets.

[e] $184/\sqrt{f}$ for square and rectangular tubular sections.

[f] b/t, where b = distance between nearest lines of fasteners or welds or between roots of flanges of rolled sections.

[g] $238/\sqrt{F_y}$ for square and rectangular tubular sections of uniform thickness.

[h] h/t, where h = clear distance between flanges measured in plane of web; limits are 300 if web has bearing stiffeners and intermediate stiffeners and 260 if only bearing stiffeners.

[i] h/t, where h = clear distance between flanges and $F_b = 0.6F_y$ if beam is laterally supported or value by Eqs. (6-5) or (6-6) if not; larger values permitted with reduced allowable stress or web stiffeners.

[j] d/t, where d = depth of beam; for members under combined bending and axial compression, use $d/t = (640/\sqrt{F_y})(1 - 3.74f_a/F_y)$ if $f_a/F_y \leqslant 0.16$ and $257/\sqrt{F_y}$ if $f_a/F_y > 0.16$.

taken as zero for tees and angles. Values of C_b for various loading conditions are given in Ref. 1, and Eq. (6–3) can also be used. The effective length coefficient is determined as for columns. For example, $K = 1$ if the beam is free to turn about the y axis at each end and 0.5 if y-axis rotation is prevented. Similarly, $K = 0.5$ if the beam is free to rotate about the y axis at each end and has lateral support at midspan.

For bending about the major (x) axis, the AISI allowable stress for laterally unsupported symmetrical channels and I sections symmetrical about the plane of the web, given here in slightly different form, is

$$\frac{F_b}{F_y} = \frac{2}{3} - 0.0188 \frac{F_y}{E} \frac{L^2 S_{xc}}{dI_{yc} C_b} \qquad 3.55C_b \frac{E}{F_y} \leqslant \frac{L^2 S_{xc}}{dI_{yc}} \leqslant 17.77C_b \frac{E}{F_y} \tag{6–2a}$$

$$F_b = 174{,}690 \frac{dI_{yc} C_b}{L^2 S_{xc}} \qquad 17.77C_b \frac{E}{F_y} \leqslant \frac{L^2 S_{xc}}{dI_{yc}} \tag{6–2b}$$

where L = unbraced length of beam
d = depth of beam
S_{xc} = x-axis section modulus of compression flange
I_{yc} = y-axis moment of inertia of compression portion of cross section

The value of C_b is given by

$$C_b = 1.75 + 1.05 \frac{M_1}{M_2} + 0.3 \left(\frac{M_1}{M_2}\right)^2 \leqslant 2.3 \tag{6–3a}$$

$C_b = 1$ if the moment at any point between M_1 and M_2 exceeds M_2 (6–3b)

M_1 in these equations is the smaller and M_2 the larger bending moment at the ends of the unbraced length of the beam, with M_1/M_2 to be taken negative if the unbraced length bends in single curvature.

The AISI allowable lateral-buckling stress for Z sections is

$$\frac{F_b}{F_y} = \frac{2}{3} - 0.0375 \frac{F_y}{E} \frac{L^2 S_{xc}}{dI_{yc} C_b} \qquad 1.78C_b \frac{E}{F_y} \leqslant \frac{L^2 S_{xc}}{dI_{yc}} \leqslant 8.88C_b \frac{E}{F_y} \tag{6–4a}$$

$$F_b = 87{,}350 \frac{dI_{yc} C_b}{L^2 S_{xc}} \qquad 8.88C_b \frac{E}{F_y} \leqslant \frac{L^2 S_{xc}}{dI_{yc}} \tag{6–4b}$$

Equation (6–2b) can be derived directly from Eq. (6–1). Since the torsion constant J is proportional to the cube of the element thickness, the second term under the radical in Eq. (6–1) is small, compared to the first, for light-gage sections and can be neglected. Since $C_w = d^2 I_y/4$ for I sections, $r_{eq}^2 = C_b(d/S_x)(I_y/2)$ if $J = 0$. Using $I_y/2 = I_{yc}$, substituting the result into the Euler formula $F = \pi^2 E/(L/r)^2$, and using a factor of safety of 1.67, we obtain Eq. (6–2b).

The AISC allowable stress for laterally unsupported noncompact I- and channel-shaped sections, given here in slightly different form, is the larger value given by Eqs. (6–5) or Eq. (6–6) but not to exceed $0.6F_y$:

$$\frac{F_b}{F_y} = \frac{2}{3} - \frac{F_y}{C_b}\left(\frac{L/r_T}{1237}\right)^2 \qquad 319\sqrt{\frac{C_b}{F_y}} \leqslant \frac{L}{r_t} \leqslant 714\sqrt{\frac{C_b}{F_y}} \tag{6–5a}$$

$$F_b = \frac{170{,}000C_b}{(L/r_T)^2} \qquad 714\sqrt{\frac{C_b}{F_y}} \leqslant \frac{L}{r_T} \tag{6–5b}$$

$$F_b = \frac{12{,}000C_b}{Ld/A_f} \tag{6–6}$$

where L = distance between lateral supports
r_T = radius of gyration of a T consisting of the compression flange and the adjacent third of the part of the web in compression
A_f = area of compression flange

Equations (6–5) and (6–6) are based indirectly on Eq. (6–1). Equations (6–5) give approximately the stress that would be obtained by treating the beam as a column with r_{eq} determined by Eq. (6–1) with $J = 0$, while Eq. (6–6) gives the stress that would be determined in the same way with $C_w = 0$. The larger value is to be taken since each is an underestimate because it neglects one element of the buckling resistance.

Lateral-torsional buckling of box beams is not usually a problem because of their superior torsional stiffness. The allowable stress $F_b = 0.6F_y$ for noncompact box sections is applicable according to the AISI if

$$\frac{L}{b} \leqslant \frac{2500}{F_y} \tag{6–7}$$

where b is the width of the section, and according to the AISC if

$$\frac{d}{b} \leqslant 6 \tag{6–8}$$

The AISC stress $F_b = 0.66F_y$ for compact sections is permitted only if the laterally unsupported length is such as to prevent lateral buckling before the plastic moment can be developed. Since this requires larger extreme-fiber strains than for noncompact sections, the distance between lateral supports must be smaller. According to the AISC, lateral support is needed at intervals L given by

$$L \leqslant \frac{76b}{\sqrt{F_y}} \leqslant \frac{20{,}000}{F_y d/A_f} \qquad \text{for I shapes} \qquad (6\text{–}9)$$

$$\frac{L}{b} = \frac{1950 + 1200M_1/M_2}{F_y} \qquad \text{for box shapes} \qquad (6\text{–}10)$$

but need not be less than $1200/F_y$ for box shapes.

Shear. The AISI allowable shear F_v on the gross area of unstiffened beam webs is given by

$$F_v = \frac{152\sqrt{F_y}}{h/t} \leqslant 0.4F_y \qquad \frac{h}{t} \leqslant \frac{547}{\sqrt{F_y}} \qquad (6\text{–}11a)$$

$$F_v = \frac{83{,}200}{(h/t)^2} \qquad \frac{h}{t} > \frac{547}{\sqrt{F_y}} \qquad (6\text{–}11b)$$

Furthermore, the web must satisfy the interaction formula

$$\left(\frac{f_v}{F_v}\right)^2 + \left(\frac{f_b}{F_b}\right)^2 \leqslant 1 \qquad (6\text{–}11c)$$

where, for beams with stiffened compression flanges,

$$F_b = 0.6F_y \left[1.21 - 0.00034\sqrt{F_y}\left(\frac{h}{t}\right)\right] \leqslant 0.6F_y$$

and for beams with unstiffened compression flanges,

$$F_b = 0.6F_y \left[1.26 - 0.00051\sqrt{F_y}\left(\frac{h}{t}\right)\right] \leqslant 0.6F_y$$

The AISC allowable shear on the gross area of unstiffened beam webs is $0.4F_y$ if the web slenderness h/t does not exceed $380/\sqrt{F_y}$.

Deflection. It is difficult to prescribe tolerable deflection limits because they depend so much on judgment and service requirements. A frequently quoted limit prohibits deflection exceeding $\frac{1}{360}$ of the span. This value is prescribed by the AISC for beams supporting plastered ceilings. Other limits suggested in the commentary to the AISC specification are given in terms of the span-depth ratio of the beam. Span-depth ratios $L/d = 800/F_y$ and $1000/F_y$ are suggested for fully stressed beams in floors and for purlins, except those in flat roofs, respectively. The relation between span-depth ratio and deflection-span ratio for a uniformly loaded beam is given by[1]

$$\frac{\delta}{L} = \frac{5}{24} \frac{f_b}{E} \frac{L}{d} \tag{6-12}$$

Thus, the deflection-span ratios for the L/d limits quoted above are $\frac{1}{275}$ and $\frac{1}{220}$, respectively, for fully stressed compact sections ($f_b = 0.66F_y$). Larger deflections may be acceptable in certain components of bins and their supporting structures. For example, a deflection-span limit of $\frac{1}{180}$ gives a deflection of 1 in. in a 15-ft span, which, unless it is visually unacceptable or interferes in some way with the operation, does not seem excessive, particularly if it is due to a maximum design load that has a small probability of occurrence.

Equation (6-12) shows that the value of L/d for an understressed beam which will have the same deflection as a fully stressed beam can be determined by multiplying the specified L/d for the fully stressed beam by the inverse ratio of the stresses. Thus, the allowable L/d for an A36 beam of compact section stressed to only 20 ksi is $24/20 = 1.2$ times the value for a fully stressed beam.

6-3. Compression Members

Short centrally loaded compression members fail by general yielding or by local buckling. Longer members which do not fail by local buckling fail by overall buckling in a manner which depends on the cross-sectional shape. There are three categories of overall buckling:

(a) Compression members with cross sections symmetrical about both principal axes can fail by bend buckling about either principal axis or by torsional buckling (Figs. 6-1a, b, and c). Each of these modes is independent of the others.

(b) Compression members with one axis of symmetry can fail by bend buckling about the principal axis perpendicular to the axis of symmetry (Fig. 6-1d) or by a combination of twisting and bending about the other principal axis (Fig. 6-1e).

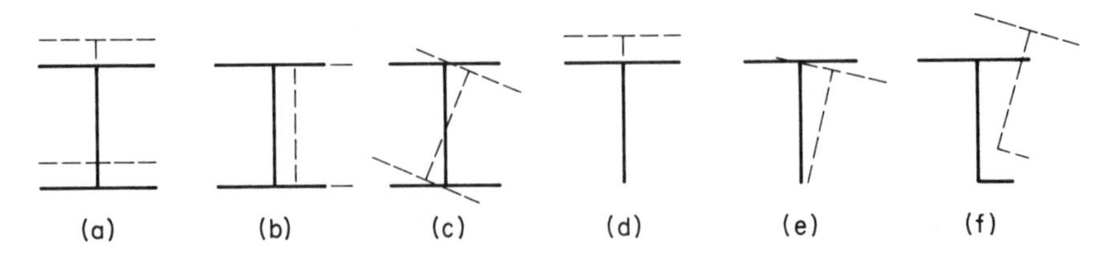

Figure 6-1

(c) Compression members with no axis of symmetry fail by a combination of the twist buckling and the two bending modes (Fig. 6–1f).

Whether a column fails by one of the three buckling modes in (a) or the two in (b) depends on the relative stiffnesses in these modes and on restraints that may prevent buckling in one or more modes. For example, attached construction may prevent a compression member from twisting but allow it to bend.

All closed sections, such as tubular members, are so torsionally stiff as to always fail in bend buckling. Members with doubly symmetrical cross sections, such as the I, or point symmetrical cross sections like the Z, also fail in the weaker bending mode rather than by twisting.

Members with one axis of symmetry which are only slightly weaker in the twist-bend buckling mode are the standard rolled channel and the built-up I with unequal flanges. Therefore, they can be designed in the usual way for bend buckling.[4] However, thin-walled channels such as are used in cold-formed construction may be much weaker in twist-bend buckling. The principal thin-walled sections that need to be checked for possible failure in this mode are angles, hat sections, C-sections, and channels. The charts of Fig. 6–2 enable one to determine whether these members will fail by bend buckling or twist-bend buckling. If a proposed member will fail by twist-bend buckling, it may be wise to redesign it rather than to make the more difficult check for this mode.

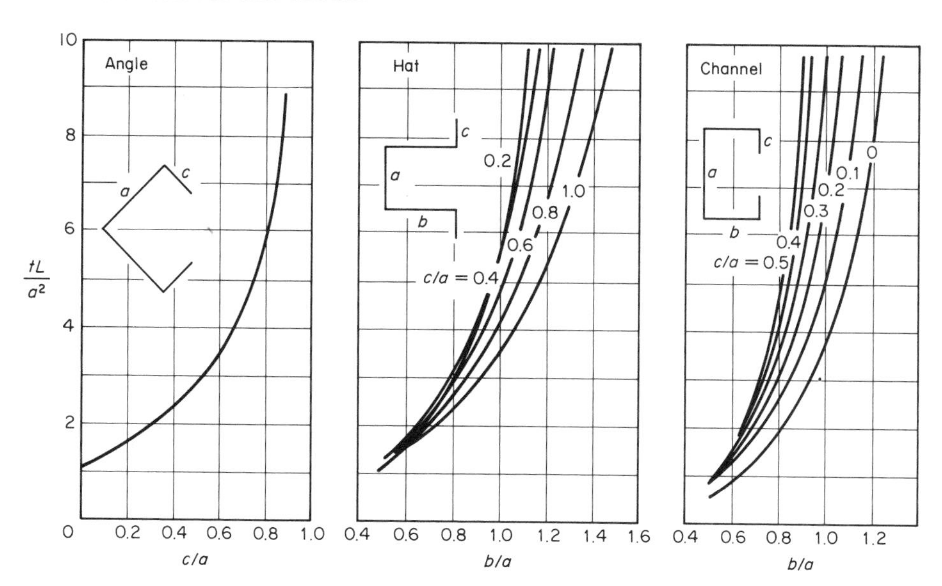

Figure 6–2 Type of buckling for angle, hat, or channel compression members. Members which plot to left of pertinent curve buckle in simple flexure; members which plot to right buckle in combined torsion and flexure. (From Ref. 5.)

Allowable stresses. The AISC allowable stress F_a for centrally loaded columns is given by

$$\frac{F_a}{F_y} = \frac{1 - \dfrac{1}{2}\left(\dfrac{KL/r}{C_c}\right)^2}{\dfrac{5}{3} + \dfrac{3}{8}\dfrac{KL/r}{C_c} - \dfrac{1}{8}\left(\dfrac{KL/r}{C_c}\right)^3} \qquad \frac{KL}{r} \leqslant C_c \tag{6–13a}$$

$$F_a = \frac{149{,}000}{(KL/r)^2} \qquad \frac{KL}{r} \geqslant C_c \tag{6–13b}$$

where K is the effective length coefficient, r is the radius of gyration, and $C_c = \pi\sqrt{2E/F_y}$. The denominator in Eq. (6–13a) is the factor of safety, which varies from 1.67 at $KL/r = 0$ to 1.92 at $KL/r = C_c$. Equation (6–13b) is the Euler buckling stress divided by a factor of safety of 1.92. These equations are derived from bend-buckling formulas. The AISC does not cover twist-bend buckling.

Theoretical values of K for various end conditions are shown in Fig. 6–3. More complete tabulations are available for other situations, such as in trusses and building frames.[4,6]

The AISI allowable stress for members that are not susceptible to torsional buckling, or which are braced against torsional buckling, is

$$F_a = 0.522QF_y - \left(\frac{QF_y\,KL/r}{1494}\right)^2 \qquad \frac{KL}{r} \leqslant \frac{C_c}{\sqrt{Q}} \tag{6–14a}$$

$$F_a = \frac{151{,}900}{(KL/r)^2} \qquad \frac{KL}{r} \geqslant \frac{C_c}{\sqrt{Q}} \tag{6–14b}$$

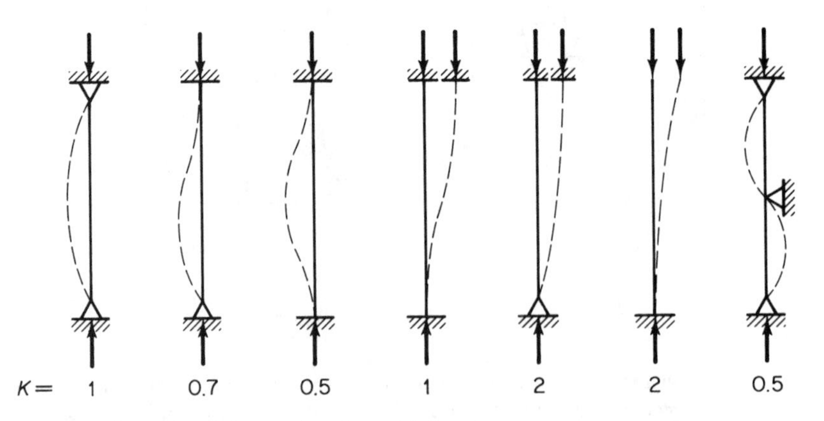

Figure 6–3 Effective length coefficients for compression members.

Q in these formulas is a factor which accounts for local buckling of elements of the cross section whose slenderness limits exceed the values that enable them to reach yield stress (Table 6–2). If all elements meet the requirements of Table 6–2, $Q = 1$. The evaluation of Q for other cases is explained in Sec. 6–9. Larger allowable stresses are permitted if $Q = 1$ and the thickness of the steel is 0.09 in. or more; the AISC formula, Eq. (6–13a), is specified for this case instead of Eq. (6–14a).

Local buckling. The AISC allowable stresses according to Eqs. (6–13) apply only if the limits on element slenderness in Table 6–2 are satisfied. However, larger slendernesses are permitted if F_y is replaced by QF_y in Eq. (6–13a) and in the formula for C_c. The AISI values in Table 6–2 are those for which $Q = 1$ in Eqs. (6–14). Some of the entries in Table 6–2 are the same as in Table 6–1; they are repeated here for convenience.

Twist-bend buckling. Twist-bend buckling for members whose cross sections have one axis of symmetry can be investigated by computing the

TABLE 6–2. WIDTH-THICKNESS LIMITS FOR COMPRESSION MEMBERS[a–d]

Type of element	AISI	AISC
Unstiffened		
Single angle	$\frac{63.3}{\sqrt{F_y}}$	$\frac{76}{\sqrt{F_y}}$
Stem of tee	$\frac{63.3}{\sqrt{F_y}}$	$\frac{127}{\sqrt{F_y}}$
Other	$\frac{63.3}{\sqrt{F_y}}$	$\frac{95}{\sqrt{F_y}}$
Stiffened		
Flanges of square and rectangular sections of uniform thickness	$\frac{184}{\sqrt{f}}$	$\frac{238}{\sqrt{F_y}}$
Other	$\frac{171}{\sqrt{f}}$	$\frac{253}{\sqrt{F_y}}$

[a] f (service-load stress) and F_y in ksi.

[b] AISI limits are values of w/t, where w = flat width of element excluding fillets.

[c] AISC limits are values of b/t, where b = full nominal dimension of angle legs, flanges of channels and zees, webs of tees, and half the nominal width of flanges of Is and tees.

[d] The AISI does not distinguish between types of unstiffened elements.

equivalent radius of gyration r_{tb} given by (the y axis is taken as the axis of symmetry)[1]

$$\frac{1}{r_{tb}^2} = \frac{1}{2r_y^2} + \frac{1}{2r_t^2} + \sqrt{\left(\frac{1}{2r_y^2} - \frac{1}{2r_t^2}\right)^2 + \left(\frac{y_0}{r_y r_t r_p}\right)^2} \qquad (6\text{–}15)$$

in which

$$r_t = \sqrt{\frac{C_w + 0.04J(KL)^2}{I_{ps}}}$$

I_{ps} = polar moment of inertia referred to shear center $= I_x + I_y + Ay_0^2$
A = cross-sectional area
K = effective length coefficient
y_0 = distance from center of gravity to shear center
$r_p = \sqrt{I_{ps}/A}$
r_y = radius of gyration for y axis (axis of symmetry)

C_w and J in this formula are defined in connection with Eq. (6–1). The same effective length coefficient must be used to determine r_t and the equivalent slenderness ratio KL/r_{tb}. If KL/r_{tb} is larger than K_xL/r_x, the member fails by twist-bend buckling. The effective length coefficient K_x for bend buckling about the x axis may differ from the value K for twist-bend buckling. The allowable stress is determined by substituting the larger effective slenderness ratio in a column formula. Equations (6–13) may be used for the AISC specification and Eqs. (6–14) for the AISI. The AISI formulas for allowable stress in twist-bend buckling give the same result as the procedure just described if KL/r_{tb} is used in Eqs. (6–14). However, it is then necessary to compute the allowable stress for bend buckling about the axis perpendicular to the axis of symmetry to determine which buckling mode controls. Using the equivalent radius of gyration, one need compute only the allowable stress corresponding to the smaller of r_{tb} and the radius of gyration for the axis perpendicular to the axis of symmetry and then calculate the corresponding allowable stress.

Example 6-1

Determine the allowable load P for a compression member consisting of a WT8 × 18 10 ft long. The ends are free to rotate about the x and y axes ($K = 1$) and free to warp ($K = 1$). A36 steel, AISC specification.

The following properties of the section are from the AISC manual (the torsion constant J is one-half that of the W16 × 36 from which the WT8 is cut). The warping constant C_w can be taken as zero for tees.[1,4]

$$A = 5.28 \text{ in.}^2, \quad t_w = 0.295 \text{ in.}, \quad I_x = 30.6 \text{ in.}^4, \quad I_y = 12.2 \text{ in.}^4,$$

$$r_x = 2.41 \text{ in.}, \quad r_y = 1.52 \text{ in.}, \quad J = \frac{0.54}{2} = 0.27 \text{ in.}^4, \quad C_w = 0$$

The shear center of a tee is at the intersection of the flange and stem centerlines. Therefore,

$$y_0 = 1.88 - \frac{0.295}{2} = 1.74 \text{ in.}$$

$$I_{ps} = 30.6 + 12.2 + 5.28 \times 1.74^2 = 58.8 \text{ in.}^4$$

$$r_p = \sqrt{\frac{58.8}{5.28}} = 3.34 \text{ in.}$$

$$r_t = \sqrt{\frac{0 + 0.04 \times 0.27 \times 120^2}{58.8}} = 1.63 \text{ in.}$$

$$\frac{1}{2r_y^2} \pm \frac{1}{2r_t^2} = \frac{0.5}{1.52^2} \pm \frac{0.5}{1.63^2} = 0.405, 0.0282$$

$$\frac{y_0}{r_y r_p r_t} = \frac{1.74}{1.52 \times 3.34 \times 1.63} = 0.210$$

Then from Eq. (6–15),

$$\frac{1}{r_{tb}^2} = 0.405 + \sqrt{0.0282^2 + 0.210^2} = 0.617 \qquad r_{tb} = 1.27 \text{ in.}$$

Since r_{tb} is less than r_x, twist-bend buckling controls:

$$\frac{L}{r_{tb}} = \frac{120}{1.27} = 94.5$$

Equation (6–13a) gives $F_a = 13.66$ ksi

$$P = 13.66 \times 5.30 = 72.4 \text{ kips}$$

6-4. Combined Bending and Axial Compression

The AISC formula for members subjected to bending and axial compression is

$$\frac{f_a}{F_a} + \frac{C_{mx}}{1 - f_a/F'_{ex}} \frac{f_{bx}}{F_{bx}} + \frac{C_{my}}{1 - f_a/F'_{ey}} \frac{f_{by}}{F_{by}} \leqslant 1 \qquad (6\text{–}16a)$$

$$\frac{f_a}{0.6F_y} + \frac{f_{bx}}{F_{bx}} + \frac{f_{by}}{F_{by}} \leqslant 1 \qquad (6\text{–}16b)$$

where x, y = axes of bending
$f_a = P/A$ = stress due to axial force P
f_b = bending compressive stress at the section being checked
F_a = allowable compressive stress for P alone
F_b = allowable compressive stress for bending alone
$F'_e = 149{,}000/(Kl/r)^2$

C_m = 0.85 for members in frames subject to sidesway
C_m = 0.85 for transversely loaded members with rotational end restraint, in frames not subject to sidesway*
C_m = 1 for transversely loaded members without rotational end restraint, in frames not subject to sidesway*
$C_m = 0.6 - 0.4M_1/M_2 \geqslant 0.4$ for members in frames not subject to sidesway and with no transverse load between supports in the plane of bending, where M_1/M_2 is the ratio of the smaller end moment to the larger and is positive if the member is bent in reverse curvature

The factor $1 - f_a/F'_e$ in the second and third terms of Eq. (6–16a) accounts for the moments $P\delta$ of the axial force P, which result from the deflection δ. Two cases must be recognized:

(a) If there are transverse loads and end moments M_1 and M_2 ($M_2 > M_1$), Eq. (6–16a) must be computed at the section between supports where the moment is a maximum, while f_b in Eq. (6–16b) is computed for the support at which M_2 acts. In computing F_b, C_b in Eqs. (6–5) and (6–6) must be taken equal to unity, because C_m and C_b perform the same function (accounting for the variation in moment along the member) and are in fact reciprocal.[1]

(b) If there are no transverse loads, f_b in both Eqs. (6–16) is computed for the larger end moment (M_2).

The secondary moment $P\delta$ tends to be small compared to the primary moment if P is relatively small, and for cases where f_a/F'_e is less than 0.15, the following simpler formula may be used:

$$\frac{f_a}{F_a} + \frac{f_{bx}}{F_{bx}} + \frac{f_{by}}{F_{by}} \leqslant 1 \qquad (6\text{–}17)$$

Equations (6–16) and (6–17) are also used in the AISI specification except that they are restricted to doubly symmetric shapes or any shape not subject to twist-bend buckling, and F_a in Eq. (6–17) is multiplied by a factor Q defined in Sec. 6–9.

6–5. Curved Members

Curved members subjected to bending and axial compression occur as rafters in domed roofs. Since they usually have reaction components directed along the chord, they can usually be analyzed as arches.

*According to the specification, a value determined by analysis may be used in lieu of the specified value for this case.

Buckling in plane of arch. Formulas for buckling of an arch in its plane have been derived for arches having various forms and load distributions. However, since buckling in the plane of a curved rafter is restrained by the roof shell (except for snap-through buckling), they will not be given here. These buckling modes are discussed in more detail in Sec. 7–9.

Lateral buckling. If the roof plate of supported roofs is not fastened to the rafters, the possibility of lateral-torsional buckling of the rafter must be considered. Tests discussed in Sec. 7–5 show that friction between a beam flange and the load-carrying elements it supports is very effective in resisting lateral-torsional buckling, and it is reasonable to expect this to be true for arched beams as well. However, lateral-torsional buckling due to compression directed along the chord, which can result from conveyor or other loads on the compression ring, is possible when there is no roof load to ensure contact of the rafter and the roof plate (Sec. 7–5). This mode of collapse has been investigated for a pin-ended, circular-arc member.[7] The buckling force is

$$P_{cr} = k \frac{EI_y}{(r\alpha/2)^2} \tag{6-18}$$

where r, α (radians), and P are defined in Fig. 6–4. The coefficient k depends on the lateral-bending stiffness EI_y, the St. Venant torsional stiffness GJ, and the warping stiffness EC_w. Charts are given in Ref. 7. The following formula gives good results for cross sections whose warping stiffness is small compared to the St. Venant torsional stiffness[8]:

$$P_{cr} = \frac{\pi^2 EI_y}{8r^2\,[1 - \cos(\alpha/2)]} \tag{6-19}$$

Equation (6–19) gives conservative results for cross sections, such as the I, whose warping stiffness may contribute significantly to resistance to lateral-torsional buckling. It is subject to the usual restriction that buckling occurs while member stresses are at or below the proportional limit. The authors suggest that inelastic buckling be treated as for straight compression members. This is readily accomplished by using an equivalent slenderness ratio derived by equating $F_{cr} = P_{cr}/A$ to the Euler critical stress $\pi^2 E/(L/r_y)^2_{eq}$.

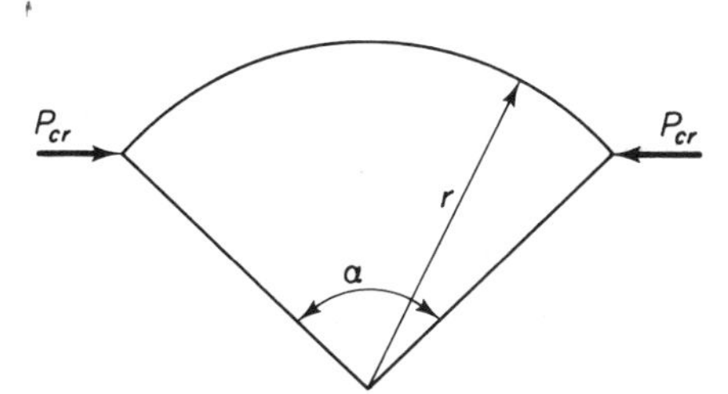

Figure 6–4

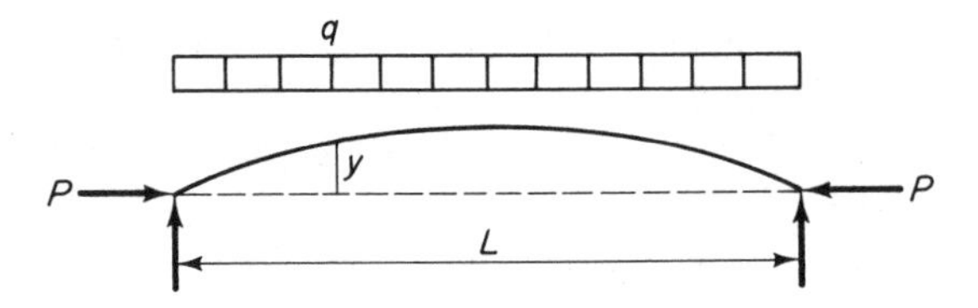

Figure 6–5

The result is

$$\left(\frac{L}{r_y}\right)_{eq} = \frac{r}{r_y}\sqrt{8\left(1 - \cos\frac{\alpha}{2}\right)} \tag{6–20}$$

where r_y = weak-axis radius of gyration of the member cross section. Allowable stresses can be determined by using $(L/r_y)_{eq}$ in Eqs. (6–13) or (6–14), as applicable, with $K = 1$.

Combined bending and axial compression. Equation (6–16a) should not be used for curved members subjected to bending and axial compression (Fig. 6–5). It is incorrect because F_a is evaluated for a column, that is, for buckling under the effect of P alone. The maximum stress in the curved member of Fig. 6–5 is given with good approximation by

$$f = \frac{P}{A} + \frac{M_q - Py}{S_x}\,\frac{1}{1 - P/P_E} \tag{6–21}$$

where M_q = moment due to transverse load
$P_E = \pi^2 EI_x/L^2$
L = chord length
x = cross-sectional principal axis normal to plane of buckling

The factor $1/(1 - P/P_E)$ amplifies the moment $M_q - Py$, which is computed for the undeformed member, to account for the second-order moment due to deflection. It is an approximation because the true amplification factor depends on the shape of the member and the nature of the transverse load.

For allowable-stress design Eq. (6–21) can be written in the form

$$f = f_a + \frac{f_b}{1 - f_a/F'_E} \leqslant F \tag{6–22}$$

where F is the allowable combined stress and F'_E is an allowable Euler stress as in the AISC formula. Consistency requires the same factor of safety to be used in determining these two values.[1] However, the value of F'_E specified by the AISC for Eq. (6–16a) is computed for a factor of safety of 1.92, while the basic value of F is $0.6F_y$.

The superior bending strength of compact sections cannot be taken into account with the combined stress expressed as in Eq. (6–22). It would be incorrect to use $F = 0.66F_y$ because in the limiting case where f_b is zero the

formula would give an incorrect allowable value for f_a for which the compact section has no advantage. To cover this situation, we can divide through by F and then arbitrarily use $F = 0.6F_y$ in the first term and $F = F_b$ in the second. This gives

$$\frac{f_a}{0.6F_y} + \frac{f_b}{F_b}\,\frac{1}{1 - f_a/F'_E} \leqslant 1 \tag{6-23}$$

where $F_b = 0.66F_y$ if the section is compact and $0.6F_y$ if it is not. As already mentioned, lateral buckling need not be considered in determining F_b.

6-6. Buckling of Flat Plates

The stress at which a flat rectangular plate, loaded in its plane in various ways along its edges, buckles out of its plane is given by

$$F_{cr} = \frac{k\pi^2 E}{12(1 - \mu^2)(b/t)^2} \tag{6-24}$$

where k = coefficient dependent on the nature of the loading, the plate aspect ratio a/b, and the type of edge support
μ = Poisson's ratio
a, b = lengths of edges
t = plate thickness

Since this equation contains the modulus of elasticity E, it holds only for buckling at stresses in the elastic range.

Plates in compression. For the case of plates in compression, b in Eq. (6-24) is the length of the loaded edges. Values of k for various types of edge support are given in Fig. 6-6. Except for case e in Fig. 6-6, the plate

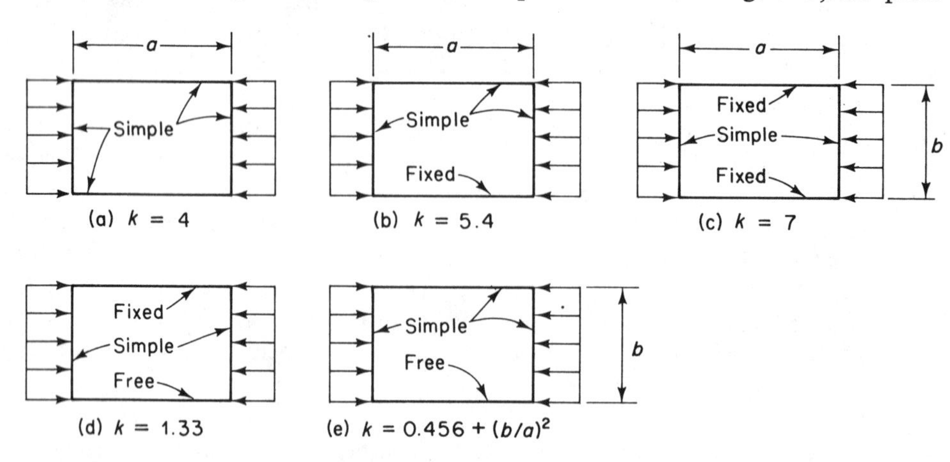

Figure 6-6 Plate-buckling coefficients.

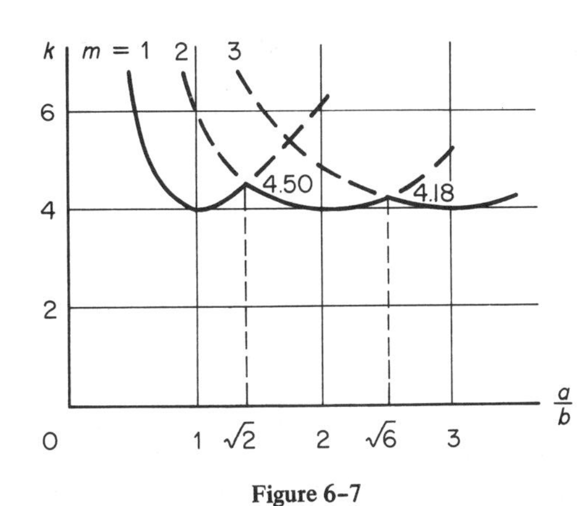

Figure 6–7

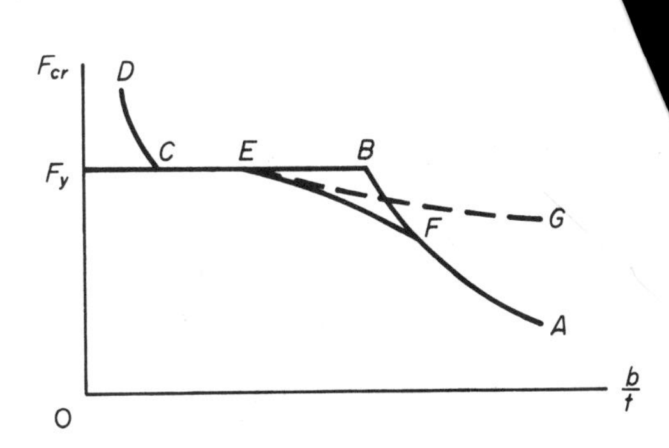

Figure 6–8

buckles in one or more half-waves in the direction of the compression and one half-wave transversely; the plate of case e buckles in only one half-wave in both directions. Figure 6–7 shows the variation of k with a/b for case a for one, two, and three longitudinal half-waves. The curves are tangent to the line $k = 4$ for these and all other values of m, so that $k = 4$ is an acceptable value for all aspect ratios except the very unusual case of a plate of aspect ratio less than about 0.7. Values of k for the other conditions of support, except e, in Fig. 6–6 are obtained similarly. For case e, k approaches a limiting value 0.456 with increasing a/b, but for a plate as short as $a = 5b$, is only 9% larger than the limiting value. Therefore, $k = 0.456$ is a good approximation in most cases.

The variation of critical stress with b/t is shown in Fig. 6–8. *ABCD* represents a plate made of steel with a flat-top yield and no residual stresses or eccentricities of the edge stress; *AB* is given by Eq. (6–24), while *CD* indicates the strain hardening that can develop in a stiff plate. However, *AFECD* is a more realistic representation, where inelastic buckling begins at *F*. Therefore, in practical cases Eq. (6–24) gives only the branch *AF* of the local-buckling curve.

The limiting values of b/t in Table 6–2 and the flange b/t for $F_b = 0.6F_y$ in Table 6–1 correspond approximately to point *E* in Fig. 6–8; in other words, the plate slenderness is such as to allow the yield stress to be reached without buckling.* Therefore, in these cases investigation of the local-buckling stress is unnecessary. Furthermore, where slenderer plate elements are used, so that buckling occurs at stresses represented by *AFE*, the postbuckling strength, shown by *EG* in Fig. 6–8, is taken into account in the AISC and AISI specifications, so that computation of a plate-buckling stress is again unnecessary.

*Values for $F_b = 0.66F_y$ in Table 6–1 allow the plate to be strained beyond the beginning of yield without buckling, so that the plastic bending moment can be attained.

6-7. Postbuckling Strength of Compressed Plates

If a compressed plate supported on the four edges buckles at a stress smaller than the yield stress, further shortening increases the stress in the vicinity of the edges, where the plate is forced to remain straight or nearly so, but produces little or no increase in the interior portion because here such shortening is accommodated more easily by an increase in amplitude of the buckle. The result is a nonuniform distribution of stress. Procedures for computing the resultant force are based on a uniform distribution of the edge stress f_e over two edge strips each of width $b_e/2$, where b_e is the *effective width.*

The stress in a compressed plate supported at only one unloaded edge (case e of Fig. 6–6) is smallest at the unsupported edge, and the effective width is assumed to be a strip b_e adjacent to the supported edge.

Various formulas for computing the effective width are discussed in Ref. 1. The following have been shown to be in good agreement with results of tests [9,10]:

1. Plates supported on four edges (stiffened element):

$$\frac{b_e}{b} = \sqrt{\frac{F_{\mathrm{cr}}}{f_e}}\left(1 - 0.22\sqrt{\frac{F_{\mathrm{cr}}}{f_e}}\right) \tag{6-25}$$

2. Plates free on one unloaded edge (unstiffened element):

$$\frac{b_e}{b} = 1.19\sqrt{\frac{F_{\mathrm{cr}}}{f_e}}\left(1 - 0.298\sqrt{\frac{F_{\mathrm{cr}}}{f_e}}\right) \tag{6-26}$$

where b_e = effective width
b = flat width (width excluding fillets)
f_e = compressive stress at supported edge of element

The ultimate strength of the element is assumed to be reached when $f_e = F_y$. However, the formulas are also valid for $f_e < F_y$.

Substituting F_{cr} from Eq. (6–24) into Eqs. (6–25) and (6–26), with E = 29,500 ksi and $\mu = 0.3$, gives the following:

1. For stiffened elements:

$$\frac{b_e}{t} = 163.3\sqrt{\frac{k}{f_e}}\left(1 - \frac{35.9}{b/t}\sqrt{\frac{k}{f_e}}\right) \quad \text{for } \frac{b}{t} \geqslant 110.0\sqrt{\frac{k}{f_e}} \tag{6-27}$$

2. For unstiffened elements:

$$\frac{b_e}{t} = 194.3\sqrt{\frac{k}{f_e}}\left(1 - \frac{48.7}{b/t}\sqrt{\frac{k}{f_e}}\right) \quad \text{for } \frac{b}{t} \geqslant 97.2\sqrt{\frac{k}{f_e}} \tag{6-28}$$

The limiting values of b/t in these equations are found by equating b_e to b; that is, they give the values of b/t below which the element is fully effective.

Values of k for a stiffened element vary from 4 if the unloaded edges are hinged to 7 if they are fixed and for an unstiffened element from 0.456 if the supported unloaded edge is hinged to 1.33 if it is fixed (Fig. 6–6). Intermediate values for rotational restraint from adjoining elements, such as webs, can be computed by formulas given in Ref. 4. Of course, conservative values of the effective width are obtained by assuming hinged edges.

To obtain effective width formulas in terms of a service (working-load) stress f at a factor of safety of 1.65, f_e in Eqs. (6–27) and (6–28) is replaced by $1.65f$. Then, with $k = 4$, Eq. (6–27) gives

$$\frac{b_e}{t} = \frac{253}{\sqrt{f}}\left(1 - \frac{55.3}{(b/t)\sqrt{f}}\right) \quad \text{for } \frac{b}{t} \geqslant \frac{171}{\sqrt{f}} \tag{6–29}$$

This is the AISI formula for safe-load determination of all stiffened elements except flanges of square and rectangular tubes, for which the coefficients 55.3 and 171 are replaced with 50.3 and 184.

The AISC effective width formula for stiffened elements in members other than square and rectangular tubes of uniform thickness is

$$\frac{b_e}{t} = \frac{253}{\sqrt{f}}\left(1 - \frac{44.3}{(b/t)\sqrt{f}}\right) \quad \text{for } \frac{b}{t} \geqslant \frac{196}{\sqrt{f}} \tag{6–30}$$

The coefficients 44.3 and 196 are replaced by 50.3 and 184, respectively, for flanges of uniformly thick square and rectangular sections.

Instead of 0.456, $k = 0.5$ has been suggested as a minimum for unstiffened elements.[3] Experimentally determined values ranged from 0.560 to 1.04 in 24 tests reported in Ref. 9. With $k = 0.5$ and $f_e = 1.65f$, Eq. (6–28) gives

$$\frac{b_e}{t} = \frac{107.0}{\sqrt{f}}\left(1 - \frac{26.8}{(b/t)\sqrt{f}}\right) \quad \text{for } \frac{b}{t} \geqslant \frac{53.5}{\sqrt{f}} \tag{6–31}$$

for unstiffened elements. Effective width formulas for unstiffened elements are not used in the AISC (1978) and AISI (1980) specifications. Instead, the gross width of such elements is used in determining section properties, and allowable stresses, based on the local-buckling stress, are specified. The following are the AISI (1980) allowable stresses:

$$F_c = 0.6F_y \qquad \frac{b}{t} \leqslant \frac{63.3}{\sqrt{F_y}} \tag{6–32a}$$

$$F_c = F_y\left(0.767 - 0.00264\sqrt{F_y}\,\frac{b}{t}\right) \qquad \frac{63.3}{\sqrt{F_y}} < \frac{b}{t} \leqslant \frac{144}{\sqrt{F_y}} \tag{6–32b}$$

$$F_c = \frac{8000}{(b/t)^2} \qquad \frac{144}{\sqrt{F_y}} < \frac{b}{t} \leqslant 25 \tag{6-32c}$$

$$F_c = 19.8 - 0.28\frac{b}{t} \qquad 25 < \frac{b}{t} \leqslant 60 \tag{6-32d}$$

For angles with $25 < b/t \leqslant 60$, use Eq. (6–32c).

This procedure may lead to significant errors, both conservative and unconservative, especially for compression members.[11]

6-8. Design of Beams Based on Postbuckling Strength of Elements

Two procedures are available for the design of beams with compression elements that require consideration of postbuckling strength:

1. According to the AISC (1978) and AISI (1980) specifications, the section modulus is computed for a reduced cross section based on the effective widths of all the stiffened elements that are in compression and the full widths of all the unstiffened elements that are in compression together with the full widths of all other elements of the cross section. If there are unstiffened elements in compression, the smallest of the allowable stresses for these elements or, if the member is not supported laterally, the allowable lateral-torsional buckling stress, if it is the smallest, must be used to determine the allowable moment.

If the neutral axis is closer to the compression flange than to the tension flange and there are stiffened elements in compression, the analysis must be by trial and error because the compressive stress that determines their effective widths is unknown at the start.

2. The section modulus is computed for a reduced cross section based on the effective widths of all elements in compression together with the other elements of the cross section, and the allowable bending moment is determined with no reduction in allowable stress for unstiffened elements but with any reduction required to allow for lateral-torsional buckling. As in procedure 1, the analysis must be by trial and error if the neutral axis is closer to the compression flange than to the tension flange. This procedure is not presently permitted by the AISC and AISI specifications.

Beam deflections at service loads will be overestimated slightly if they are based on section properties for safe-load determination, because service-load effective widths are larger than safe-load widths. If more accurate values are desired, section properties for computing deflection should be based on effective widths given by formulas obtained by multiplying the numerical coefficients in Eqs. (6–29)–(6–31), including the coefficients of the limiting values of b/t, by $\sqrt{1.65}$.

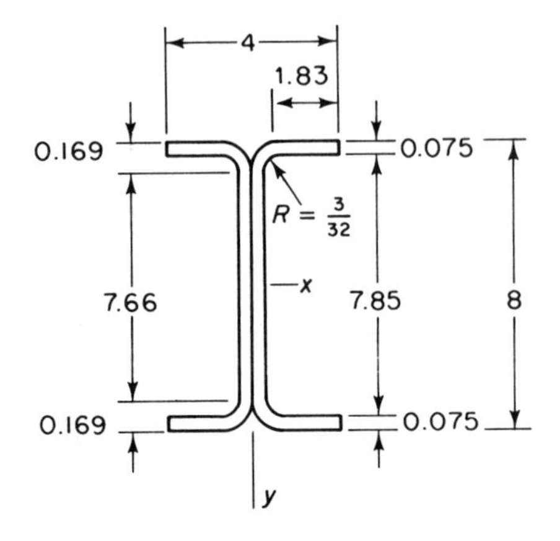

Figure 6-9

Example 6-2

Determine the allowable bending moment for a laterally supported beam with the cross section shown in Fig. 6-9. A446 Grade A steel, $F_y = 33$ ksi.

Procedure 1 (AISI specification)

Since there are no stiffened elements in compression, the entire cross section is effective. The following cross-section properties are for a section with square corners using centerline lengths of the elements:

$$\frac{I}{t} = 2 \times 4 \times 3.9625^2 + \frac{2 \times 7.85^3}{12} = 206.2 \text{ in.}^3$$

$$I = 206.2 \times 0.075 = 15.5 \text{ in.}^4 \qquad S = \frac{15.5}{4} = 3.88 \text{ in.}^3$$

(If corner radii are accounted for, $S = 3.80$ in.^{3})

$$\frac{b}{t} = \frac{1.83}{0.075} = 24.4 \qquad \frac{144}{\sqrt{F_y}} = 25.1$$

Therefore, Eq. (6-32b) applies, and

$$F_c = 33(0.767 - 0.00264 \times 24.4\sqrt{33}) = 13.1 \text{ ksi}$$

$$M = 13.1 \times 3.88 = 50.8 \text{ in.-kips}$$

Procedure 2

If the compression flange is not fully effective, the compressive bending stress will control because the neutral axis will be closer to the tension flange than to the compression flange. Therefore, $F_c = 20$ ksi and, since $b/t = 1.83/0.075 = 24.4$ exceeds $53.5/\sqrt{20}$, Eq. (6-31) applies:

$$\frac{b_e}{t} = \frac{107.0}{\sqrt{20}}\left(1 - \frac{26.8}{24.4\sqrt{20}}\right)$$

$$= 18.0 \qquad b_e = 1.35 \text{ in.}$$

compression flange width = 4 − 2(1.83 − 1.35) = 3.04 in.

$$\bar{y} = \frac{3.9625(4 - 3.04)}{4 + 3.04 + 2 \times 7.85} = 0.168 \text{ in.}$$

$$\frac{I}{t} = 3.04 \times 4.13^2 + 4 \times 3.79^2 +$$

$$2\left(\frac{7.85^3}{12} + 7.85 \times 0.168^2\right) = 190.3 \text{ in.}^3$$

$$I = 190.3 \times 0.075 = 14.28 \text{ in.}^4$$

$$S = \frac{14.28}{4.168} = 3.43 \text{ in.}^3$$

(If corner radii are accounted for, $S = 3.36$ in.^{3})

$$M = 20 \times 3.43 = 68.6 \text{ in.-kips}$$

Procedure 2 gives a 35% larger allowable moment.

6-9. Design of Compression Members Based on Postbuckling Strength of Elements

Two procedures are available for the design of compression members that are not subject to twist buckling or that are braced against twisting.

1. According to the AISC (1978) and AISI (1980) specifications, the allowable stress for such members depends on the slenderness ratio computed for the radius of gyration of the gross (unreduced) cross-sectional area and on a factor $Q = Q_s Q_a$ [Eqs. (6-14)] with

$$Q_s = \frac{F_c}{F}$$

$$Q_a = \frac{A_e}{A_g}$$

where F_c = allowable stress for the unstiffened element with the largest b/t ratio
F = basic allowable stress
A_e = area of cross section based on the gross width of the unstiffened elements and the effective widths of the stiffened elements, with the effective widths computed by Eq. (6-29) using F for f if the member has no unstiffened elements and F_c if it does
A_g = gross cross-sectional area

The resulting allowable stress is multiplied by the *gross* area to obtain the allowable load P.

Compression members of singly symmetric or nonsymmetric cross sections which are subject to twist buckling can be conservatively proportioned by this procedure.[3] The equivalent radius of gyration of Eq. (6–15) can be used for singly symmetric sections to compute the slenderness ratio for Eqs. (6–14). (See Example 6–1.)

2. The following procedure, which is not a part of the AISC or AISI specifications, has been shown to be more accurate and consistent than the one just outlined.[9,11] Both the cross-sectional area and the radius of gyration are computed for the reduced cross section formed by using the effective widths of both the stiffened and the unstiffened elements and the gross widths of all others. The allowable stress is given by Eqs. (6–13), or Eqs. (6–14) with $Q = 1$, and the allowable load P is equal to this allowable stress times the *effective* area.

Since the effective widths depend on the axial stress P/A_e and A_e in turn depends on the effective widths, the solution must be by successive approximations.

Example 6–3

Determine the allowable load P for a 15-ft pin-ended column with the cross section shown in Fig. 6–10. The column is supported against buckling about the y axis. A446 Grade D steel, $F_y = 50$ ksi

Procedure 1 (AISI specification)

Flange:

$$\frac{b}{t} = \frac{1.68}{0.135} = 12.4$$

$$\frac{63.3}{\sqrt{50}} = 8.95 \qquad \frac{144}{\sqrt{50}} = 20.4 \qquad \text{Eq. (6–32b) applies}$$

$$F_c = 50(0.767 - 0.00264 \times 12.4\sqrt{50}) = 26.8 \text{ ksi}$$

$$F = 0.6 \times 50 = 30 \text{ ksi} \qquad Q_s = \frac{26.8}{30} = 0.893$$

Web:

$$\frac{b}{t} = \frac{7.36}{0.135} = 54.5 \qquad \frac{171}{\sqrt{26.8}} = 33.0 \qquad \text{Eq. (6–29) applies}$$

$$\frac{b_e}{t} = \frac{253}{\sqrt{26.8}}\left(1 - \frac{55.3}{54.5\sqrt{26.8}}\right) = 39.3 \qquad b_e = 5.30$$

$$\text{ineffective web area} = 2 \times 0.135(7.36 - 5.30) = 0.56 \text{ in.}^2$$

$$A_g = 2 \times 0.135(4 + 7.73) = 3.17 \text{ in.}^2$$

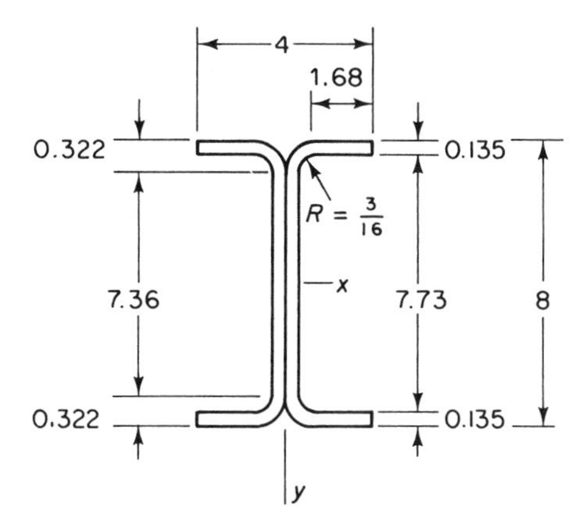

Figure 6-10

$$A_e = 3.17 - 0.56 = 2.61 \text{ in.}^2 \qquad Q_a = \frac{2.61}{3.17} = 0.823$$

$$Q = Q_a Q_s = 0.823 \times 0.893 = 0.735$$

$$\frac{I_x}{t} = 2 \times 4 \times 3.93^2 + \frac{7.73^3}{12} - \frac{(7.36 - 5.30)^3}{12} = 161.3 \text{ in.}^3$$

$$I_x = 21.8 \text{ in.}^4$$

$$r_x = \sqrt{\frac{21.8}{2.61}} = 2.89 \qquad \frac{KL}{r_x} = \frac{180}{2.89} = 62.3$$

$$F_a = 0.522 \times 0.735 \times 50 - \left(\frac{0.735 \times 50 \times 62.3}{1494}\right)^2$$

$$= 16.8 \text{ ksi} \qquad [\text{Eq. (6-14a)}]$$

$$P = F_a A_g = 16.8 \times 3.17 = 53.3 \text{ kips}$$

Procedure 2

An estimate of F_a, which is needed to determine the effective widths, can be obtained from an estimate of L/r. For cross sections of this type, r_x will range from about $0.35d$ to $0.4d$. Using the average value gives $r_x = 0.375 \times 8 = 3$ in. Then $KL/r_x = 180/3 = 60$ and, with $Q = 1$, Eq. (6-14a) gives $F_a = 22.1$ ksi

Flange: [Eq. (6-31)]

$$\frac{53.5}{\sqrt{22.1}} = 11.4 \qquad \frac{b}{t} = \frac{1.68}{0.135} = 12.4$$

$$\frac{b_e}{t} = \frac{107.0}{\sqrt{22.1}}\left(1 - \frac{26.8}{12.4\sqrt{22.1}}\right) = 12.3 \qquad b_e = 1.66$$

Since b_e is only 0.02 in. less than the flat width 1.68 in., the flange can be considered fully effective.

Web:

$$\frac{b}{t} = \frac{7.36}{0.135} = 54.5 \qquad \frac{171}{\sqrt{22.1}} = 36.4$$

$$\frac{b_e}{t} = \frac{253}{\sqrt{22.1}}\left(1 - \frac{55.3}{54.5\sqrt{22.1}}\right) = 42.2 \qquad b_e = 5.70 \text{ in.}$$

ineffective web area = 2 X 0.135(7.36 − 5.70) = 0.45 in.^{2}

$$A_g = 2 \times 0.135(4 + 7.73) = 3.17 \text{ in.}^2 \qquad A_e = 3.17 - 0.45 = 2.72 \text{ in.}^2$$

$$\frac{I_x}{t} = 2 \times 4 \times 3.93^2 + \frac{7.73^3}{12} - \frac{(7.36 - 5.70)^3}{12} = 161.7 \text{ in.}^3 \qquad I_x = 21.8 \text{ in.}^4$$

$$r_x = \sqrt{\frac{21.8}{2.72}} = 2.83 \text{ in.} \qquad \frac{KL}{r_x} = \frac{180}{2.83} = 63.6$$

$$F_a = 0.522 \times 50 - \left(\frac{50 \times 63.6}{1494}\right)^2 = 21.6 \text{ ksi}$$

This is close enough to the assumed value of 22.1 ksi (the effective width of web is increased by only 0.03 in. with $f = 21.6$ ksi).

$$P = F_a A_e = 21.6 \times 2.72 = 58.8 \text{ kips}$$

Procedure 2 gives a 10% larger allowable load.

6-10. Stiffened Flat Plates in Compression

The buckling strength of flat plates can be increased by using stiffeners. Stiffeners for axially compressed plates may be longitudinal, transverse, or a combination of the two. Such plates may be used as load-bearing skirts for rectangular bins or as load-bearing panels in the support structure and as flange and/or web plates in box-type beams and columns. Transverse stiffeners alone are generally not efficient, because the half-wavelength of the buckle tends to be about equal to the width of the plate, so that the spacing of transverse stiffeners must be less than the width of the plate if they are to increase the buckling strength.

Buckling of axially compressed stiffened plates is discussed in Refs. 4, 6, and 12. There are two buckling modes: overall buckling and local buckling of the plate elements between stiffeners. Charts for the overall-buckling coefficient k in Eq. (6-24), with b equal to the distance between stiffeners, were developed by Seide and Stein.[13] Figure 6-11 shows (in the solid lines below $k = 4$) k for $A_s/bt = 0.2$ and $EI_s/Db = 20$ for plates with one ($n = 2$) and three ($n = 4$) longitudinal stiffeners, where

a = length of plate

b = distance between stiffeners

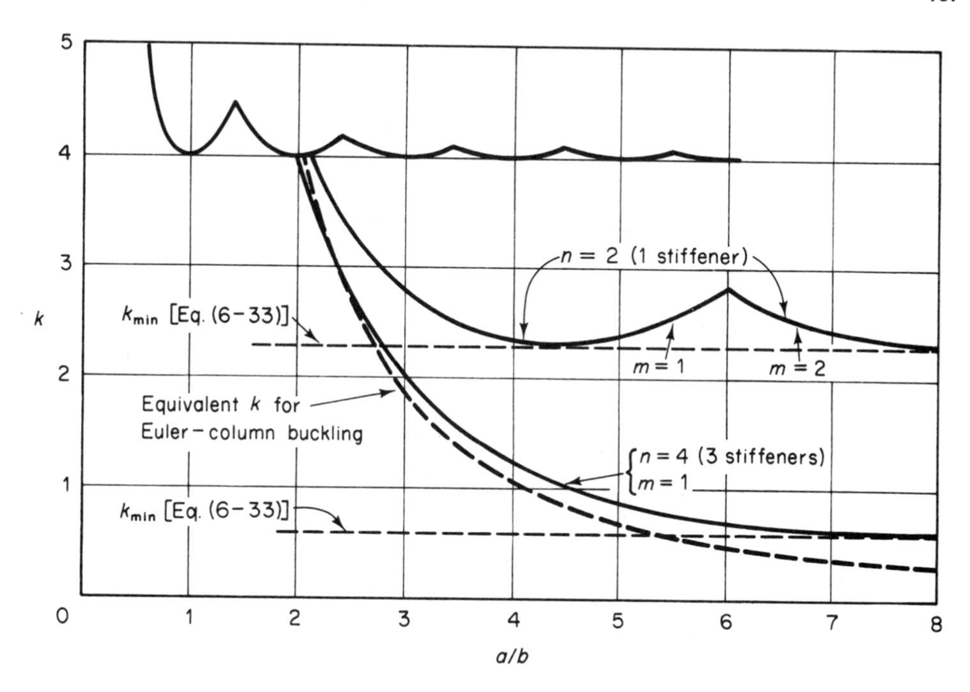

Figure 6-11 Buckling coefficient k for stiffened plate. $A_s/bt = 0.2$, $EI_s/DB = 20$. (From Ref. 13.)

t = thickness of plate

A_s = cross-sectional area of one stiffener and an effective width of plate

I_s = moment of inertia of one stiffener and an effective width of plate

$$D = \frac{Et^3}{12(1 - \nu^2)} = \text{flexural stiffness of plate}$$

m = number of longitudinal half-waves

n = number of panels into which the plate is divided

If the stiffeners are symmetrical about the plate, that is, identical stiffeners on both sides, the full width b is effective in computing A_s and I_s. However, Sharp[14] has shown, based on later work by Seide, that it is sufficiently accurate to also use the full width b for stiffeners on only one side.

Also shown in Fig. 6-11, above and tangent to the line $k = 4$, are the k curves for a longitudinally compressed plate with no stiffeners (Fig. 6-7). This shows that, except for the slightly larger values at the intersections of consecutive curves, k for a stiffened plate cannot exceed 4 because the buckling strength of the plate cannot exceed the local-buckling strengths of the plate segments.

The curve in Fig. 6–11 for $n = 2$ shows that k has a minimum value for each value of m. Sharp[14] has shown that this minimum value is given by

$$k_{\min} = \frac{2}{n^2} \frac{1 + \sqrt{1 + EI_s/Db}}{1 + A_s/bt} \qquad (6\text{–}33)$$

Values of $k_{\min}$ according to Eq. (6–33) are shown in Fig. 6–11. It will be noted that it would be much too conservative to use these values for plates that buckle in one longitudinal half-wave ($m = 1$) if a/b is less than the value at the vertex of the curve for $m = 1$. An approximate solution for this case can be obtained by computing F_{cr} for a pin-ended column consisting of one stiffener and the width b of plate. To compare values of F_{cr} computed in this way with the theoretically correct ones, an equivalent k can be found by equating F_{cr} by Eq. (6–24) to $F_{cr} = \pi^2 E/(L/r)^2$, where $r^2 = I_s/A_s$. The result is shown in Fig. 6–11. This approximate solution gives conservative results, especially for the single-stiffener case in the vicinity of $a/b = 3$, because it neglects the restraint against buckling owing to the transverse curvature of the plate.

Figure 6–11 shows that the elastic-buckling strength of a uniformly compressed stiffened plate, simply supported on all four edges, can be determined conservatively by taking the larger of F_{cr} computed by considering one stiffener and the width of plate between stiffeners as a column and the value by Eq. (6–24) with $k_{\min}$ from Eq. (6–33), provided this larger value does not exceed the local-buckling stress computed with $k = 4$. Instead of computing the three values of F_{cr}, the corresponding equivalent column-slenderness ratios can be compared to determine which is applicable. Equating $\pi^2 E/(L/r)^2$ to F_{cr} by Eq. (6–24) with $k_{\min}$ from Eq. (6–33) gives

$$\left(\frac{L}{r}\right)_{eq} = 2.3n\frac{b}{t}\sqrt{\frac{1 + A_s/bt}{1 + \sqrt{1 + 10.7 I_s/bt^3}}} \qquad (6\text{–}34)$$

In the same way, with $k = 4$ for local buckling of the plate,

$$\left(\frac{L}{r}\right)_{eq} = 1.65\frac{b}{t} \qquad (6\text{–}35)$$

The buckling stress is then determined by a column formula using the *smaller* of L/r for the stiffener-plate combination and $(L/r)_{eq}$ by Eq. (6–34), unless this smaller value is smaller than $(L/r)_{eq}$ by Eq. (6–35) in which case the latter must be used.

Figure 6–11 shows that, for the number of stiffeners and the range of a/b likely to be encountered, buckling of longitudinally stiffened plates in bin-supporting structures will be in one half-wave in each direction if the edges are hinged. This is the case for which one stiffener and the associated width of plate can be considered as a column. Furthermore, the postbuckling strength of such a column can be determined by using an effective width of plate if b/t exceeds the limiting value for which the plate is fully effective.

Horne and Narayanan developed a design procedure, considering one stiffener and an effective width of plate as a column, which takes into account local imperfections (waviness) of the plate, overall crookedness of the panel, and eccentricity of the load.[15] They also carried out a large number of tests on stiffened panels, the results of which were in satisfactory agreement with the predictions of the design procedure. Tests were made on panels of $\frac{3}{8}$-in. plate with a nominal yield stress of 36 ksi and $\frac{1}{4}$-in. plate with a nominal yield of 50 ksi. The stiffeners were spaced 18 in., which gave b/t values of 48 and 70. The panels were 72 in. long. The effect of imperfections was investigated by testing similar specimens, some of which were nominally straight while the others had certain imperfections. Results of tests on 31 panels are reported in Ref. 15. Results are given as ratios of the collapse load to the squash load, the latter being defined as the yield stress of the stiffener times the area of the stiffener plus the yield stress of the plate times the area of the plate. In general, the imperfections had little effect, and in two cases the ratio of collapse load to squash load was larger for the imperfect panel than for the corresponding nominally straight panel.

The ratios of the AISI allowable loads [Eqs. (6–14a) and (6–29)] to the nominal squash loads, the ratios of the test loads to the squash loads, and the corresponding factors of safety of the AISI loads for a few of the tests described are as follows:

(a) For $\frac{3}{8}$-in. panels with $6 \times \frac{5}{8}$ in. stiffeners ($L/r = 39$): AISI, 0.452; four tests, 0.75–0.85; FS, 1.66–1.88

(b) For $\frac{1}{4}$-in. panels with $6 \times \frac{3}{8}$ in. stiffeners ($L/r = 40$): AISI, 0.348; four tests, 0.55–0.64; FS, 1.58–1.84

(c) For $\frac{1}{4}$-in. panels with $3 \times \frac{1}{2}$ in. stiffeners ($L/r = 87$): AISI, 0.264; four tests, 0.44–0.48; FS, 1.67–1.82.

These factors of safety are consistent and adequate, which suggests that the AISI column formulas can be used with confidence for axially compressed plates with welded stiffeners. The AISC allowable loads are somewhat higher and give factors of safety for these tests ranging from 1.45 to 1.72.

The stiffeners of the test panels are unstiffened elements and would be limited to $b/t \leqslant 63.3/\sqrt{F_y}$ by the AISI specification and $85/\sqrt{F_y}$ by the AISC (Table 6–2). The $6 \times \frac{3}{8}$ in. stiffeners of one of the sets of $\frac{1}{4}$-in. test panels give $b/t = 16$, which is considerably in excess of both these limits for $F_y = 50$ ksi (9 for AISI and 13.4 for AISC). Therefore, it would appear that the AISC limit is adequate.

Eccentric load. Tests on one each of the panels described in (b) and (c), loaded eccentrically, are reported in Ref. 15. The eccentricity in each case was 0.31 in. toward the plate. The ratios of the AISI allowable loads, by Eq. (6–16a), to the squash loads were 0.32 and 0.21 for panels (a) and (b), respectively. The corresponding test values were 0.56 and 0.34, which give factors of safety of 1.75 and 1.64. The 0.31-in. eccentricity reduced the allowable load *P* from 117 to 108 kips for the type (b) panel and from 79 to 62 kips for type (c).

Example 6-4

Design a load-bearing skirt for a square bin 12 × 12 ft in cross section. The skirt is 8 ft high to the junction of the hopper and bin wall. A full load, including the weight of the bin and hopper, is 480,000 lb. Use A36 steel and AISI specification where applicable. Assume the skirt to be hinged and detailed to eliminate eccentricity of the load. (Since there will be some rotational restraint, some eccentricity of load can be tolerated.)

The load on one wall is 480/4 = 120 kips. Assume a stiffener cross section at 10% of the plate area. Try $\frac{3}{16}$-in. plate:

$$f = \frac{120}{1.1 \times 144 \times 0.1875} = 4.04 \text{ ksi}$$

This stress is low enough to be on the Euler curve. Therefore, from Eq. (6–14b),

$$\frac{KL}{r} = \sqrt{\frac{151{,}900}{4.04}} = 194$$

$$\frac{b}{t} = \frac{194}{1.65} = 118 \qquad b = 118 \times 0.1875 = 22 \text{ in.}$$

Try $2\frac{1}{2} \times \frac{1}{4}$ in. stiffeners 24 in. on centers on one side of the plate:

$$A = 24 \times 0.1875 + 2.5 \times 0.25 - 4.5 + 0.625 = 5.125 \text{ in.}^2$$

$$\bar{y} = 0.625 \times \frac{1.344}{5.125} = 0.164 \text{ in.}$$

$$I = 4.5 \times 0.164^2 + 0.25 \times \frac{2.5^3}{12} + 0.625 \times 1.18^2 = 1.32 \text{ in.}^4$$

$$r = \sqrt{\frac{1.32}{5.125}} = 0.508 \text{ in.} \qquad \frac{L}{r} = \frac{96}{0.508} = 189$$

From Eq. (6–29) with $f = 0.6F_y = 22$ ksi and $b/t = 24/0.1875 = 128$,

$$\frac{b_e}{t} = \frac{253}{\sqrt{22}}\left(1 - \frac{55.3}{128\sqrt{22}}\right) = 49.0 \qquad b_e = 9.18 \text{ in.}$$

$$A_{\text{eff}} = 9.18 \times 0.1875 + 2.5 \times 0.25 = 2.35 \text{ in.}^2$$

$$Q_a = \frac{2.35}{5.125} = 0.459$$

From Eqs. (6–14),

$$\frac{C_c}{\sqrt{Q}} = \pi \sqrt{\frac{2 \times 29{,}500}{0.459 \times 36}} = 188 < 189$$

$$F_a = \frac{151{,}900}{189^2} = 4.25 \text{ ksi}$$

F_a is the allowable stress on the gross, not the effective, area. Assuming that the skirts will be groove-welded at the corners, using a backup strip, there are five stiffeners on one face of the skirt. But since one-half of the effective width of the 24 in. of plate between the end stiffener and the corner is at the stiffener and the other half at the corner, the entire width of plate can be included in the gross area. Therefore,

$$P = 4.25(144 \times 0.1875 + 5 \times 0.625) = 128 > 120 \text{ kips}$$

6-11. Axially Compressed Plates with Longitudinal and Transverse Stiffeners

Although transverse stiffeners alone are generally not efficient in increasing the buckling strength of an axially compressed plate, it may sometimes be advantageous to reinforce a long, longitudinally stiffened plate with one or more transverse stiffeners.

The required moment of inertia I_t of one transverse stiffener for a plate with both longitudinal and transverse stiffeners is given by[16]

$$I_t = \frac{0.02Pd^3}{El}\left(2 - \frac{1}{N_t}\right)\left(1 + \frac{1}{N_l}\right) \tag{6–36}$$

where P = critical (ultimate) load for the panel
- d = width of panel (length of transverse stiffeners)
- l = length of panel (distance between transverse stiffeners)
- N_t = number of transverse stiffeners
- N_l = number of longitudinal stiffeners

It is important to note that P is *not* the allowable (service) load but the critical load for the panel.

Transverse stiffeners sized according to Eq. (6–36) are stiff enough to enforce nodes in the buckled shape, so that the plate can be designed as a longitudinally stiffened plate of length l. The width of plate effective with the transverse stiffener is uncertain but can be assumed to be such that the center of gravity of the stiffener-plate combination is at the face of the plate adjacent to the stiffener; thus, the stiffener can be sized by taking I_t about the axis at the juncture of the stiffener and the plate and neglecting the moment of inertia of the plate itself about that axis.

Another formula for transverse stiffener requirements is given in Ref. 6. The formula is derived in the same way, but the result is given in terms of the stiffness EI of one longitudinal stiffener and the corresponding strip of attached plate. To use this formula correctly, the tangent modulus must be used to compute EI in the inelastic range of column buckling. To do otherwise is to assume the Euler buckling load to be valid for the entire range of column slenderness ratios, which can result in a considerable overestimate of the transverse stiffener requirement. Therefore, the authors consider Eq. (6–36) to be the better choice, since inelastic buckling can be accounted for by basing P on the SSRC column formula [Eq. (6–13a) with the denominator omitted] or other applicable formula and using the Euler formula instead of Eq. (6–13b).

6–12. Effective Widths of Wide Flanges in Tension

In beams with flanges that are wide relative to the span and depth of the beam, bending stresses are not uniformly distributed over the width of the flange. Analysis of such beams is based on an effective width over which the stress can be assumed to be uniform. Effective widths of plates in compression are discussed in Sec. 6–7.

Effective widths of flange plates in tension are given in Table 6–3. These widths depend on the pattern of loading and support, the ratio of span L to flange width b, and the ratio of the flange moment of inertia I_f to the web moment of inertia I_w both about the neutral axis of the beam. The variation with moments of inertia is negligible for $0.25 < I_f/I_w < 3$.

The effective width at the critical section of the beam is not appreciably affected by a lengthwise taper of the flange. Also, effective widths are practically the same for I-beams, T-beams, U-beams, and box beams.

The use of Table 6–3 is illustrated in Example 7–3.

6–13. Local Buckling of Axially Compressed Cylindrical Shells

The theoretical, elastic local-buckling stress of an axially compressed cylindrical shell is given by[12]

$$f_{cr} = \frac{Et}{r\sqrt{3(1 - \nu^2)}} = \frac{0.6Et}{r} \tag{6–37}$$

where t = thickness and r = radius. Because they are sensitive to surface imperfections, axially compressed cylindrical shells usually buckle at stresses smaller than the theoretical value. The reduction increases, relatively, with increase in the ratio r/t, and shells with large r/t may buckle at stresses considerably smaller than the theoretical. In general, tubes produced in manufacturing plants by extrusion, piercing, forming and welding, etc., are likely to have truer surfaces than cylindrical shells fabricated in a structural fabri-

TABLE 6-3. RATIO OF EFFECTIVE WIDTH w_e TO TOTAL WIDTH w OF WIDE FLANGES IN TENSION[a]

	L/b									
Case[b]	2	2.5	3	4	5	6	8	10	15	20
1	0.757	0.801	0.830	0.870	0.895	0.913	0.934	0.946	–	–
2	0.632	0.685	0.724	0.780	0.815	0.842	0.876	0.829	–	–
3	–	0.609	0.650	0.710	0.751	0.784	0.826	0.858	–	–
4	0.571	0.638	0.686	0.757	0.801	0.830	0.870	0.895	0.936	0.946
5	0.495	0.560	0.610	0.686	0.740	0.778	0.826	0.855	0.910	0.930
6	0.690	0.772	0.830	0.897	0.936	0.957	0.977	0.985	0.991	0.995

[a]From Ref. 17.

[b]Cases:

1. Cantilever beam, concentrated load at end.
2. Cantilever beam, uniformly distributed load.
3. Cantilever beam, load increasing uniformly from zero at free end to maximum at fixed end.
4. Simply supported beam, concentrated load at center.
5. Simply supported beam, concentrated load at quarter point.
6. Simply supported beam, uniformly distributed load.

cating shop. Consequently, fabricated shells may have lower local-buckling strengths. Furthermore, there may be wide variations in the buckling strengths of apparently identical cylinders. Therefore, Eq. (6–37) is usually written in the form

$$f_{cr} = \frac{CEt}{r} \tag{6–38}$$

for which various formulas or curves to determine C have been developed by analyses of cylinders with waveform surface to represent imperfections, or by tests, or both. Some formulas give average values of C, while others give lower bounds. Some are based on test results of shells of a specific material or of specimens of, say, fabricating-shop quality, while others are based only on variation in geometry.

Steinhardt and Schulz[18] made a study of the results of tests by 12 investigators, or teams of investigators, which covered a range of radius/thickness (r/t) from 70 to 2800 and of length/radius (L/r) from 0.05 to 3 and included specimens of both fabricating-shop and manufacturing-plant quality. The test results were divided into 23 groups according to their r/t values. Thus, cylinders for which $175 < r/t < 225$ were classified as $r/t = 200$, those with $225 < r/t < 275$ as $r/t = 250$, etc. The following average-value formula was derived by minimizing the sum of the squares of the differences between the average values of all the test results for each r/t group and the corresponding predicted values:

$$C = 0.6\left[\left(0.03\frac{r}{t}\right)^{-1/8} - 0.39\right] \tag{6–39}$$

The following additional formula was developed to give a curve above which 90% of the averages of the test results for each r/t group lie:

$$C = 0.6\left[\left(\frac{r}{t}\right)^{-1/8} - 0.27\right] \tag{6–40}$$

Equations (6–39) and (6–40) are compared in Fig. 6–12 with the test results from which they were derived. In general, individual results are not shown; instead the range for each r/t group is indicated by a solid line. The exceptions are the results of tests of fabricated specimens from Ref. 19, which were made subsequent to the Steinhardt-Schulz investigation, and of some fabricated specimens from Ref. 20, which were not included in the study. They are discussed later.

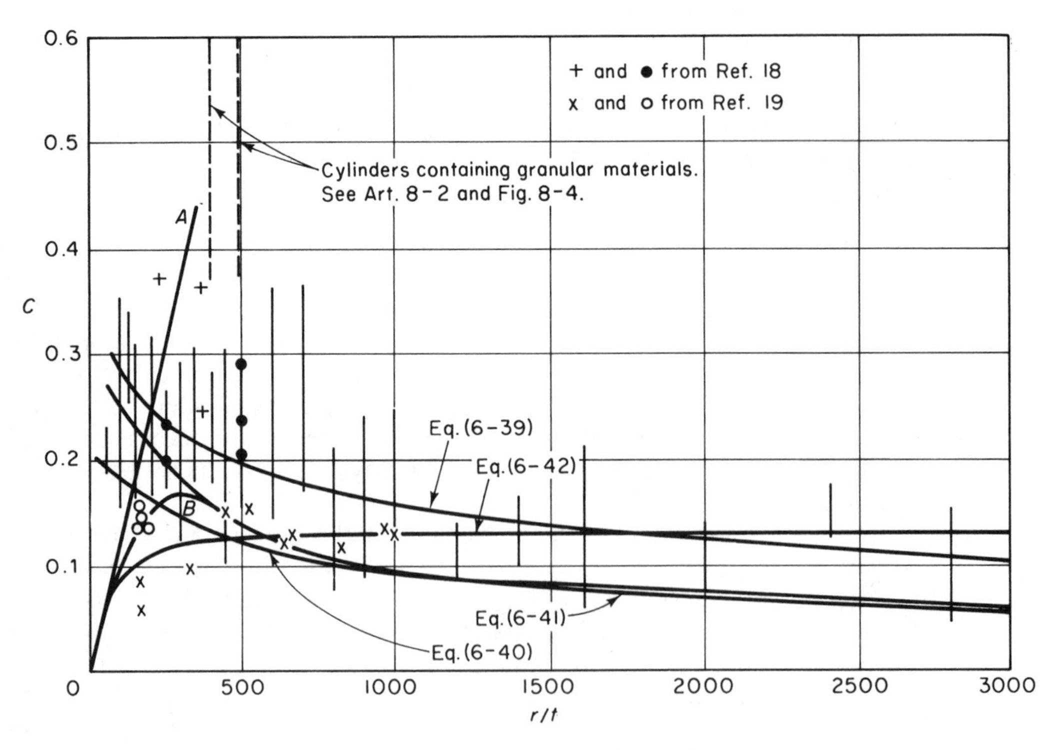

Figure 6–12 Results of tests on axially compressed steel cylindrical shells. (From Ref. 18.)

The following formula by Pflüger,[21] which is based on the results of an initial-imperfection analysis, compared favorably as an average-value formula with the test results of three of the investigations studied by Steinhardt and Schulz:

$$C = \frac{0.6}{\sqrt{1 + 0.01r/t}}$$

Formulas similar in form are recommended by the European Convention for Constructional Steelwork (ECCS).[22-24] They are given here in somewhat different form to be consistent in notation with other formulas in this section. Values of C for elastic buckling are

$$C = \frac{0.374}{\sqrt{1 + 0.01r/t}} \qquad \text{for } \frac{r}{t} \leqslant 212 \qquad (6\text{–}41\text{a})$$

$$C = \frac{0.315}{\sqrt{0.1 + 0.01r/t}} \qquad \text{for } \frac{r}{t} \geqslant 212 \qquad (6\text{–}41\text{b})$$

provided the resulting value of f_{cr} by Eq. (6–38) is less than $3/8F_y$. If f_{cr} exceeds $3/8F_y$, buckling is assumed to be inelastic [Eq. (6–43)]. These equations were obtained by dividing lower-bound formulas, which were based on hundreds of experimental values from 18 publications, by a partial factor of safety of 4/3 to allow for imperfection sensitivity. The graph is shown in Fig. 6–12. Some of the test points fall below this curve. Furthermore, ordinates to the lower-bound curve to which the safety factor was applied would be 4/3 times the ordinates to the Eq. (6–41) curve, so that additional test points would lie below the lower-bound curve. This is because Saal[23] discarded some of the test results used by Steinhardt and Schulz because he felt that the corresponding test cylinders and/or test conditions did not measure up to the requirements, discussed later in this article, that determine the applicability of the ECCS formula.

The following formula by Boardman, which gives f_{cr} directly rather than through an equation for C, was based on test results from Ref. 20 and another series of tests by the same investigators:

$$f_{cr} = 4000 \frac{t}{r} \left(1 - \frac{100}{3} \frac{t}{r}\right) \qquad (6\text{–}42)$$

The plot of the equivalent C is shown in Fig. 6–12. Instead of decreasing with increasing r/t, as is suggested by the test results, C increases. It will also be noted that it falls off rapidly with decrease in r/t below about 300, which is as it should be for inelastic buckling, as is shown by the line OB which is discussed later in this section. Since the axial compressive stress must exceed the proportional limit of the material for buckling to be inelastic, theoretically this segment of the curve represents a specific steel. The formula was derived for mild steels.

Miller[19] suggests that $C = 0.125$ be used for elastic buckling of fabricated shells because of the limited amount of test data; this is nearly the same as C by Eq. (6–42) for r/t greater than about 500. He also suggests that buckling be assumed elastic only if f_{cr} by Eq. (6–38) is less than $0.55F_y$. Thus, for $F_y = 36$ ksi, buckling would be elastic for $r/t > 190$. Miller also suggests formulas for inelastic buckling.

Many other formulas for local buckling of axially compressed shells have been developed, some of which are discussed in Ref. 6.

Inelastic buckling. For a sharply yielding material, Eq. (6–38) holds for geometrically perfect cylinders free of residual stress for all values of r/t for which the buckling stress is less than the yield stress. The limiting value of C is given by $C = F_y r/Et$, which plots as the straight line OA in Fig. 6–12. The complete C diagram consists of this line to its intersection with an elastic-buckling curve together with that portion of the latter that lies to the right.

Values of C for elastoplastic buckling of cylinders with residual stresses or of gradually yielding steel are given by a curve from the origin. The ECCS formula for the elastoplastic buckling stress $f_{cr(i)}$ is

$$f_{cr(i)} = F_y \left[1 - 0.347\left(\frac{F_y}{f_{cr}}\right)^{0.6}\right] \qquad \text{for } f_{cr} > \frac{3}{8}F_y \qquad (6\text{–}43)$$

where f_{cr} is the elastic-buckling stress by Eqs. (6–41). This formula is shown in Ref. 24 to be a good lower bound on test results. With $F_y = 36$ ksi, it gives the line OB in Fig. 6–12.

The partial safety factor 4/3 to cover imperfections which is included in Eqs. (6–41) is not needed for a thick cylinder since such a cylinder fails by yielding rather than buckling. Therefore, Eq. (6–43) is formulated so as to gradually reduce this safety factor from 4/3 for $f_{cr} = \frac{3}{8}F_y$ to 1 for $f_{cr} = F_y$ at $r/t = 0$. (The critical stress above which buckling is assumed to be inelastic is $\frac{1}{2}F_y$; the imperfection safety factor reduces it to $\frac{3}{8}F_y$.)

Steinhardt and Schulz derive an inelastic-buckling formula which gives a curve similar to the ECCS curve but much steeper, and suggest that a straight line such as OA in Fig. 6–12, based on a proportional limit rather than the yield stress, might be used instead. However, they note that their formula gives an "idealized" limit and that additional test data would probably suggest a more gradual transition.

Range of applicability of ECCS formulas. Because of the imperfection sensitivity of axially compressed cylindrical shells, maximum values of amplitudes of inward waviness are specified as limits of applicability of the ECCS formulas. The three restrictive amplitudes are (1) deviation w from straight-

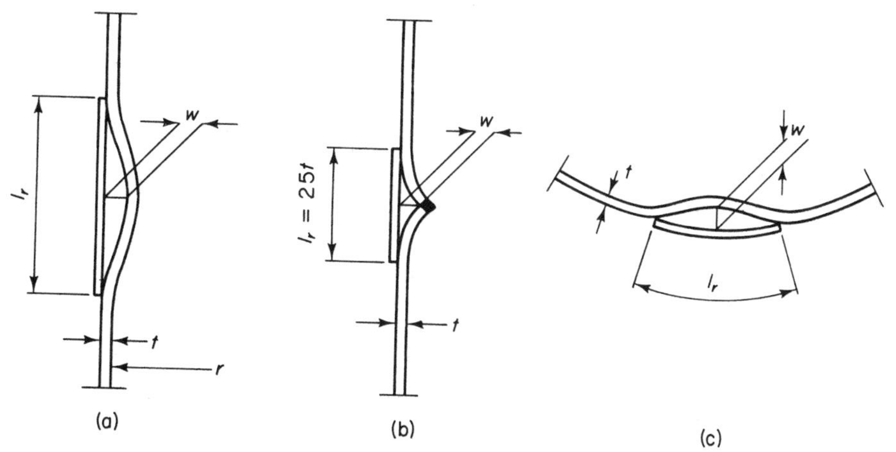

Figure 6–13 Allowable imperfections in cylindrical shells by ECCS. (From Ref. 22.)

ness of a meridian between circumferential joints (Fig. 6–13a), at circumferential welds (Fig. 6–13b), and from circularity as measured from a circular template (Fig. 6–13c). In all three cases $w \leqslant 0.01l_r$, where $l_r = 4\sqrt{rt}$ but not more than 95% of the distance between welds for Figs. 6–13a and c and $25t$ for Fig. 6–13b. If w exceeds $0.01l_r$, values of C by Eqs. (6–41) are to be reduced by 50% for $w = 0.02l_r$ and interpolated linearly for $0.01l_r \leqslant w \leqslant 0.02l_r$. Furthermore, radial or circumferential displacements of the ends of the cylinder must be prevented, and provision must be made for reasonably uniform delivery of the compressive forces at the ends to avoid concentrations of load.

Length effect. Axially compressed cylindrical shells may fail by column buckling, or by interacting local and overall buckling, if L/r exceeds a critical value. This is not a problem in the case of cylindrical bins, but it must be considered for cylindrical columns. According to the ECCS, the critical length is given by

$$\frac{L}{r} = 0.95\sqrt{\frac{r}{t}}$$

Axially compressed cylinders with L/r less than this value fail by local buckling, while those with larger L/r should be investigated for interacting overall and local buckling. This can be done by using a column-buckling formula such as the AISC formula [Eq. (6–13a)] with f_{cr} for local buckling substituted for F_y.

Length also affects the local-buckling strength if the cylinder is very short; the strength increases, relative to the value by Eq. (6–38), as L/r decreases. Tests reported in Ref. 18 show that this effect begins at about $L/r = 0.8$, and formulas to compute the increase are given.

Discussion of test results of Fig. 6-12. Although fabricated cylinders are generally believed to be weaker in axial compression than manufactured ones, the test results from Refs. 19 and 20 shown in Fig. 6–12 suggest that this may not be so. Two of the cylinders from Ref. 19 developed buckling strengths noticeably above the band of tests studied by Steinhardt and Schulz. The cylinder with $r/t = 243$ was made of cold-rolled steel of half-hard temper with a yield stress of 79.0 ksi (2% offset) and a proportional limit of 34.0 ksi. It buckled at 45.5 ksi, which is above the proportional limit but well below the yield stress. The two cylinders with $r/t = 372$ were made of SAE 1075 steel with a yield stress of 66.1 ksi and a proportional limit of 29.6 ksi. They were identical in geometry ($r = 7.5$ in., $t = 0.0202$ in., $L = 9.3$ in.), yet one carried 50% more load than the other. The stronger of the two buckled at 28.3 ksi, which is below the proportional limit and only 43% of the yield stress. Therefore, the superior performance of this cylinder and the one with $r/t = 243$ is probably due to their having been exceptionally good in respect to surface flaws and other dimensional defects rather than to properties of the steel. Steinhardt and Schulz report a number of tests on cylinders made to very close tolerances, with r/t ranging from 100 to 1000, which gave buckling strengths as high as 79% of the theoretical value ($C = 0.47$) and which averaged $C = 0.38$. These specimens were considered to be unrepresentative and were not used in deriving Eqs. (6–39) and (6–40).

The other five specimens from Ref. 19 were made of ASTM A366 steel with yield stress ranging from 28.7 to 41.4 ksi. Three buckled at stresses below the proportional limit. The other two buckled at stresses above the proportional limit but well below the yield stress.

The fabricated cylinders from Ref. 20 denoted in Fig. 6–12 by x were made of $\frac{1}{32}$-in. plate with yield stress ranging from 25.6 to 45.8 ksi. All were 30 in. high and had radii of 5, 10, 15, 20, 25, and 30 in. with r/t from 166 to 990. The three of these in the low r/t range are well below the lower boundary of the band of the other test results. They were among a group of specimens for which shims were used to fill the gaps caused by uneven bearing of the cylinders. They were not used in determining Eqs. (6–41) and (6–43) because they were considered to be unrepresentative of cylinders conforming to the restrictions on the applicability of the ECCS formulas.

The four cylinders denoted by circles in Fig. 6–12 were 72 in. long, 40 in. in diameter, and $\frac{1}{4}$ in. thick with yield stresses ranging from 35.6 to 37.8 ksi. Two of them had a butt-welded girth seam at midheight and two a riveted lap girth seam. The average strength of the two lap-jointed cylinders was 11% less than the average of the two with butt joints.

Factor of safety. It is clear that there is no one factor of safety that is appropriate for the large number of formulas that have been developed to predict local buckling of circular, cylindrical shells since some are average-value formulas, some lower-bound formulas, etc. The ECCS suggests a factor of safety (load factor) of 1.5 for Eqs. (6–41) and (6–43). With the partial factor of safety 4/3 in Eqs. (6–41) and one varying from 4/3 to 1 in Eq. (6–43), this gives a factor of safety on the lower bound of test results increasing from 1.5 (at $r/t = 0$) to 2 for inelastic buckling and 2 for the entire elastic-buckling range. Factors of safety for average-value formulas would of course have to be larger.

The Steinhardt-Schulz 50% curve [Eq. (6–39)] with a factor of safety 2.25 gives very nearly the same results (±10%) as the ECCS Eq. (6–41) with a factor of safety of 1.5.

6-14. Buckling of Cylindrical Shells in Bending

Tests show that for cylindrical shells which buckle elastically the ratio of the buckling stress in flexure to the buckling stress in uniform axial compression ranges from about 1.2 for $r/t = 200$ to about 1.6 for $r/t = 1000$. The ECCS formula[22] for the coefficient C_b to be used in place of C in Eq. (6–38) to determine the lower-bound bend-buckling stress is

$$C_b = 0.0849 + 0.811C \tag{6–44}$$

where C is the coefficient for axial compression [Eqs. (6–41)]. This formula also includes the imperfection-sensitivity safety factor 4/3 that is used in Eqs. (6–41). The same factor of safety, 1.5, as for axial compression is suggested. Inelastic buckling is covered by Eq. (6–43) using f_{cr} based on C_b.

There are few tests on bend buckling of fabricated tubes. Tests on two cylinders 5 ft in diameter by 6 ft long, one with a 0.202-in. wall ($r/t = 149$) and the other 0.135 in. ($r/t = 222$), are reported in Ref. 25. The tubes were made from plate cold-formed in rollers to the desired curvatures and welded with full-penetration submerged-arc groove welds. The ratios of the test buckling stresses to the predictions by the ECCS formula were 1.14 for the cylinder with $r/t = 150$ and 1.03 for the one with $r/t = 222$.

6-15. Buckling of Axially Compressed, Longitudinally Stiffened Cylindrical Shells

The following formula for the buckling stress of axially compressed, longitudinally stiffened cylinders was developed by Shang et al.[26]:

$$f_{cr} = \frac{k_1 k_2 \pi^2 E}{12(1 - \nu^2)(b/t)^2} \tag{6–45}$$

where $k_1 = 1.3 + 0.24b^2/rt$
$k_2 = 1 - 1.75\ [(500 - \gamma)/k_1\gamma]$ if $\gamma < 500$
$k_2 = 1$ if $\gamma \geqslant 500$
b = circumferential distance between stiffeners

r, t = radius, thickness of cylinder
$\gamma = \{I_s/[bt^3/12(1-\nu^2)]\} = 10.9I_s/bt^3$ with $\nu = 0.3$
I_s = moment of inertia of stiffener, including an effective width of cylinder

With $E = 30{,}000$ ksi and $\nu = 0.3$, Eq. (6–45) gives

$$f_{cr} = \frac{27{,}000k_1k_2}{(b/t)^2} \tag{6–46}$$

Substituting the value of k_1 into Eq. (6–46) gives

$$f_{cr} = k_2\left(\frac{35{,}000}{(b/t)^2} + \frac{6480}{r/t}\right) \tag{6–47}$$

The buckling strength of the stiffened cylinder is equal to f_{cr} times the cross-sectional area of the shell and stiffeners. In addition, the buckling stress of a single stiffener should be at least equal to f_{cr}. The investigators made this check by considering a stiffener and its effective width of shell as a pin-ended column of the same length as the cylinder.

The test cylinders were in three groups with respect to r/t: two 9 in. in diameter with $r/t = 300$ and 562 and one 15 in. in diameter with $r/t = 1070$. They were fabricated from steel sheet which had an 0.00008-in. electroplated tin coating to facilitate soldering the webs of the T-shaped stiffeners to the outer surfaces of the cylinders. Four stiffener sizes were used, with the number ranging from 9 to 36 per cylinder. Three of each type were tested as well as 23 unstiffened cylinders.

An effective width of shell consisting of two strips of width $0.76\sqrt{rt}$, one on each side of the stiffener web, was considered to be a part of the stiffener cross section. Values of γ ranged from 92 to 8727. In general, cylinders with γ greater than about 500 buckled antisymmetrically, with the stiffeners remaining straight but somewhat twisted, so that each panel buckled as a simply supported panel with f_{cr} independent of I_s. Cylinders with γ less than about 500 buckled symmetrically, accompanied in some cases with buckling of the stiffeners. In all cases there was a sudden reduction in load at buckling.

After the skin buckled, the stiffeners began to twist or to bow radially or circumferentially as the ultimate load was approached. In general, cylinders with $r/t = 300$ had no postbuckling strength. Those with $r/t = 562$ and 1070 had postbuckling strengths ranging from 1.05 to as much as 3.4 times the buckling load. Postbuckling strength increased with an increase in number of stiffeners of a given size and with an increase in size of stiffeners of a given number.

The ratios of the average buckling stress of each group of three test cylinders to the value predicted by Eq. (6–47) ranged from 0.89 to 1.32. This ratio was less than 1 for only two groups. The average of test value to predicted value for the entire group of 54 stiffened cylinders was 1.09.

Values of b^2/rt ranged from 8.7 to 500, and the investigators claim validity for their procedure only between these limits. However, the graph of k_1 is practically linear for $70 < b^2/rt < 500$, and the good fit of test values of k_1 with the corresponding straight line suggests that the formula can be used for $b^2/rt > 500$.

No limit on the applicability of Eq. (6–47) with respect to inelastic behavior is given. Eleven of the 18 stiffened cylinders with $r/t = 300$ had buckling strengths (with no postbuckling strength) in excess of 20 ksi; 1 of the 11 buckled at 30.5 and another at 35.5 ksi. The coupon yield strengths of the unstiffened cylinders with $r/t = 300$ ranged from 23 to 48 ksi. The stiffened cylinders were made of the same steel but with no coupon tests, so that no correlation between buckling strengths and yield strengths can be made.

A procedure for computing the ultimate strength of axially compressed, longitudinally stiffened cylinders was developed by Walker and Sridharan.[27] The collapse load of a broad-paneled cylinder, which fails by local buckling with no deflection of the stiffeners, is taken to be the sum of the yield strengths of the stiffeners and the ultimate strengths of the plate panels. The strength of the curved plate is computed by assuming it to be flat and using an effective width analysis. The collapse load of a narrow-paneled cylinder is computed by considering the stiffeners, together with the half panels on either side, as columns. Circumferential continuity is accounted for by treating these columns as supported by an elastic medium. The stiffness of the medium depends on the number n_w of circumferential waves in the buckling mode of the cylinder, which is determined by computing the cylinder buckling stress for various values of n_w. The smallest value of the buckling stress determines the correct value of n_w. An effective width of panel is taken as part of the column cross section. The analysis involves a step-by-step procedure which can be programmed for a desk computer. Predicted loads were in good agreement with test results for both the broad-paneled cylinders and the narrow-paneled cylinders. Experimental values of the ratios of the average stress f_{av} on the gross area of the shell plus stiffeners to the compressive yield stress F_{yc} are compared with the predicted values in Fig. 6–14.

There is no sharp boundary between broad-paneled and narrow-paneled cylinders, and there is a region where there may be a significant interaction between the two buckling modes. On the basis of experimental evidence, Walker and Sridharan describe a method for determining this interactive region.

Factor of safety. A comparison of the buckling behavior of stiffened and unstiffened cylindrical shells is shown in Fig. 6–15. Unstiffened shells usually suffer a sudden drop in carrying capacity upon buckling. Broad-paneled stiffened shells may also lose carrying capacity at first buckling or may experience a succession of small losses and recoveries. However, these losses are smaller than for unstiffened shells, and the load begins to increase

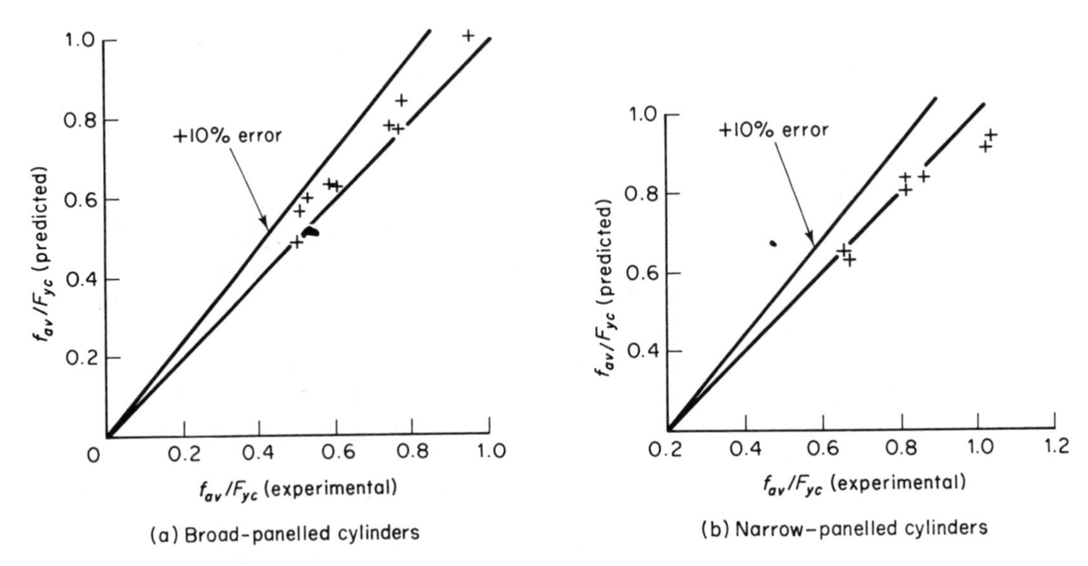

Figure 6-14 Longitudinally stiffened cylindrical shells in axial compression—comparison of theory with tests. (From Ref. 27.)

immediately with further strain. Broad-paneled shells may also show only a loss of stiffness as the first buckle occurs, with no drop in load. Narrow-paneled cylinders usually show the same type of behavior as columns. These differences in behavior suggest that the factor of safety on collapse of stiffened shells may be taken somewhat smaller than the factor of safety against buckling of unstiffened shells.

Because of the wide range of the differences between the buckling load and the collapse load, which were mentioned earlier in discussing the tests by Shang et al., it is clear that using a uniform factor of safety in design based on only the buckling load will give factors of safety on collapse that vary considerably. A better procedure would be to design for a uniform factor of safety on collapse together with a smaller uniform factor of safety on buckling. However, it should be noted that the theoretical, overall buckling stress of an imperfection-free stiffened cylinder may exceed the buckling stress of an imperfect one by a considerable margin, as is the case for unstiffened cylinders. For example, the ratios of the theoretical elastic buckling stresses of the 14 cylinders reported in Ref. 27 to the yield stress ranged from 1.33 to 7.35, while the collapse (not the initial-buckling) stresses ranged from 0.5 to 1.04 times the yield stress. Equation (6–45) accounts for imperfections through the coefficients k_1 and k_2 which were developed from an analysis of test results. Therefore, safety against initial buckling should be checked by f_{cr} according to Eq. (6–47), or similarly based data or formulas, unless an elastic buckling analysis of a cylinder with specific imperfections is made.

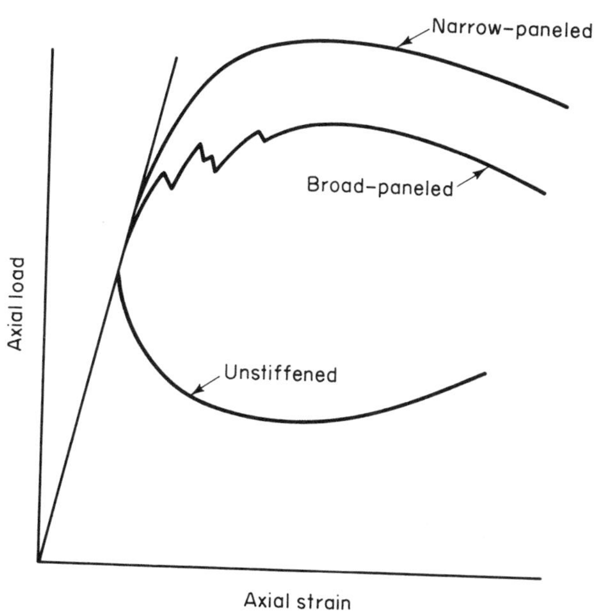

Figure 6-15 Load-strain relationships for axially compressed cylindrical shells. (From Ref. 27.)

Based on the ECCS-recommended factors of safety on lower-bound buckling loads of unstiffened shells and the equivalent factors of safety on average values, which are discussed in Sec. 6-13, the authors suggest that stiffened skirts for skirt-supported bins be designed for a factor of safety ranging from 1.75 to 2 on collapse, together with a factor of safety of 1.2 to 1.3 on initial buckling. If design is based only on the initial-buckling strength, a factor of safety of 1.5 to 1.75 is suggested. The lower factors of safety are intended for bins which are expected to be of a quality of workmanship comparable to the ECCS standard (Fig. 6-13).

Somewhat smaller factors of safety would be appropriate for stiffened bin cylinders because of the stabilizing effect of the stored solid (Sec. 6-16).

6-16. Buckling of Circular Cylinders under Axial Compression and Internal Pressure

Circular cylinders are more resistant to buckling due to axial compression if they are under internal pressure. The lateral pressure diminishes the effects of imperfections, and under very large pressure the buckling stress approaches the theoretical value for a perfect cylinder. A number of investigations of this phenomenon have been made, and various empirical formulas based on results of tests have been developed.

The following ECCS lower-bound formula[22] was developed by curve-fitting to test results reported in Ref. 28. It is given here in somewhat different form to be consistent with formulas in Sec. 6–13 for axially compressed cylinders:

$$f_{cr(p)} = C_p E \frac{t}{r} \leqslant \frac{3}{8} F_y \tag{6–48}$$

$$C_p = C + (0.45 - C) \frac{\rho}{\rho + 0.007} \tag{6–49}$$

where $f_{cr(p)}$ = critical axial stress for axially compressed cylinder under internal pressure p
C = coefficient from Eq. (6–41) for axially compressed cylinder with no internal pressure
p = internal pressure
$\rho = (p/E)(r/t)^{3/2}$

When $f_{cr(p)} > \frac{3}{8} F_y$, the inelastic critical stress $f_{cr(p)i}$ is computed by Eq. (6–43) with $f_{cr(p)}$ from Eq. (6–48) substituted for f_{cr}. The partial safety factor 4/3 which is included in Eq. (6–41) to allow for imperfection sensitivity is included in Eq. (6–49). These equations, with the imperfection safety factor omitted, are in good agreement with the tests reported in Ref. 28 except when p is small ($\rho < 0.01$). For $p = 0$, results are as much as 40% less than the curve values.[29] Of course, this is on the safe side.

Factor of safety. Together with the imperfection safety factor, a factor of safety (load factor) of 1.5 gives a factor of safety of 2 with respect to the lower-bound curve in the elastic range. The partial safety factor decreases in the inelastic range and becomes unity at $R/t = 0$. Therefore, a factor of safety of 1.5 for the entire range of R/t gives factors of safety on the lower bound of test results ranging from 1.5 (at $R/t = 0$) to 2 for inelastic buckling, and 2 throughout the elastic-buckling range. However, ECCS suggests that a smaller factor of safety may be justified for walls of bins because the internal pressure in the tests was produced by compressed air, and the resistance of granular material to inward movements of the wall is different from that for a gas. A uniform factor of safety 1.25 applied to $f_{cr(p)}$ by Eq. (6–49), combined with the imperfection safety factor 4/3, gives a uniform factor of safety of 1.67 with respect to the lower bound of test results. However, the factor of safety varies from 1.25 (at $r/t = 0$) to 1.67 in the inelastic range if a uniform factor of safety 1.25 is applied to $f_{cr(p)i}$ because of the varying imperfection safety factor. This is probably too low in the range of small r/t. A more nearly uniform factor of safety results if the factor of safety FS = $1.67 - 0.156 F_y/f_{cr}$ is applied to $f_{cr(p)i}$. Since $f_{cr} = \infty$ at $r/t = 0$ and $\frac{3}{8} F_y$ at the beginning of the elastic-buckling range, this gives factors of safety of 1.67 and 1.25, respectively, at these two points, which combined with the variable imperfection safety factor gives a nearly uniform factor of safety of 1.67 for the inelastic range.

6-17. Laterally Loaded Plates

Walls of rectangular bins are usually continuous over supporting stiffeners or beams. Since the plates are attached to the supporting members, the edges of a plate panel are not free to approach one another when the panel deflects under pressure from the contained material, so that tensile reactions develop in addition to the bending stresses. These tensile reactions are significant only if the deflection of the plate is large compared to its thickness. Plate theory which accounts for the tensile reactions is called the large-deflection theory (Sec. 6-19).

Allowable stresses. Improper use of standard-specification allowable stresses in designing plate components of bins can lead to inconsistent and, in some cases, very conservative results. For example, since they are not ordinarily encountered in steel frames for buildings, transversely loaded plates are not covered by the AISC specification. The most nearly applicable AISC allowable stress would be $0.75F_y$, which is specified for rectangular cross sections bent about the weak axis. This allowable stress takes into account the larger plastic-moment capacity of rectangular sections compared with I sections, for which the allowable stress is $0.6F_y$ or, for compact sections, $0.66F_y$. Furthermore, an allowance can be made for the fact that the ratio of the plastic-moment strength to the yield strength of a fixed-end (or continuous) beam is larger than that of a simply supported beam. Thus, the maximum moments by elastic analysis are $wL^2/8$ for a one-way, simply supported plate and $wL^2/12$ for one with fixed ends. But the moment redistribution that develops in the fixed-end plate with increase in load beyond the elastic limit results in moments, when the plastic strength is reached, of $wL^2/16$ at each end and at midspan. The AISC specification makes some provision for this in beams by allowing the calculated elastic negative moment due to gravity loads to be reduced by 10%, provided the maximum positive moment is increased by 10% of the average negative end moments. For the equal negative moments which develop when the plastic bending strength in continuous spans is reached, the same result could be obtained by using a 10% larger allowable stress with the calculated elastic moment, i.e., $0.75F_y$ for the simple-span maximum moment $wL^2/8$ and $0.82F_y$ for the fixed-end maximum moment $wL^2/12$.

Two limiting loads are important in developing design criteria for transversely loaded plates: (1) the load at which the maximum extreme-fiber bending stress reaches the yield point and (2) the load at which the fully plastic strength of the plate is attained. An upper bound to the plastic-moment strength can be found by the yield-line theory. Although upper bounds are, by definition, larger than the true plastic-moment strengths, they can be used to make some relative evaluations of factors of safety and the corresponding allowable stresses.

Small-deflection theory. The limiting load p per unit surface area of a plate is given by

$$p = \frac{F_y}{k(b/t)^2} \tag{6-50}$$

For a uniformly loaded rectangular plate simply supported on the four edges, $k = k_{ys}$ for extreme-fiber yield and $k = k_{ps}$ for the fully plastic strength, where

$$k_{ys} = \frac{0.75}{1 + 1.61(b/a)^3} \tag{6-51}$$

$$k_{ps} = \frac{[\sqrt{3 + (b/a)^2} - b/a]^2}{6} \tag{6-52}$$

in which a = length of long side and b = length of short side. The corresponding values for a plate rotationally fixed at the four edges are

$$k_{yf} = \frac{0.50}{1 + 0.623(b/a)^6} \tag{6-53}$$

$$k_{pf} = \frac{[\sqrt{3 + (b/a)^2} - b/a]^2}{12} \tag{6-54}$$

Equations (6–51) and (6–53) are from Ref. 33 and Eqs. (6–52) and (6–54) from Ref. 30. Values of k_p are based on upper bounds for p.

Values of k by the preceding equations for several values of a/b are given in Table 6–4. The values of k_{ps} and k_{pf} for $a/b = \infty$ also give lower-bound values of p; that is, they give the true values. The ratios p_p/p_y given in Table 6–4 are a measure of the reserve strength which exists after the extreme-fiber yield stress is reached. For the simply supported plate this ratio ranges from 1.5 for infinite a/b to something over 2 for plates on the order of $a/b = 2$ and for the fixed-edge plate from 2 for infinite a/b to approximately 4 for a/b = 1.5. Of course, these are upper-bound values for all a/b except $a/b = \infty$, but it is reasonable to assume that the true values will not fall below the values for $a/b = \infty$. Table 6–4 shows that for design based on elastic theory a larger allowable stress is justified for plates continuous (or rotationally fixed) at the edges than for simply supported plates. This is analogous to the AISC provision, discussed earlier in this section, that permits a reduction in the elastic moments at the supports of continuous beams for the purpose of computing the section modulus.

Based on the preceding discussion, it is suggested that laterally loaded plates be designed for an allowable stress $0.75F_y$ if they are simply supported

TABLE 6-4. COEFFICIENT k IN EQ. (6-50)

	a/b							
	1	1.5	2	3	4	5	10	∞
k_{ys}	0.287	0.508	0.624	0.707	0.730	0.741	0.746	0.750
k_{ps}	0.167	0.236	0.282	0.341	0.375	0.397	0.446	0.500
p_{ps}/p_{ys}	1.72	2.15	2.21	2.08	1.95	1.87	1.67	1.50
k_{yf}	0.308	0.474	0.495	0.500	0.500	0.500	0.500	0.500
k_{pf}	0.083	0.118	0.141	0.171	0.188	0.198	0.223	0.250
p_{pf}/p_{yf}	3.69	4.02	3.50	2.93	2.67	2.52	2.24	2.00

and $0.9F_y$ if they are fixed or continuous at the edges. This gives factors of safety with respect to yield of 1.33 for simply supported plates and 1.11 for plates continuous at the edges. Factors of safety with respect to plastic strength are greater than 2 for both types of support. In judging the low factors of safety with respect to yield, it should be noted that they are with respect to yielding at only two points (top and bottom at the middle) of a simply supported plate and four points (top and bottom at each long edge) of a plate continuous or rotationally fixed at the edges. However, these suggested allowable stresses do not provide for additional stresses from other sources, such as in-plane bending (Sec. 8-12).

Equation (6-50) can be written in the form

$$\frac{b}{t} = \sqrt{\frac{F}{k_y p}} \tag{6-55}$$

where F is the allowable stress and k_y is determined by Eq. (6-51) or (6-53). If the preceding recommendations are adopted, $F = 0.75F_y$ for simply supported plates and $0.9F_y$ for plates continuous or rotationally fixed at the edges.

Where deflection is to be limited, deflection relative to span is usually of interest. The following formula gives the value of b/t required to yield a specified deflection-span ratio δ/b:

$$\frac{b}{t} = \sqrt[3]{k \frac{\delta}{b} \frac{E}{p}} \tag{6-56}$$

where $k = k_{\delta s} = 7.03[1 + 2.21(b/a)^3]$ for simply supported edges
$k = k_{\delta f} = 35.2[1 + 1.056(b/a)^5]$ for rotationally fixed edges

It will be noted in Table 6-4 that values of k_{ys} for a/b greater than about 3 and k_{yf} for a/b greater than about 1.5 are nearly constant. Therefore, such plates can be designed for one-way bending using the values of k for

$a/b = \infty$. With these values Eqs. (6–55) and (6–56) give the following: For simply supported plates with $a/b \geqslant 3$,

$$\frac{b}{t} = 1.15\left(\frac{F}{p}\right)^{1/2} \qquad (6\text{–}57)$$

$$\frac{b}{t} = 1.92\left(\frac{\delta}{b}\frac{E}{p}\right)^{1/3} \qquad (6\text{–}58)$$

For plates with rotationally fixed edges with $a/b \geqslant 1.5$,

$$\frac{b}{t} = 1.41\left(\frac{F}{p}\right)^{1/2} \qquad (6\text{–}59)$$

$$\frac{b}{t} = 3.28\left(\frac{\delta}{b}\frac{E}{p}\right)^{1/3} \qquad (6\text{–}60)$$

Example 6–5

Determine the required thickness of an A36-steel plate 4 × 8 ft, simply supported on the four edges, which carries a uniformly distributed load of 650 psf. The deflection limit $\delta/b = \frac{1}{100}$.

Since $a/b = 2$, use Eqs. (6–55) and (6–56). Then with $p = 650/144 = 4.51$ psi and using the authors' suggested allowable stress $F = 0.75F_y = 27{,}000$ psi, Eqs. (6–51) and (6–55) give

$$k_{ys} = \frac{0.75}{1 + 1.61 \times 0.5^3} = 0.624$$

$$\frac{b}{t} = \sqrt{\frac{27{,}000}{0.624 \times 4.51}} = 98$$

From Eq. (6–56),

$$k_{\delta s} = 7.03(1 + 2.21 \times 0.5^3) = 8.97$$

$$\frac{b}{t} = \left(8.97 \times \frac{1}{100}\frac{30 \times 10^6}{4.51}\right)^{1/3} = 84$$

Deflection controls. Use $t = 48/84 = 0.57 = \frac{9}{16}$ in.

Example 6–6

This example is the same as Example 6–5 except that there is no deflection limit. For this case, $b/t = 98$ as calculated in Example 6–5. Then $t = 48/98 = 0.490 = \frac{1}{2}$ in. With $b/t = 48/0.5 = 96$, Eq. (6–56) gives

$$\frac{\delta}{b} = \frac{96^3 \times 4.51}{8.97 \times 30 \times 10^6} = 0.0148 = \frac{1}{67}$$

$$\delta = 0.0148 \times 48 = 0.71 \text{ in.}$$

Example 6-7

This example is the same as Example 6-5 except that the four edges are fixed. Since $a/b = 2$, the simpler formulas Eqs. (6-59) and (6-60) can be used. Then, with the authors' suggested allowable stress for fixed edges, $F = 0.9F_y = 32{,}400$ psi and

$$\frac{b}{t} = 1.41\sqrt{\frac{32{,}400}{4.51}} = 120$$

$$\frac{b}{t} = 3.28\left(\frac{1}{100}\,\frac{30 \times 10^6}{4.51}\right)^{1/3} = 133$$

Stress controls. Use $t = 48/120 = 0.400 = \frac{7}{16}$ in.

6-18. Plates Under Nonuniform Lateral Load

Tables of coefficients, based on solutions by the small-deflection theory, for computing the bending moments in plates with triangularly distributed transverse load are available.[31] However, a good approximation to the maximum moments can be obtained by using an equivalent uniform load to determine the plate thickness by Eq. (6-55) or, for large a/b, Eq. (6-57) or (6-59). Furthermore, trapezoidal load distributions can be handled by superimposing a uniformly distributed load and the equivalent uniform load for a triangular distribution.

Two cases of triangular loading must be considered: (1) variation along the width (Fig. 6-16a) and (2) variation along the length (Fig. 6-16b). The maximum bending stress in a simply supported plate is parallel to the short side for both Case 1 and Case 2, at a point beyond the plate midpoint toward the p_{max} edge. The equivalent uniform load given here is in terms of

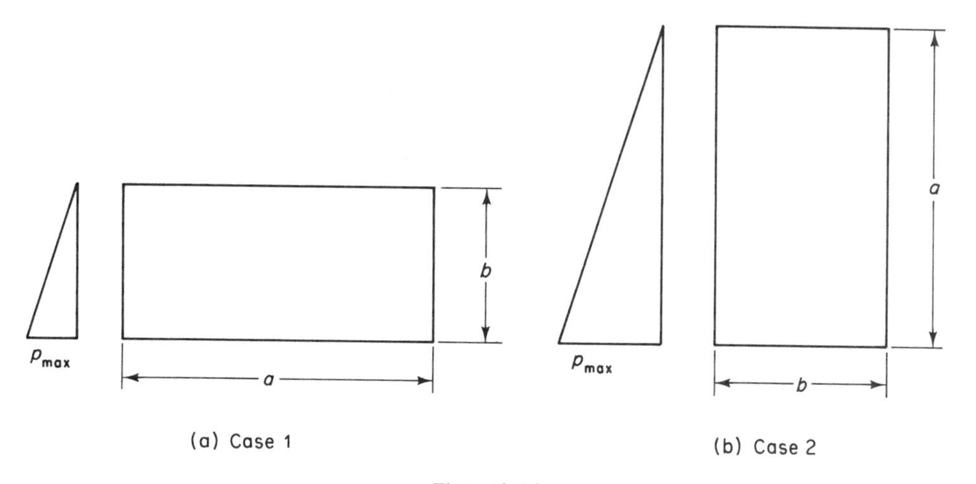

Figure 6-16

the maximum stress at the midpoint of a uniformly loaded, simply supported plate. The maximum stress in a plate fixed at the edges is at the midpoint of the p_{max} edge for Case 1. It is also at this point for Case 2 if $a/b < 1.4$, but if $a/b > 1.4$, it is on the long edge, beyond the midpoint toward the p_{max} edge. The equivalent uniform load given here is in terms of the long-edge midpoint moment of a uniformly loaded plate with all edges fixed.

Note that in each of the following cases a is the long dimension and b is the short one.

Case 1 with Simply Supported Edges

The equivalent uniform load is $p_{eq} = 0.55p_{max}$. Bending moments by this equivalent load vary from 0.99 to 1.07 times the theoretically correct values given in Ref. 32.

Use p_{eq} to compute b/t by Eq. (6–57) if $a/b > 3$ and by Eq. (6–55) with k_{ys} from Eq. (6–51) if $a/b \leqslant 3$.

Case 2 with Simply Supported Edges

The equivalent uniform loads are

$$p_{eq} = 0.55p_{max} \qquad \text{for } 1 \leqslant \frac{a}{b} \leqslant 2$$

$$p_{eq} = \left(0.45 + 0.05\,\frac{a}{b}\right)p_{max} \qquad \text{for } 2 < \frac{a}{b} \leqslant 6$$

$$p_{eq} = 0.75p_{max} \qquad \text{for } \frac{a}{b} > 6$$

Bending moments by these equivalent loads vary from 0.96 to 1.07 times the values given in Ref. 32.

Use p_{eq} to compute b/t by Eq. (6–57) if $a/b > 3$ and by Eq. (6–55) with k_{ys} from Eq. (6–51) if $a/b \leqslant 3$.

Case 1 with All Edges Fixed

The equivalent uniform load is $p_{eq} = 0.62p_{max}$. Bending moments by this equivalent load vary from 0.96 to 1.07 times the theoretical values given in Ref. 31.

Use p_{eq} to compute b/t by Eq. (6–59) if $a/b > 2$ and by Eq. (6–55) with k_{yf} from Eq. (6–53) if $a/b \leqslant 2$.

Case 2 with All Edges Fixed

The equivalent uniform loads are

$$p_{eq} = 0.62p_{max} \qquad \text{for } \frac{a}{b} > 2$$

$$p_{eq} = \left[1.40 - 2.33\frac{b}{a} + 1.55\left(\frac{b}{a}\right)^2\right]p_{max} \qquad \text{for } 1 \leqslant \frac{a}{b} \leqslant 2$$

Bending moments by these equivalent uniform loads vary from 0.96 to 1.03 times the theoretical values given in Ref. 31.

Use p_{eq} to compute b/t by Eq. (6–59) if $a/b > 2$ and by Eq. (6–55) with k_{yf} from Eq. (6–53) if $a/b \leqslant 2$.

Case 1 with Unloaded Edge Hinged, Other Three Fixed

This case represents a panel whose unloaded edge is not rotationally restrained by continuity with an adjoining panel. The equivalent uniform loads are

$$p_{eq} = \left(0.85 - 0.2\frac{b}{a}\right)p_{max} \qquad 1 \leqslant \frac{a}{b} \leqslant 2$$

$$p_{eq} = 0.75p_{max} \qquad \text{for } \frac{a}{b} > 2$$

Bending moments by these equivalent uniform loads vary from 0.97 to 1.10 times the values given in Ref. 31.

Use p_{eq} to compute b/t by Eq. (6–59) if $a/b \geqslant 2$ and by Eq. (6–55) with k_{yf} from Eq. (6–53) if $1 < a/b < 2$.

Case 2 with Unloaded Edge Hinged, Other Three Fixed

The equivalent uniform load is

$$p_{eq} = \left[1.32 - 2.22\frac{b}{a} + 1.55\left(\frac{b}{a}\right)^2\right]p_{max} \qquad \text{for } 1 \leqslant \frac{a}{b} \leqslant 2$$

$$p_{eq} = 0.60p_{max} \qquad \text{for } \frac{a}{b} > 2$$

Bending moments by these equivalent uniform loads vary from 0.96 to 0.99 times the values in Ref. 31.

Use p_{eq} to compute b/t by Eq. (6–59) if $a/b \geqslant 2$ and by Eq. (6–55) with k_{yf} from Eq. (6–53) if $1 < a/b < 2$.

Deflection. The preceding equivalent uniform loads give conservative estimates of deflection when used in formulas for uniformly distributed load. Deflections for the simply supported plate range from 0.98 to 1.13 times the theoretically correct values in Ref. 32 and from 0.93 to 1.13 times values in Ref. 31 for the plate with top edge hinged and the others fixed. Deflections for the plate with all edges fixed range from 1.23 to 1.32 for Case 1 and from 1.12 to 1.23 for Case 2 times values in Ref. 31. Therefore, Eqs. (6–56), (6–58), and (6–60) can be used to determine b/t for deflection control. (It should be noted that b/t varies with $\delta^{1/3}$, so that a 30% change in δ results in only a 9% change in b/t.)

Trapezoidal load. A conservative solution is obtained by superimposing the maximum moments for a uniformly distributed load and a triangularly distributed one. The solution is conservative because the maximum moments do not occur at the same points for the two distributions. An example is given below.

Allowable stresses. It was shown in Sec. 6–17 that consistent allowable stresses for design according to elastically computed bending moments depend on the relative values of the fully plastic load and the load at which the maximum extreme-fiber stress reaches the yield point. For the simply supported plate with large a/b this ratio is the same for trapezoidal load as for uniform load, namely, 1.5. The failure mechanism for a fixed-edge, triangularly loaded plate with large a/b is the same as for a triangularly loaded beam with fixed ends, for which it can be shown that the ratio of the plastic strength to the yield strength is 2.34. The corresponding ratio for the uniformly loaded fixed-end beam is 2.00. Since this ratio increases with decreasing a/b, the same allowable stresses that were suggested for uniformly loaded plates, namely $0.75F_y$ for simply supported edges and $0.9F_y$ for fixed edges, can be justified for triangularly loaded plates. However, these values must be reduced if there are additional stresses from other sources (Sec. 8–12).

Example 6–8

Determine the required thickness of an A36 steel plate spanning 12 ft in one direction and continuous over supports 5 ft on centers in the other. Load perpendicular to the plate varies from zero at one end of the 12-ft span to 260 psf at the other. Assume the edge where $p = 0$ to be simply supported and the other three fixed. The deflection limit $\delta/b = 1/150$ ($\delta = 60/150 = 0.4$ in.).

This is a Case 2 plate. Since $a/b > 2$, $p_{eq} = 0.6p_{max}$, and b/t can be determined by Eqs. (6–59) and (6–60). Using the authors' suggested allowable stress for this case, $F = 0.9F_y = 32{,}400$ psi, we get

$$p_{eq} = 0.6 \times 260 = 156 \text{ psf} = 1.08 \text{ psi}$$

$$\frac{b}{t} = 1.41\left(\frac{32{,}400}{1.08}\right)^{1/2} = 244$$

$$\frac{b}{t} = 3.28\left(\frac{1}{150}\,\frac{30 \times 10^6}{1.08}\right)^{1/3} = 187$$

Deflection controls, and $t = 60/187 = 0.32$. Use $\frac{5}{16}$-in. plate.

Example 6-9

Determine the required thickness of the plate in Example 6-8 if the load varies linearly from 460 to 780 psf over the 12-ft span and both ends of this span are fixed.

The equivalent uniform load for the triangular portion is $0.60p_{max}$. Therefore, the equivalent uniform load for the trapezoidal distribution is $p_{eq} = 0.60 \times 320 + 460 = 658$ psf = 4.57 psi. Then from Eqs. (6-59) and (6-60),

$$\frac{b}{t} = 1.41\left(\frac{32{,}400}{4.57}\right)^{1/2} = 119$$

$$\frac{b}{t} = 3.28\left(\frac{1}{150}\,\frac{30 \times 10^6}{4.57}\right)^{1/3} = 116$$

and $t = 60/116 = 0.52$ in. Use $\frac{1}{2}$-in. plate.

6-19. Laterally Loaded Plates—Large-Deflection Theory

The maximum stress in a uniformly loaded plate with its four edges simply supported and held against translation occurs at the center and is nearly independent of a/b for $2 < a/b < \infty$. If the four edges are held against rotation and translation, the maximum stress occurs at the center of the long edge and is nearly independent of a/b for $1.5 < a/b < \infty$. Therefore, simply supported plates with $a/b \geqslant 2$ and fixed-edge plates with $a/b \geqslant 1.5$ can be analyzed for one-way bending.

Formulas based on linearly elastic one-way bending which are derived in Ref. 32 were used to compute the values in Tables 6-5 and 6-6. The stress f is the sum of a uniform tension f_a and the tension due to bending. Tables 6-5 and 6-6 can be used to determine p for a given b/t or the b/t required to support a given load p, as well as the maximum deflection δ. The deflection-span ratio δ/b as a function of p and b/t is given in Figs. 6-17 and 6-18.

Values of pE/f^2 in Tables 6-5 and 6-6 show that, for given values of b/t and f, a plate with simply supported edges carries a larger load than one with fixed edges except for plates with very small b/t. This apparent incon-

TABLE 6-5. UNIFORMLY LOADED RECTANGULAR PLATES WITH SIMPLY SUPPORTED EDGES: LARGE-DEFLECTION THEORY: $a/b \geqslant 2$

$\frac{b}{t}\sqrt{\frac{f}{E}}$	$\frac{pE}{f^2}$	$\sqrt{\frac{p}{E}}\left(\frac{b}{t}\right)^2$	$\frac{f_a}{f}$	$\frac{\delta}{t}$
1.5	0.702	1.89	0.159	0.363
1.75	0.594	2.36	0.199	0.472
2	0.535	2.93	0.239	0.591
2.25	0.504	3.59	0.279	0.713
2.5	0.488	4.37	0.315	0.848
2.75	0.479	5.23	0.352	0.980
3	0.475	6.20	0.386	1.12
3.5	0.469	8.39	0.446	1.40
4	0.462	10.9	0.497	1.70
4.5	0.455	13.7	0.542	1.99
5	0.444	16.7	0.580	2.28
5.5	0.432	19.9	0.613	2.56
6	0.419	23.3	0.641	2.85
7	0.393	30.7	0.685	3.44
8	0.368	38.8	0.719	4.02
9	0.342	47.4	0.748	4.59
10	0.323	56.8	0.770	5.19
11	0.303	66.6	0.789	5.77
12	0.286	77.0	0.804	6.36
13	0.271	88.0	0.818	6.99
14	0.257	99.4	0.830	7.55
15	0.244	111.1	0.841	8.13
16	0.232	123.3	0.850	8.71
17	0.222	136.2	0.857	9.32
18	0.212	149.2	0.865	9.90

sistency results from the fact that a simply supported plate develops larger deflection and a larger membrane tension, as shown by the relative values of f_a/f and δ/t in Tables 6–5 and 6–6, which enables it to carry a larger part of the load by the more efficient membrane action. In the postyield range, however, load beyond the load which produces plastic hinges at the edges of a plate with fixed edges is carried by simple-support action, so that the simply supported plate may have lost much of its advantage when the fully plastic conditions are reached.

It should be noted that the relatively large capacity of the plate with simple supports can be attained only if the edges have very little rotational restraint. Furthermore, the simple-support solution does not apply in any situation where the plate is continuous at the supports, which is the usual case. On the other hand, the fixed-edge case applies to a single panel as well

TABLE 6-6. UNIFORMLY LOADED RECTANGULAR PLATES WITH MOMENT-RESISTANT EDGES: LARGE-DEFLECTION THEORY: $a/b \geqslant 1.5$

$\frac{b}{t}\sqrt{\frac{f}{E}}$	$\frac{pE}{f^2}$	$\sqrt{\frac{p}{E}}\left(\frac{b}{t}\right)^2$	$\frac{f_a}{f}$	$\frac{\delta}{t}$
1.50	0.880	2.11	0.019	0.116
1.75	0.645	2.46	0.025	0.169
2	0.497	2.82	0.032	0.223
2.25	0.393	3.17	0.040	0.273
2.5	0.321	3.54	0.047	0.329
2.75	0.268	3.92	0.055	0.393
3	0.229	4.31	0.062	0.458
3.5	0.175	5.12	0.078	0.588
4	0.141	6.01	0.090	0.737
4.5	0.118	6.96	0.102	0.882
5	0.101	7.95	0.112	1.02
5.5	0.0890	9.02	0.120	1.17
6	0.0794	10.1	0.127	1.21
7	0.0655	12.5	0.137	1.59
8	0.0558	15.1	0.144	1.88
9	0.0486	17.9	0.147	2.18
10	0.0431	20.8	0.150	2.47
11	0.0386	23.8	0.154	2.74
12	0.0350	26.9	0.156	3.01
13	0.0320	30.2	0.157	3.29
14	0.0295	33.7	0.158	3.59
15	0.0274	37.2	0.159	3.87
16	0.0254	40.8	0.159	4.15
17	0.0239	44.7	0.159	4.45
18	0.0224	48.5	0.160	4.72

as to one which is continuous in the direction of the length or width, or both, provided the discontinuous edges are fixed.

Because of the relatively large uniform tension f_a in the simply supported plate, which is shown by the values of f_a/f in Table 6–5, the rectangular-section shape factor for bending, 1.5, is decreased and becomes unity for $f_a/f = 1$. On the other hand, the relatively small membrane tension for the plate with fixed edges give it practically all the advantage of the shape factor. Therefore, the authors suggest the following allowable stresses for use with the elastic-analysis data of Tables 6–5 and 6–6: $F = 0.65F_y$ for simply supported plates, rather than $0.75F_y$ which was suggested for the small-deflection analysis, and $F = 0.9F_y$ for plates with fixed edges. However, these allowable values do not provide for additional stresses from other sources, such as in-plane bending (Sec. 8–12).

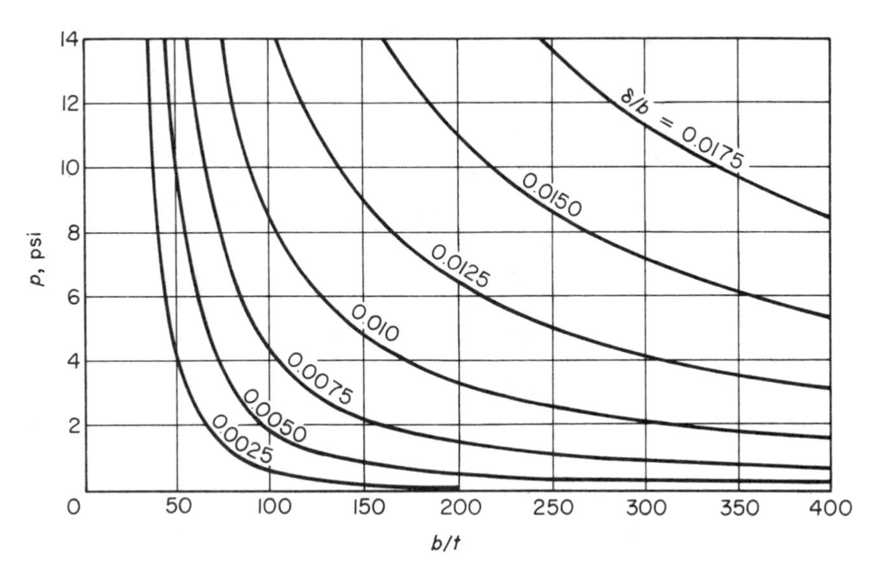

Figure 6-17 Deflection-span ratio for laterally loaded flat plates with simply supported edges: $a/b \geqslant 2$.

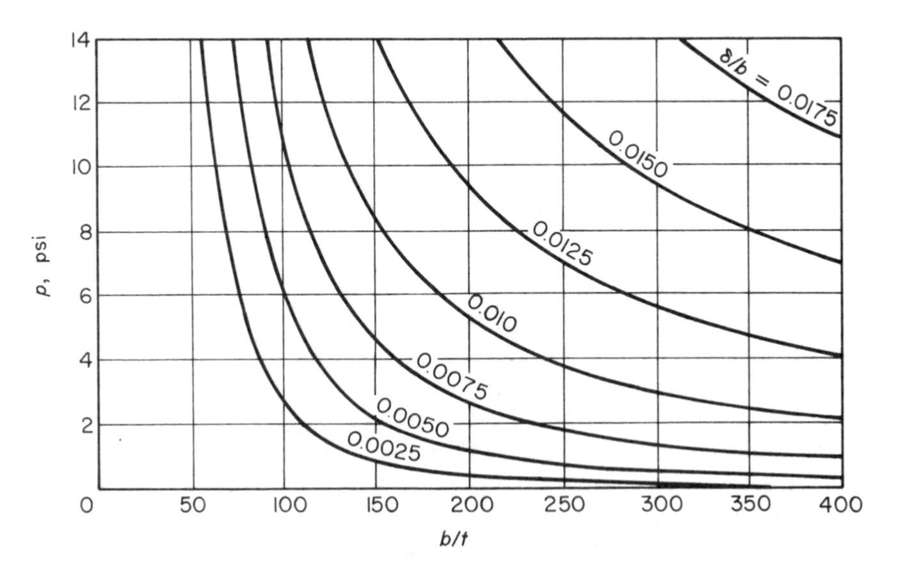

Figure 6-18 Deflection-span ratio for laterally loaded flat plates with clamped edges: $a/b \geqslant 2$.

Example 6-10

This example is the same as Example 6-5 except use the large-deflection analysis. Since $a/b = 2$, Table 6-5 can be used. Then with $f = F = 0.65 \times 36{,}000 = 23{,}400$ psi,

$$\frac{pE}{f^2} = \frac{4.51 \times 30 \times 10^6}{23{,}400^2} = 0.247$$

From Table 6-5,

$$\frac{b}{t}\sqrt{\frac{f}{E}} = 14.8 \qquad \frac{b}{t} = 14.8\sqrt{\frac{30 \times 10^6}{23{,}400}} = 530$$

With the prescribed deflection limit $\delta/b = 1/100$, Fig. 6-17 gives, for $p = 4.51$ psi, $b/t = 160$. Therefore, deflection controls.

$$t = \frac{48}{160} = 0.30 \text{ in. Try } \tfrac{9}{32} \text{ in.} = 0.281 \text{ in.:}$$

$$\frac{b}{t} = \frac{48}{0.281} = 171 \qquad \left(\frac{b}{t}\right)^2\sqrt{\frac{p}{E}} = (171)^2\sqrt{\frac{4.51}{30 \times 10^6}} = 11.34$$

Then from Table 6-5,

$$\frac{\delta}{t} = 1.74 \qquad \delta = 1.74 \times 0.281 = 0.490 \text{ in.} \qquad \frac{\delta}{b} = \frac{0.490}{48} = \frac{1}{98}$$

If the value of f_a is desired, entering Table 6-5 with $\sqrt{p/E}(b/t)^2 = 11.34$ as already calculated gives

$$\frac{b}{t}\sqrt{\frac{f}{E}} = 4.08 \qquad f = \left(\frac{4.08}{171}\right)^2 \times 30 \times 10^6 = 17{,}100 \text{ psi}$$

$$\frac{f_a}{f} = 0.504 \qquad f_a = 0.504 \times 17{,}100 = 8620 \text{ psi}$$

Example 6-11

This example is the same as Example 6-10 except that there is no deflection limit. In this case, $b/t = 530$ as calculated in Example 6-10, which gives

$$t = \frac{48}{530} = 0.091$$

Use $\frac{1}{8}$ in. = 0.125 in. To determine the deflection and the tension f_a, compute

$$\frac{b}{t} = \frac{48}{0.125} = 384 \qquad \left(\frac{b}{t}\right)^2\sqrt{\frac{p}{E}} = (384)^2\sqrt{\frac{4.51}{30 \times 10^6}} = 57.2$$

Then from Table 6–5,

$$\frac{\delta}{t} = 5.21 \qquad \frac{b}{t}\sqrt{\frac{f}{E}} = 10.04 \qquad \frac{f_a}{f} = 0.771$$

Therefore,

$$\delta = 5.21 \times 0.125 = 0.652 \text{ in.} \qquad \frac{\delta}{b} = \frac{0.652}{48} = \frac{1}{74}$$

$$f = \left(\frac{10.04}{384}\right)^2 \times 30 \times 10^6 = 20{,}500 \text{ psi}$$

$$f_a = 0.771 \times 20{,}500 = 15{,}800 \text{ psi}$$

Example 6–12

This example is the same as Example 6–7 except use the large-deflection analysis. Since $a/b = 2$, Table 6–6 can be used. Then with $f = F = 0.9F_y = 32{,}400$ psi,

$$\frac{pE}{f^2} = \frac{4.51 \times 30 \times 10^6}{32{,}400^2} = 0.129$$

From Table 6–6,

$$\frac{b}{t}\sqrt{\frac{f}{E}} = 4.26 \qquad \frac{b}{t} = 4.26\sqrt{\frac{30 \times 10^6}{32{,}400}} = 130$$

$$t = \frac{48}{130} = 0.369$$

Use $\frac{3}{8}$ in. Since the selected thickness is practically equal to the required thickness, Table 6–6 can be entered with the preceding value of $(b/t)\sqrt{f/E}$ to determine δ/t and f_a/f:

$$\frac{\delta}{t} = 0.812 \qquad \delta = 0.812 \times 0.375 = 0.305 \text{ in.} \qquad \frac{\delta}{b} = \frac{0.305}{48} = \frac{1}{157}$$

$$\frac{f_a}{f} = 0.096 \qquad f_a = 0.096 \times 32{,}400 = 3110 \text{ psi}$$

Nonuniform load. So far as the authors are aware, analyses of nonuniformly loaded plates by the large-deflection theory have not been made. To make some allowance for triangular and trapezoidal distributions, the authors suggest that the equivalent uniform load for each of these cases be taken as the average of the maximum load intensity and the equivalent uniform load by the small-deflection theory.

Example 6-13

This example is the same as Example 6–8 except use the large-deflection analysis. Since $a/b > 2$, Table 6–6 can be used. With $f = F = 0.9F_y = 0.9 \times 36{,}000 = 32{,}400$ psi and $p = (156 + 260)/2 = 208$ psf $= 1.44$ psi,

$$\frac{pE}{f^2} = \frac{1.44 \times 30 \times 10^6}{32{,}400^2} = 0.0412$$

From Table 6–6,

$$\frac{b}{t}\sqrt{\frac{f}{E}} = 9.42 \qquad \frac{b}{t} = 9.42\sqrt{\frac{30 \times 10^6}{32{,}400}} = 287$$

For the prescribed deflection limit $\delta/b = 1/150$, Fig. 6–18 gives, for $p = 1.44$ psi, $b/t = 240$. Therefore, deflection controls.

$$t = \frac{60}{240} = 0.25. \text{ Try } \tfrac{1}{4} \text{ in.}$$

$$\frac{b}{t} = \frac{60}{0.25} = 240 \qquad \sqrt{\frac{p}{E}}\left(\frac{b}{t}\right)^2 = \sqrt{\frac{1.44}{30 \times 10^6}} \times 240^2 = 12.62$$

From Table 6–6,

$$\frac{\delta}{t} = 1.60 \qquad \delta = 1.60 \times 0.25 = 0.40 \text{ in.} \qquad \frac{\delta}{b} = \frac{0.40}{60} = \frac{1}{150}$$

The large-deflection theory gives $t = \frac{1}{4}$ in. and the small-deflection theory $\frac{5}{16}$ in.

Example 6-14

This example is the same as Example 6–9 except use the large-deflection analysis. Then with $f = F = 0.9F_y = 32{,}400$ psi and $p = 460 + (0.60 \times 320 + 320)/2 = 719$ psf $= 4.99$ psi,

$$\frac{pE}{f^2} = \frac{4.99 \times 30 \times 10^6}{32{,}400^2} = 0.143$$

From Table 6–6

$$\frac{b}{t}\sqrt{\frac{f}{E}} = 3.97 \qquad \frac{b}{t} = 3.97\sqrt{\frac{30 \times 10^6}{32{,}400}} = 121$$

From Fig. 6–18 for $p = 4.99$ psi and $\delta/b = \frac{1}{150}$, $b/t = 130$.

$$t = \frac{60}{121} = 0.50. \text{ Use } \tfrac{1}{2} \text{ in.}$$

$$\frac{b}{t} = \frac{60}{0.5} = 120 \qquad \sqrt{\frac{p}{E}}\left(\frac{b}{t}\right)^2 = \sqrt{\frac{4.99}{30 \times 10^6}} \times 120^2 = 5.87$$

From Table 6-6,

$$\frac{\delta}{t} = 0.714 \qquad \delta = 0.714 \times 0.5 = 0.357 \text{ in.} \qquad \frac{\delta}{b} = \frac{0.357}{60} = \frac{1}{168}$$

6-20. Rings

Rings are used in some types of roofs for cylindrical bins. They may also be needed at the juncture of the roof and the cylinder and of the cylinder and hopper.

Rafters and trusses of supported roofs are interconnected at the crown by a ring. Such a ring is acted upon by vertical forces if it is supported, as in Figs. 7-2a and b, and by radial thrusts if it is not supported, as in Fig. 7-2c. It is also subjected to torsional moments if the rafter-to-ring connections are moment-resistant.

Supported rings. Rings may be supported at two points, as in Figs. 7-2a and b, or at four by using a second truss perpendicular to the first, as noted in Fig. 7-2b. Roof plates for small-diameter bins are connected directly to the ring, while larger bins require rafters.

If a supported ring is free to twist (roll) at its supports, it is statically determinate for symmetrically distributed loads. The maximum moments, which are at the supports, are negative. For a uniformly distributed load q per unit of circumference, the support moment M_x for a ring supported at two diametrically opposite points is

$$M_x = qR^2 \tag{6-61}$$

and for a ring supported at four equally spaced points

$$M_x = qR^2\left(1 - \frac{\pi}{4}\right) = 0.215qR^2 \tag{6-62}$$

where R is the radius of the ring. There are also twisting (torsional) moments. They are zero at the supports. The maximum torsional moments M_t for a ring on two supports are given by

$$M_t = 0.331qR^2 \tag{6-63}$$

These moments are at points of zero bending moment, which are 39.5° on either side of each support. For a ring on four supports,

$$M_t = 0.0331qR^2 \tag{6-64}$$

which occurs at 19.2° on either side of each support, which are also points of zero bending moment.

For a ring with a symmetrical system of equal concentrated loads Q, supported at two diametrically opposite points,

$$M_x = \frac{QR}{2}\sum_{0}^{\pi} \sin\alpha \tag{6-65}$$

where α is the angle between a radius to a support and the radius to each of the loads Q on half the ring (Fig. 6-19), while for a ring on four supports (Fig. 6-20),

$$M_x = \frac{QR}{2}\sum_{0}^{\pi} \sin\alpha - \frac{nQR}{8} \tag{6-66}$$

where n is the number of loads Q on the complete (360°) ring.

A ring with concentrated loads can be analyzed as a ring loaded uniformly with $q = Q/R\phi$, where ϕ is the angle between the radii to adjacent Qs. With this value of q, for a ring on two supports Eq. (6-61) gives

$$M_x = 0.159nQR \tag{6-67}$$

and for a ring on four supports Eq. (6-62) gives

$$M_x = 0.0342nQR \tag{6-68}$$

where n is the number of loads Q on the complete (360°) ring. The corresponding torsional moments from Eqs. (6-63) and (6-64) are

$$M_t = 0.0526nQR \tag{6-69}$$

$$M_t = 0.0053nQR \tag{6-70}$$

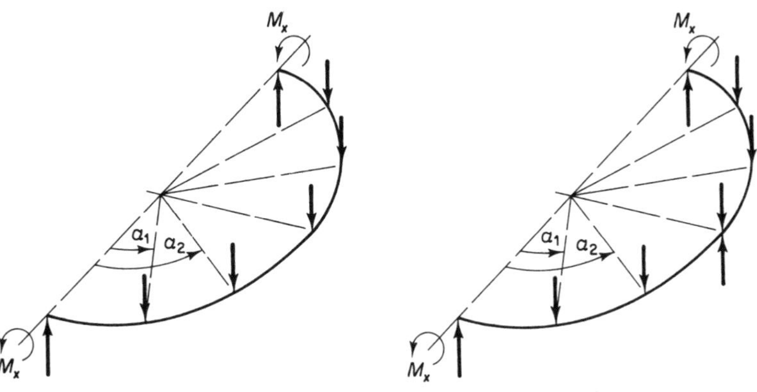

Figure 6-19

Figure 6-20

The error in these approximate formulas is about 2% for a ring on two supports with loads 30° apart and 6% for one on four supports with loads 22.5° apart and, of course, decreases with increase in n.

If a ring is not free to roll at the supports, it is statically indeterminate. Formulas for this case tend to be complicated. They involve both the bending stiffness EI and the torsional stiffness GJ, and, for open sections, the warping stiffness EC. Formulas are given in Ref. 33.

Radially loaded rings. The bending moment M, axial force P, and shear V in a ring loaded with a symmetrical system of radial loads Q are given by the following: At the load points,

$$M = \frac{QR}{2}\left(\cot\frac{\pi}{n} - \frac{n}{\pi}\right) \tag{6-71a}$$

$$P = \frac{Q}{2}\cot\frac{\pi}{n} \tag{6-71b}$$

$$V = \frac{Q}{2} \tag{6-71c}$$

and midway between load points,

$$M = \frac{QR}{2}\left(\csc\frac{\pi}{n} - \frac{n}{\pi}\right) \tag{6-72a}$$

$$P = \frac{Q}{2}\csc\frac{\pi}{n} \tag{6-72b}$$

$$V = 0 \tag{6-72c}$$

where n = number of radial loads Q on the complete ring. Q is positive directed outward, positive P produces circumferential tension, and positive moment produces tension on the inner fiber.

Buckling of radially loaded rings. A compressed ring may fail by buckling in its plane. The buckled shape is oval. The critical value of a uniform radial load q per unit length of circumference is[12]

$$q_{\text{cr}} = \frac{3EI_x}{R^3} \tag{6-73}$$

where EI_x is the bending stiffness in the plane of the ring.

Equation (6-73) should not be used by computing I_x and sizing the ring accordingly without checking the circumferential compressive stress, because the formula does not apply if the circumferential stress exceeds F_y for steels with a flat-top yield or the proportional limit for steels that yield gradually. Furthermore, if the proportional limit is exceeded, the formula is correct only if the tangent modulus E_t is substituted for E. Alternatively, the ring can be sized as for columns by converting Eq. (6–73) to a buckling-stress formula. Since the circumferential stress is qR/A, the buckling stress is given by

$$F_{cr} = \frac{q_{cr}R}{A} = \frac{3EI_x}{R^2A} = \frac{3E}{(R/r_x)^2} \tag{6–74}$$

where r_x is the radius of gyration of the ring cross section for its axis normal to the plane of the ring. In the authors' opinion, this formula should be used only for F_{cr} equal to or less than a proportional limit even for sharply yielding steels since structural members made of such steels usually exhibit a proportional limit because of residual stresses, imperfections, etc. For example, this is the case with columns, for which a proportional limit of $0.5F_y$ is assumed in the AISC and AISI specifications. Therefore, the authors suggest the following formulas for steel rings:

$$F_{cr} = F_y \left[1 - \frac{F_{y,\text{ksi}}}{400{,}000}\left(\frac{R}{r_x}\right)^2\right] \qquad \frac{R}{r_x} < \frac{370}{\sqrt{F_{y,\text{ksi}}}} \tag{6–75a}$$

$$F_{cr} = \frac{90{,}000}{(R/r_x)^2}\,\text{ksi} \qquad \frac{R}{r_x} \geqslant \frac{370}{\sqrt{F_{y,\text{ksi}}}} \tag{6–75b}$$

These formulas are derived with $E = 30{,}000$ ksi and a proportional limit of $0.66F_y$. Since they give the buckling strength, they must be divided by a factor of safety for allowable-stress design. The authors suggest a factor of safety of 1.5, which gives

$$F_a = 0.67F_y \left[1 - \frac{F_{y,\text{ksi}}}{400{,}000}\left(\frac{R}{r_x}\right)^2\right] \qquad \frac{R}{r_x} < \frac{370}{\sqrt{F_{y,\text{ksi}}}} \tag{6–76a}$$

$$F_a = \frac{60{,}000}{(R/r_x)^2}\,\text{ksi} \qquad \frac{R}{r_x} \geqslant \frac{370}{\sqrt{F_{y,\text{ksi}}}} \tag{6–76b}$$

It should be noted that the buckling considered here is in the plane of the ring. Rings can also buckle out of plane, but as they occur in bins they are supported against this form of buckling.

Rings with moment-resistant rafter connections. If interconnecting rafters are attached to a compression ring with moment-resistant connections, the rafter end moments act as twisting moments on the ring. They produce torsion and bending in the ring. The maximum torsional moments, which are adjacent to the rafter connection on each side, are given by

$$M_t = \tfrac{1}{2} M_0 \tag{6–77}$$

where M_0 is the rafter end moment (negative).

The maximum bending moment occurs midway between rafters at which point the torsional moment is zero. A formula for this moment can be derived, but it can be computed more simply and with good accuracy by assuming the concentrated torques to be distributed uniformly over the circumference of the ring. The bending moment for a uniformly distributed torque t per unit length of circumference is $M_x = tR$, which with $t = M_0/R\phi$ gives

$$M_x = \frac{M_0}{\phi} = \frac{nM_0}{2\pi} = 0.159nM_0 \tag{6–78}$$

where n is the number of rafter connections. Equation (6–78) underestimates the maximum moment by less than 3% even with rafters 45° apart.

The angle of twist of the ring is needed to compute the statically indeterminate end moment of rafters with moment-resistant connections. For a box section this angle is given by

$$\gamma = \frac{M_0 R}{EI_x}\left[\left(1 - 2.5\frac{I_x}{J}\right)\cot\frac{\phi}{2} + \left(1 + 2.5\frac{I_x}{J}\right)\frac{\phi/2}{\sin^2(\phi/2)}\right] \tag{6–79a}$$

where I_x = moment of inertia of the section about its axis in the plane of the ring, ϕ = angle between radii to adjacent rafter connections, and J = torsional constant, given by

$$J = \frac{2b^2d^2}{b/t_f + d/t_w} \tag{6–79b}$$

where b = width of the flanges of the box and d = depth of the webs (measured at the middle planes of the plates), t_f = flange thickness, and t_w = web thickness.

An approximate formula can be found by assuming the concentrated moments M_0 to be distributed uniformly on the ring. The resulting torque t per unit length of circumference is $t = M_0/R\phi$, and the angle of twist is

$$\gamma = \frac{tR^2}{EI_x} = \frac{M_0 R}{\phi EI_x} = 0.159\frac{nM_0R}{EI_x} \tag{6–80}$$

Equation (6–80) underestimates the angle of twist by less than 10% for a ring with 12 or more rafters and with $d/b < 2$ and $t_f/t_w \leqslant 2.5$. With 16 or more rafters the error is less than 10% if $d/b \leqslant 3$ and $t_f/t_w \leqslant 2.5$.

Equations (6–79a) and (6–80) cannot be used for open sections such as the channel, because of the warping during twist of such sections, which involves the warping constant C, as well as I_x and J, in the formula for γ. In any case, the channel is not a satisfactory section for a ring with moment-resistant rafters because it is torsionally weak. Of course, this does not exclude it as a ring for rafters without moment-resistant connections, as in Example 7–4.

REFERENCES

1. E. H. Gaylord and C. N. Gaylord, *Design of Steel Structures,* 2nd ed., McGraw-Hill, New York, 1972.
2. *Manual of Steel Construction,* 8th ed., American Institute of Steel Construction, Chicago, 1980.
3. *Cold-Formed Steel Design Manual,* American Iron and Steel Institute, Washington, D.C., 1977.
4. F. Bleich, *Buckling Strength of Metal Structures,* McGraw-Hill, New York, 1952.
5. A. Chajes and G. Winter, Torsional-Flexural Buckling of Thin-Walled Members, *J. Struct. Div. ASCE,* Aug. 1965.
6. Structural Stability Research Council, B. G. Johnston (ed.), *Guide to Stability Design Criteria for Metal Structures,* 3rd ed., Wiley, New York, 1978.
7. K. Klöppel and W. Protte, Ein Beitrag zum Kipp-Problem des Kreisförmig gekrummten Stabes, *Der Stahlbau,* Jan. 1961.
8. R. T. Shipp, Out-of-Plane Buckling of Ring Segments Subjected to Pull or Thrust Loads, Thesis, Ohio State University, Columbus, 1969.
9. V. Kalyanaraman, T. Pekoz, and G. Winter, Unstiffened Compression Elements, *J. Struct. Div. ASCE,* Sept. 1977.
10. G. Winter, Strength of Light-Gage Steel Compression Flanges, *Trans. ASCE,* Vol. 112, 1947.
11. G. Winter, Design of Cold-Formed Steel Structural Members, Sec. 9 in E. H. Gaylord and C. N. Gaylord (eds.), *Structural Engineering Handbook,* 2nd ed., McGraw-Hill, New York, 1979.
12. S. Timoshenko and J. M. Gere: *Theory of Elastic Stability,* 3rd ed., McGraw-Hill, New York, 1969.
13. P. Seide and M. Stein, Compressive Buckling of Simply Supported Plates with Longitudinal Stiffeners, *NACA Tech. Note 893,* National Aeronautics and Space Administration, Washington, D.C., 1949.
14. M. L. Sharp, Longitudinal Stiffeners for Compression Members, *J. Struct. Div. ASCE,* Oct. 1966.
15. M. R. Horne and R. Narayanan, An Approximate Method for the Design of Stiffened Steel Compression Panels, *Proc. Inst. Civ. Eng.,* Part 2, Sept. 1975.
16. *Alcoa Structural Handbook,* Aluminum Company of America, Alcoa, Pa., 1958.
17. F. B. Hildebrand and E. Reissner, Least-Work Analysis of the Problem of Shear Lag in Box Beams, *NACA Tech. Note 893,* National Aeronautics and Space Administration, Washington, D.C., 1943.

18. O. Steinhardt and V. Schulz, Zum Beulverhalten von Kreiszylinderschalen, *Schweiz. Bauztg.*, Jan. 7, 1971.
19. Clarence D. Miller, Buckling of Axially Compressed Cylinders, *J. Struct. Div. ASCE,* March 1977.
20. W. M. Wilson and N. M. Newmark, The Strength of Thin Cylindrical Shells as Columns, *Engineering Experiment Station Bull. 255,* University of Illinois at Urbana-Champaign, Feb. 1933.
21. A. Pflüger, Zur Praktischen Berechnung der Axial Gedrückten Kreiszylinderschale, *Der Stahlbau,* June 1963.
22. Manual on the Stability of Steel Structures, European Convention for Constructional Steelwork, *Introductory Report, Second International Colloquium on Stability of Steel Structures, Tokyo, Sept. 1976,* Liege, April 1977.
23. H. Saal, Buckling of Circular Cylindrical Shells Under Combined Axial Compression and Internal Pressure, European Convention for Constructional Steelwork, *Preliminary Report, Second International Colloquium on Stability of Steel Structures,* Liege, April 1977.
24. D. Vandepitte and J. Rathé, Buckling of Circular Cylindrical Shells Under Axial Load in the Elastic-Plastic Region, *Der Stahlbau,* Dec. 1980 (in English).
25. M. J. Stephens, G. J. Kulak, and C. J. Montgomery, Local Buckling of Thin-Walled Tubular Members, *Structural Engineering Report No. 103,* Department of Civil Engineering, University of Alberta, Edmonton, Feb. 1982.
26. J. C. Shang, W. J. Marulic, and R. G. Sturm, Buckling of Longitudinally Stiffened Cylinders, *J. Struct. Div. ASCE,* Oct. 1964.
27. A. C. Walker and S. Sridharan, Analysis of the Behavior of Axially Compressed Stringer-Stiffened Cylindrical Shells, *Proc. Inst. Civ. Eng.,* Part 2, June 1980.
28. V. I. Weingarten, E. J. Morgan, and P. Seide, Elastic Stability of Thin-Walled Cylindrical and Conical Shells Under Combined Internal Pressure and Axial Compression, *J. AIAA,* June 1965.
29. H. Saal, H. Kahmer, and A. Reif, Beullasten axial gedrückter Kreiszylinderschalen mit Innedruck, *Der Stahlbau,* Sept. 1979.
30. L. L. Jones and R. H. Wood, *Yield Line Analysis of Slabs,* American Elsevier, New York, 1967.
31. W. Fischer, *Silos und Bunker in Stahlbeton,* Veb Verlag Für Bauwesen, Berlin, 1966.
32. S. Timoshenko and S. Woinowsky-Krieger, *Theory of Plates and Shells,* 2nd ed., McGraw-Hill, New York, 1959.
33. R. J. Roark, and W. C. Young, *Formulas for Stress and Strain,* 5th ed., McGraw-Hill, New York, 1975.

7

STRUCTURAL DESIGN OF BIN ROOFS

7-1. Introduction

The structural design of steel-bin roofs is considered in this chapter. There are no codes or specifications for these structures except to the extent that standards for containment vessels for other uses, such as those of the American Petroleum Institute[1,2] (API) and the American Water Works Association[3] (AWWA), may be applicable. The AISC and AISI specifications may be used for the design of beams, columns, etc. Relevant provisions of these specifications are discussed in Chapter 6. The design of flat or curved plate elements, for some of which no standard specification exists, is also discussed in Chapter 6.

Where provisions of the specifications mentioned are used, reasons for some of the differences among them should be kept in mind. For example, oil tanks are usually located in industrial areas where the consequences of failure are confined to the owner's property, while water tanks are often located on high points in heavily populated areas. Therefore, many of the AWWA provisions involve larger factors of safety than the API standards and may be unduly conservative for bins for granular materials.

7-2. Types of Roofs

Roofs for rectangular bins may be flat or pitched, with the roof plate supported by structural members. Roofs for circular bins may be flat, conical, or spheroidal. Flat roofs must be supported by rafters or other structural members. Conical and spheroidal roofs may be self-supporting or supported by structural members. Roofs should have sufficient slope for drainage; a minimum of $\frac{3}{4}$ in. in 12 in. is required by API 650. On the other hand, there will be an upward internal pressure on the roof if its slope is greater than the angle of repose of the stored material and the bin is filled above the top of the bin wall.

A self-supporting conical roof is made with roof plates formed to produce a cone supported at its periphery by the bin shell. A ring is required at the apex if the roof segments are bolted together but is not needed for welded roofs unless there is an opening at the apex.

A self-supporting dome roof is constructed of roof plate formed to a spherical surface supported at the periphery by the bin shell. A self-supporting umbrella roof is a modified dome roof so formed from roof plate that any horizontal section is a regular polygon with as many sides as there are roof plates.

Self-supporting roofs carry load by membrane stresses, that is, by in-plane tensile, compressive, and shearing stresses. They are discussed in Sec. 7–3 for spherical domes and Sec. 7–4 for conical roofs. Self-supporting roofs with plates reinforced with radial stiffeners or a combination of radial and circumferential stiffeners may also be designed as membrane structures by using an equivalent membrane thickness.[5]

A supported conical roof consists of roof plate laid on rafters arranged in a radial pattern. One end of the rafter is supported by the bin shell, the other by a ring. Rafters may be cold-formed shapes, rolled shapes, open-web joists, or trusses or, for small-diameter bins, integral parts of the roof sheet formed by flanging a radial edge of the sheet (Fig. 7–1). The center ring may be supported on beams or trusses which are supported in turn at the bin walls (Fig. 7–2), or it may be a compression ring (Fig. 7–3).

Conical roofs for large-diameter bins may be supported by trusses and rafters arranged as shown in Fig. 7–4. One end of each truss is connected to a

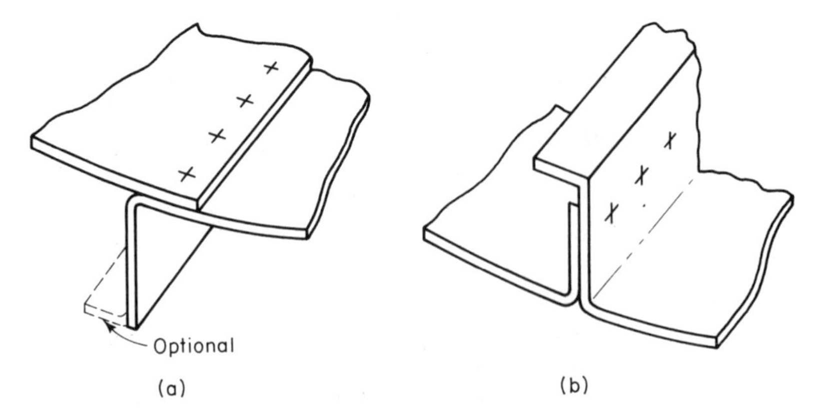

Figure 7–1 Flanged deck sheet.

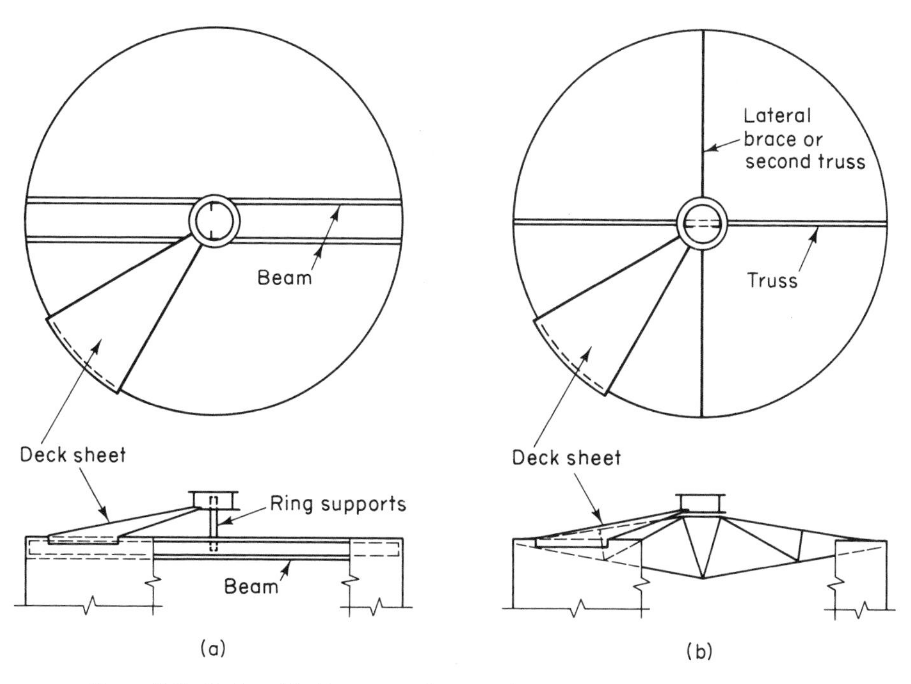

Figure 7-2 Decks with (a) supported center ring, (b) truss-supported center ring.

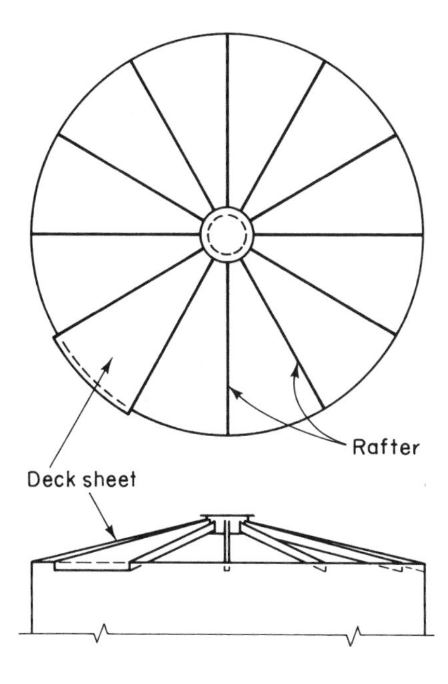

Figure 7-3 Deck with rafters supported by compression ring.

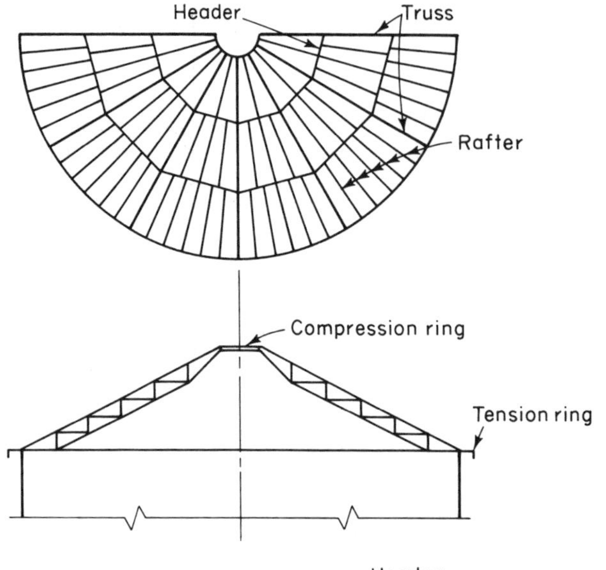

Figure 7-4 Roof framing.

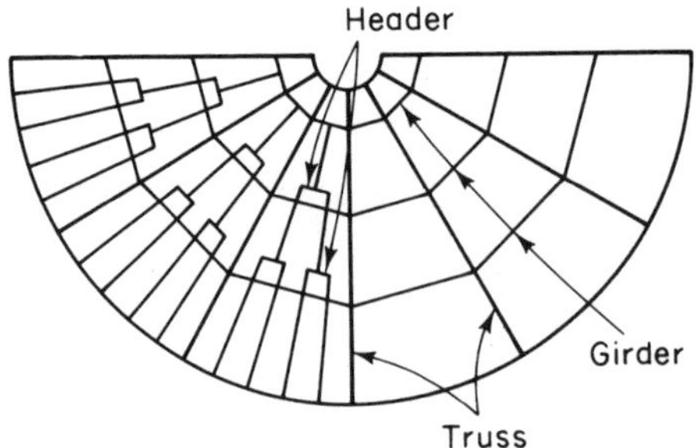

Figure 7-5 Roof framing.

tension ring at the bin wall and the other to a compression ring at the center. The rafters in Fig. 7-4 are simply supported by the headers. In Fig. 7-5 all rafters except those in the inner circle overhang girders connected to the trusses. Each pair of rafters supports a header at the free ends, with the header supporting a rafter in the adjoining circle. This arrangement gives smaller rafters because of the negative bending moments from the cantilevered ends.

Dome and umbrella roofs may be supported by a system of curved rafters arranged in the same general manner as for supported conical roofs.

7-3. Self-Supporting Spherical Domes

Formulas for the membrane forces in spherical domes are given in Fig. 7-6. Only symmetrical loadings are considered, so the principal-stress resultants are meridional (N_1) and latitudinal (N_2). Since the stresses are compressive in the upper portion, thin domes fail by buckling. The theoretical buckling

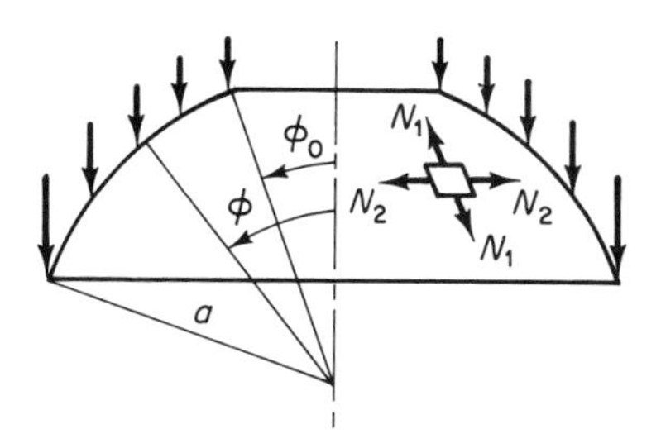

Load w per ft^2 of surface

No opening ($\phi_0 = 0$)	With opening
$N_1 = -\dfrac{wa}{1 + \cos\phi}$	$N_1 = -wa\,\dfrac{\cos\phi_0 - \cos\phi}{\sin^2\phi}$
$N_2 = -N_1 - wa\cos\phi$	$N_2 = -N_1 - wa\cos\phi$

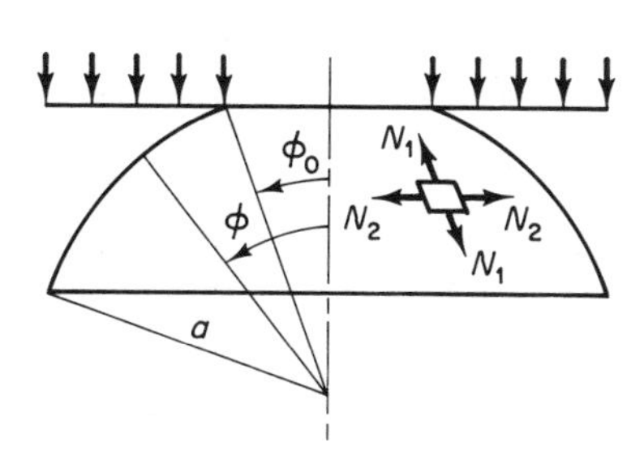

Load w per horizontal ft^2

No opening ($\phi_0 = 0$)	With opening
$N_1 = -\dfrac{wa}{2}$	$N_1 = -\dfrac{wa}{2}\left(1 - \dfrac{\sin^2\phi_0}{\sin^2\phi}\right)$
$N_2 = -\dfrac{wa}{2}\cos 2\phi$	$N_2 = -N_1 - wa\cos^2\phi$

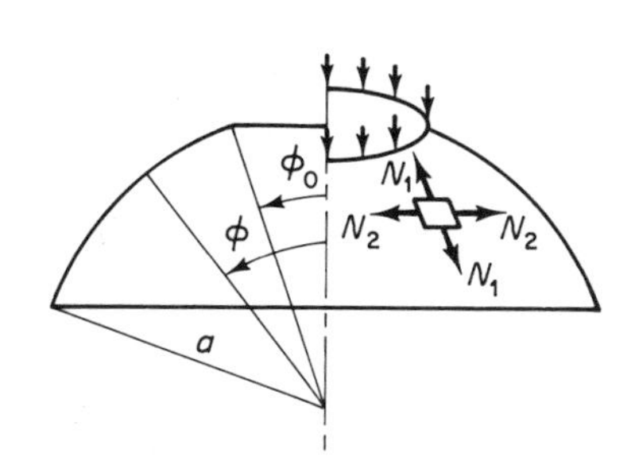

Load p per ft of opening perimeter

No opening ($\phi_0 = 0$) (= P at vertex)	With opening
$N_1 = -\dfrac{P}{2\pi a \sin^2\phi}$	$N_1 = -p\,\dfrac{\sin\phi_0}{\sin^2\phi}$
$N_2 = -N_1$	$N_2 = -N_1$

Figure 7-6 Membrane forces in spherical domes.

pressure p_{cr} per unit surface area for a spherical thin shell of radius a and thickness t is given by[4]

$$p_{cr} = \frac{2}{\sqrt{3(1 - \nu^2)}} E \left(\frac{t}{a}\right)^2 \tag{7-1}$$

With $\nu = 0.3$ this equation gives

$$p_{cr} = 1.2E\left(\frac{t}{a}\right)^2 \tag{7-2}$$

Experiments show that buckling occurs at pressures much smaller than those given by Eq. (7-2). Results of laboratory tests on spherical steel domes showed that the collapse pressure can be determined from the empirical formula[5]

$$p_{cr} = \lambda\left(1 - 0.175\frac{\phi - 20^\circ}{20^\circ}\right)\left(1 - 0.07\frac{a/t}{400}\right)0.3E\left(\frac{t}{a}\right)^2 \qquad (7\text{-}3)$$

$$\text{for } 400 \leqslant \frac{a}{t} \leqslant 2000,\ 20^\circ \leqslant \phi \leqslant 60^\circ$$

where ϕ is the base angle and λ is a factor which accounts for the rotational restraint of the tank wall on the dome. A formula for λ is given, but the following simplification is allowed by the German standard[6]: $\lambda = 1$ for $t_w/t_d \geqslant 1$, where t_d and t_w are the thicknesses of the dome and the upper course of the bin wall, respectively, while for $0 < t_w/t_d < 1$, λ is determined by interpolating linearly between 0.6 and 1.*

In 22 laboratory tests on steel domes with a/t ranging from 298 to 1238, buckling pressures ranged from 16% under to 36% over the values by Eq. (7–3). In an earlier investigation,[7] buckling pressures in 42 tests with $\lambda = 1$ and a/t from about 350 to 1900 were within 20% of the values by Eq. (7–3).

No tests were made on domes with a/t greater than 1900. However, the plot of p_{cr} versus a/t by Eq. (7–3) is nearly horizontal for a/t greater than 2000, and Klöppel and Jungbluth[7] suggest that the following equation, obtained by substituting $a/t = 2000$ in Eq. (7–3), be used:

$$p_{cr} = \lambda\left(1 - 0.175\frac{\phi - 20^\circ}{20^\circ}\right)0.195E\left(\frac{t}{a}\right)^2 \quad \text{for } \frac{a}{t} \geqslant 2000,\ 20^\circ \leqslant \phi \leqslant 60^\circ \qquad (7\text{-}4)$$

The meridional stress resultant N_1 at all points on a spherical dome with no opening is $-wa/2$ for a uniformly distributed load w (Fig. 7–6). The latitudinal stress resultant N_2 is also $-wa/2$ at the crown but is less at all other points. Since both stress resultants for a uniformly distributed load p normal to the surface are $-pa/2$ at all points, Eq. (7–4) will give a conservative value of the uniform load w by substituting w for p_{cr}. Then, with $E = 30 \times 10^6 \times 144$ psf and $p_{cr} = \eta w$, where w is the load in pounds per square foot and η is the factor of safety, Eq. (7–4) gives

$$\frac{t}{a} = \frac{1}{29{,}000}\sqrt{\frac{\eta w}{m\lambda}} \qquad (7\text{-}5)$$

where $m = 1 - 0.175(\phi - 20^\circ)/20^\circ$.

*Many of the provisions of DIN 4119, which are quoted here and in subsequent sections, are covered in the 1979 and 1980 revisions of the standard only by reference to the literature.

DIN requires a factor of safety of 2. With this value of η, Eq. (7–5) gives, with a slight rounding on the conservative side,

$$\frac{t}{a} = \frac{1}{20{,}000}\sqrt{\frac{w}{m\lambda}} \qquad (7\text{–}6)$$

If a/t turns out to be less than 2000, the result should be checked by Eq. (7–3).

Stiffened domes. Formulas for the equivalent membrane thickness of spherical domes with radial stiffeners or a combination of radial and circumferential stiffeners are given in Ref. 5.

API 650. The formula for the required thickness of a spherical dome in this standard is derived for a live load of 25 psf plus a dead load of 20 psf, both uniform on the horizontally projected area. The dead load is the weight per square foot of $\frac{1}{2}$-in. plate, which is the maximum allowable thickness of roof plate. The formula is derived by equating the membrane stresses at the crown to the allowable stress for spherical vessels under external pressure ($900{,}000t/a$ psi for $t/a \leqslant 0.00667$) of the ASME Boiler and Pressure Vessel Code.[8] Thus, with $N_1 = N_2 = wa/2$ from Fig. 7–6 we get

$$\frac{wa}{2t} = \frac{45}{144}\frac{a}{2t} = 900{,}000\frac{t}{a}$$

from which

$$\frac{t}{a} = \frac{1}{2400} \qquad (7\text{–}7)$$

API 650 gives this formula in the following form, with t in inches and a in feet:

$$t_{\text{in}} = \frac{a_{\text{ft}}}{200} \qquad (7\text{–}8)$$

The minimum thickness is $\frac{3}{16}$ in. In addition, the radius is limited by $0.8D < a < 1.2D$, where D is the tank diameter. The corresponding limits of the base angle are $38.7° > \phi > 24.6°$.

The authors prefer the DIN formula. It is not restricted to a live-load intensity of 25 psf, it enables the lighter dead weight of roof plate less than $\frac{1}{2}$ in. thick to be accounted for, and it is less restrictive in terms of base angle.

Example 7-1

Determine the required thickness of a spherical-dome roof for a bin with $D = 80$ ft, $a = 72$ ft, thickness t_w of top course of wall = $\frac{1}{4}$ in., and live load = 25 psf.

By the API, Eq. (7-8),

$$t = \frac{72}{200} = 0.36 \text{ in.}$$

Use $\frac{3}{8}$ in. plate.

By the DIN, Eq. (7-6),

$$\phi = \sin^{-1} \frac{40}{72} = 33.7^\circ$$

$$m = 1 - \frac{0.175(33.7 - 20)}{20} = 0.88$$

Assume $t_d = \frac{5}{16}$ in. Then $w = 12.8 + 25 = 37.8$ psf, $t_w/t_d = 0.8$, $\lambda = 0.6 + 0.6 \times 0.4 = 0.84$.

$$\frac{t}{a} = \frac{1}{20{,}000} \sqrt{\frac{37.8}{0.88 \times 0.84}} = \frac{1}{2797}$$

Eq. (7-6) applies since $a/t > 2000$,

$$t = \frac{72 \times 12}{2797} = 0.309$$

Use $\frac{5}{16}$ in. plate.

7-4. Self-Supporting Conical Roofs

Formulas for membrane forces in conical roofs are given in Fig. 7-7. Only symmetrical loadings are considered, so the principal-stress resultants are in the direction of the slant height (N_1) and circumferential (N_2). Since both are compressive, failure is usually by buckling.

Many theoretical and experimental investigations of buckling of conical shells under various types of load have been made.[9,10] Results of tests by eight investigators of buckling of aluminum, brass, steel, lucite, and mylar cones under pressure normal to the surface (Fig. 7-8a) are reported in Ref. 9. Results are shown in Fig. 7-9. A wide range of proportions is covered. Cones and truncated cones were tested, with the base angle ϕ ranging from 15° to 90°, the latter, of course, being a cylinder. The ratio of the average radius of curvature r_{av}, defined in Eq. (7-10), to the thickness ranged from 375 to 2318. The critical external pressure is approximately equal to a factor times the critical external pressure of a cylinder having the same wall thickness t, a radius equal to the average radius of curvature of the cone, and a length equal to the slant length l of the cone or frustrum (Fig. 7-8b). Although

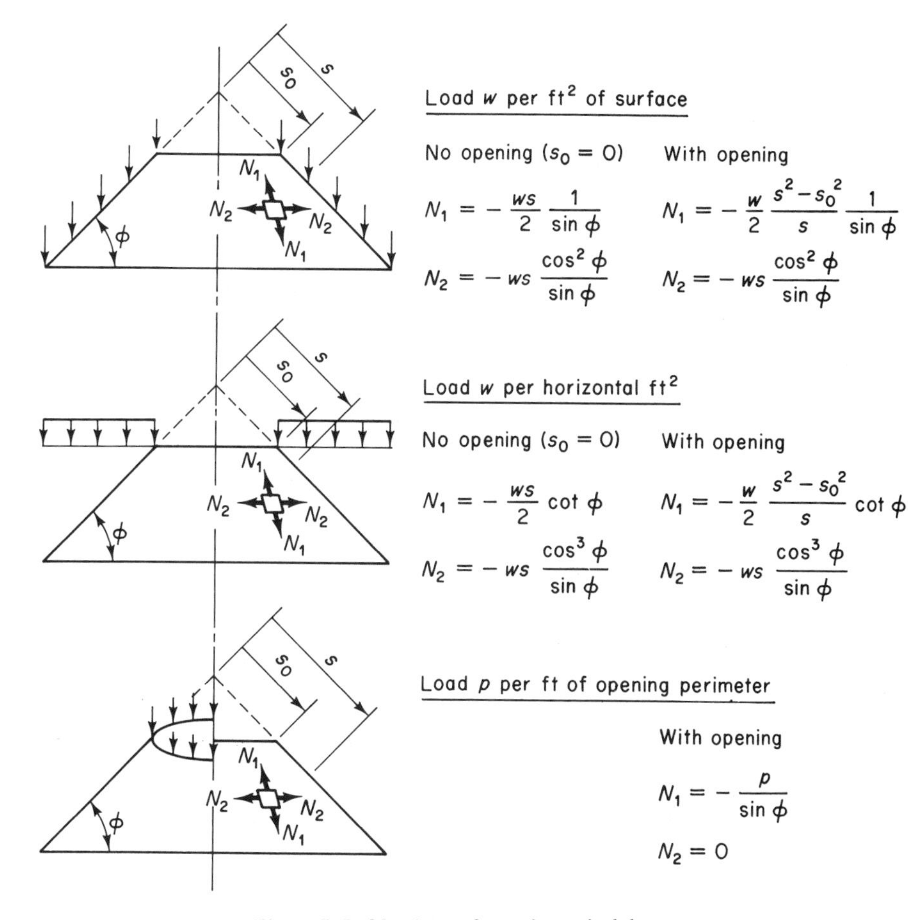

Figure 7-7 Membrane forces in conical domes.

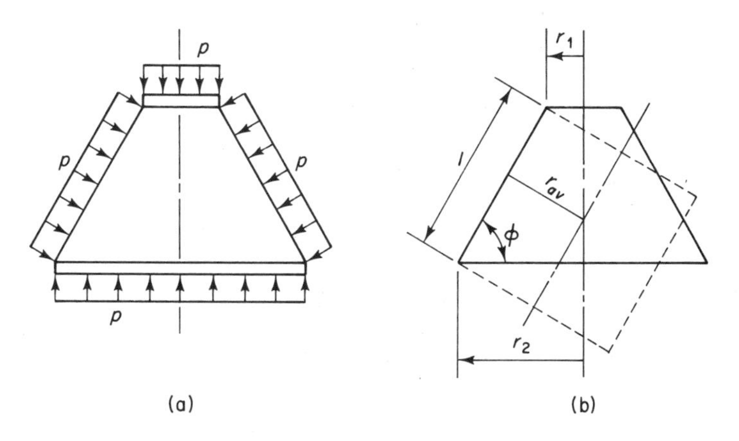

Figure 7-8

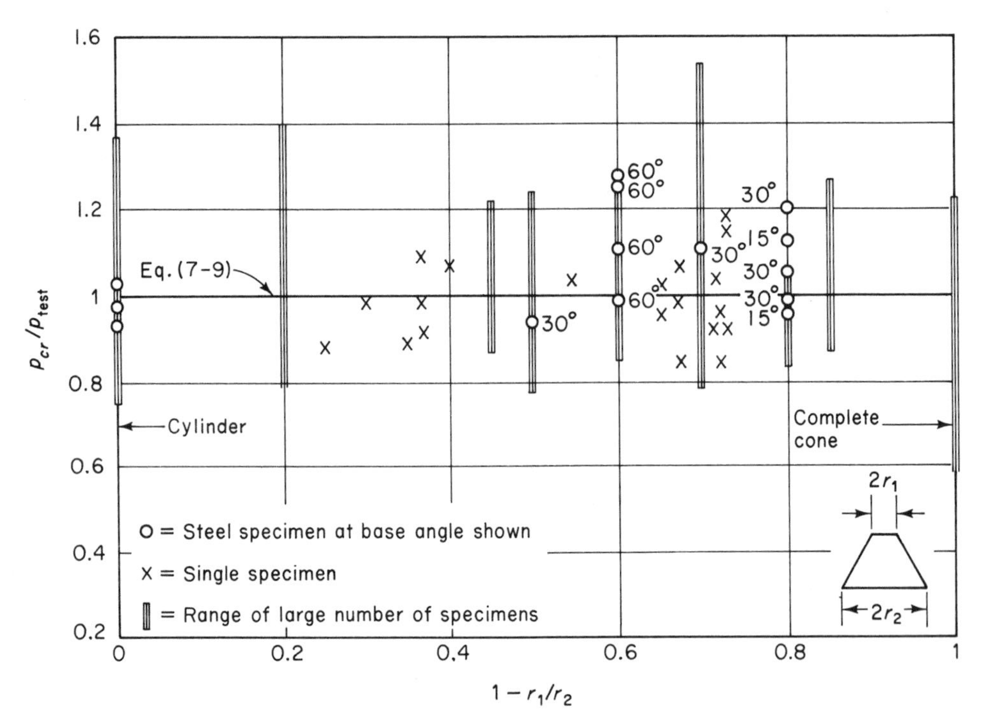

Figure 7-9 Tests on conical shells under external hydrostatic pressure. (From Ref. 9.)

there is a considerable scatter in the test results, a fairly good average fit is given by

$$p_{\text{cr}} = \frac{0.92E}{(l/r_{\text{av}})(r_{\text{av}}/t)^{5/2}} \tag{7-9}$$

The average radius of curvature is given by

$$r_{\text{av}} = \frac{r_1 + r_2}{2 \sin \phi} \tag{7-10}$$

in which r_1, r_2, and ϕ are defined in Fig. 7-8b.

The line $p_{\text{cr}}/p_{\text{test}} = 1$ in Fig. 7-9 corresponds to a fairly complex equation of which Eq. (7-9) is a conservative, simplified form. The difference is small for cones of the proportions used in bin roofs. Therefore, the line $p_{\text{cr}}/p_{\text{test}} = 1$ can also be considered to represent Eq. (7-9). Except for the tests by one investigator, $p_{\text{cr}}/p_{\text{test}}$ ranges from 0.685 to 1.53. The exception is a set of 15 tests on aluminum cones, all complete cones, for which the buckling pressures ranged from 0.58 to 1.05 times the theoretical values. On the other hand, the scatter was smaller for steel cones. For these specimens, which had base angles of 15°, 40°, and 60° and values of r_{av}/t from 405 to 2318, the experimental pressures ranged from 0.85 to 1.27 times the theoretical values. These test results are plotted individually in Fig. 7-9.

Equation (7-9) will give a conservative estimate of the buckling of a conical roof under load distributed uniformly over the horizontally projected area, which is the usual criterion for design. This is shown in Fig. 7-10, where it will be noted that the membrane forces N_1 are the same for p normal to the surface and w uniform on the projected area, while N_2 is larger for the normal pressure. On the other hand, N_1 due to the weight of the cone is smaller than N_1 for normal pressure, while the opposite is true for N_2. Thus, Eq. (7-9) would seem to be a good basis for the design of conical roofs.

With $E = 30 \times 10^6 \times 144$ psf and $w = p_{cr}/\eta$, where w is in pounds per square foot and η is the factor of safety, Eq. (7-9) gives

$$\frac{t}{r_{av}} = \frac{1}{6913}\left(\frac{\eta w l}{r_{av}}\right)^{0.4} \tag{7-11}$$

For a full cone $l/r_{av} = 2 \tan \phi$, and Eq. (7-11) gives

$$\frac{t}{r_{av}} = \frac{1}{5240}(\eta w \tan \phi)^{0.4} \tag{7-12}$$

Since the radius of an opening at the vertex of a conical roof is not likely to exceed one-tenth the radius of the bin, Eq. (7-12) may also be used for truncated-cone roofs.

If a factor of safety of 2 is used in Eq. (7-12), the ratio of experimental buckling load to allowable load ranges from 1.37 to 3.06 for all the specimens except the 15 aluminum cones. For the latter the ratios range from 1.16 to 2.10. For the steel cones the range is 1.70 to 2.54. Therefore, a factor of safety of 2 appears to be adequate, which, with a slight rounding on the safe side, reduces Eq. (7-12) to

$$\frac{t}{r_{av}} = \frac{(w \tan \phi)^{0.4}}{4000} \tag{7-13}$$

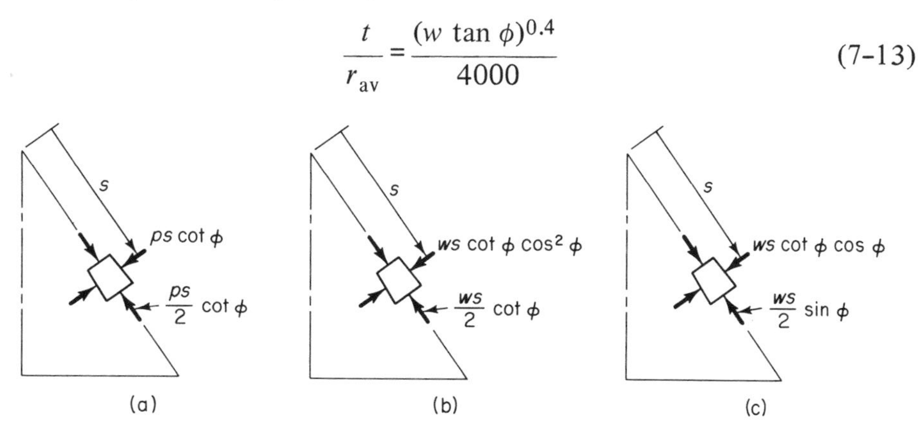

Figure 7-10 Membrane forces in cones from (a) pressure p normal to surface, (b) load w per horizontal square foot, (c) load w per square foot of surface.

This formula can be written in terms of the bin diameter D by substituting $r_{av} = D/4 \sin \phi$, which gives

$$\frac{t}{D} = \frac{(w \tan \phi)^{0.4}}{16{,}000 \sin \phi} \tag{7-14}$$

API 650. The formula for the thickness of a conical roof in this standard is derived for a live load of 25 psf plus a dead load of 20 psf. The latter is the weight of $\frac{1}{2}$-in. steel plate, which is the maximum allowable thickness of the roof. Since the membrane stress resultants are both compressive (Fig. 7-7), the ASME Boiler Code allowable stress for spherical pressure vessels, $900{,}000t/a$ psi, is used.[8] The maximum value of N_1 occurs at the base:

$$N_1 = \frac{wl \cot \phi}{2} = \frac{wD/2}{2 \sin \phi} \tag{a}$$

The radius of a sphere tangent to the cone at this point is $(D/2) \sin \phi$. Then, equating the stress due to N_1 to the allowable value, we get

$$\frac{wD/2}{(2 \sin \phi)t} = 900{,}000 \frac{t}{(D/2) \sin \phi} \tag{b}$$

With $w = 25 + 20 = 45$ psf $= 0.3125$ psi, Eq. (b) gives

$$\frac{t}{D} = \frac{1}{4800 \sin \phi} \tag{7-15}$$

In the API formula t is given in inches and the tank diameter D in feet, for which Eq. (7-15) becomes

$$t_{in} = \frac{D_{ft}}{400 \sin \phi} \tag{7-16}$$

Factors of safety based on the buckling load according to Eq. (7-9) are given in Table 7-1 for five conical roofs with thicknesses according to API 650 using a live load of 25 psf plus the weight of the roof plate. The ranges

TABLE 7-1. FACTORS OF SAFETY η OF API CONICAL ROOFS

		API		Eq. (7-14)	
D (ft)	ϕ (deg)	t (in.)	η	t (in.)	η
33	9.5	0.5	5.3	0.31	2
33	26	0.188	2.5	0.17	2
75	23	0.5	2.2	0.46	2
75	37	0.31	1.4	0.36	2
120	37	0.5	1.2	0.65	2

of thickness and base angle allowed by API are covered. It will be noted that there is a large variation in the factor of safety between the roofs for small-diameter bins and those with a large diameter. The required thicknesses according to Eq. (7–14), which gives a factor of safety of 2, are also shown. Since the cone is a single-curvature surface, an equivalent cylinder such as is used to obtain Eq. (7–9) is more logical than the equivalent sphere on which the API formula is based. Furthermore, Eq. (7–9) is consistent with test results, even though there is a large scatter, over a wide range of base angles and radius-thickness ratios. Therefore, the authors believe that a design formula based on Eq. (7–12) is to be preferred. A larger factor of safety than was used to obtain Eq. (7–14) might be advisable in some situations, particularly if the expected quality of workmanship may result in large irregularities in the roof surface.

Example 7–2

Determine the thickness required by Eq. (7–14) for a conical roof with a base angle of 15° for a bin 50 ft in diameter. The live load is 30 psf.

Try $\frac{3}{8}$-in. plate. $w = 30 + 15.3 = 45.3$ psf.

$$\frac{t}{D} = \frac{(45.3 \tan 15^\circ)^{0.4}}{16{,}000 \sin 15^\circ} = \frac{1}{1526}$$

$$t = \frac{50 \times 12}{1526} = 0.39 > \frac{3}{8}$$

Use $\frac{7}{16}$ plate or a larger base angle. Try $\phi = 16^\circ$:

$$\frac{t}{D} = \frac{(45.3 \tan 16^\circ)^{0.4}}{16{,}000 \sin 16^\circ} = \frac{1}{1581}$$

$$t = \frac{50 \times 12}{1581} = 0.379$$

Use $\frac{3}{8}$-in. plate with a 16° base angle.

7-5. Supported Roofs

The load on a cone, dome, or unbrella roof supported by rafters is not carried by the rafters alone. This is because the membrane action of the shell does not involve bending that transmits load to the rafters, and it is only after the shell has buckled and the membrane stress system is altered that the rafters begin to take load. The ultimate strength of such a roof is the sum of the local-buckling strength of the shell and the bending strength of the rafters. Tests on rafter-supported dome roofs showed that about 60% of the load was carried by the shell and 40% by the rafters.[11] The principal difficulty in an analysis that accounts for this dual supporting mechanism is the determination of the buckling strength of the shell. Since this is a local buckling of the surface between rafters, it differs from the buckling of unsupported shells which is discussed in Secs. 7–3 and 7–4. Local-buckling formulas for supported spherical domes are given in Sec. 7–9, but formulas for supported conical shells are not available.

An important consideration in the design of supported-roof rafters has to do with restraint against lateral-torsional buckling. If the rafter is not attached to the roof plate, lateral buckling due to live load and roof-plate dead load is resisted by friction, since these loads must be transmitted to the rafter by contact with the roof plate. However, lateral buckling due to rafter axial forces resulting from conveyors or other equipment supported by the compression ring is possible when there is no live load, such as snow, on the roof or when uplift on portions of the roof due to wind may result in loss of contact between roof plate and rafter. Tests by Fischer on lateral buckling of I-shaped beams which were loaded on their top flanges show that the stabilizing effect of load transmitted in this manner is substantial.[12] Ten tests were made, all on simply supported beams of IPE 120 section, which has a nominal depth of 12 cm (4.72 in.) and a flange width of 6.5 cm (2.56 in.) and weighs 0.97 kg/m (6.99 plf). One of the test arrangements is shown in Fig. 7–11. The beam was loaded symmetrically by loading both of the 2-m side spans. Although the beam was very slender (L/r_y = 345), it attained its full plastic moment, at which the failure load was 540 kp/m (1190 plf) or 2.84 times the theoretical buckling load. The AISC allowable stress, assuming no lateral support, is only 8.1 ksi, which gives an allowable load of 65 plf (97 kp/m) or only 18% of the failure load.

In another test a steel grating spanned the distance between two parallel IPE 120s. Edges of the grating were at the centers of the beam flanges so that the load was applied eccentrically to the beams. Furthermore, each beam was prebent laterally. The failure load was 510 kp/m per beam, or 2.7 times the theoretical buckling load.

In another setup two parallel beams 110 cm apart were loaded through precast planks 200 cm long which cantilevered 45 cm beyond each beam. The beam webs were set at a slope of 1/12 with the vertical in one test and 1/9 in another. The planks, which were set at the same slope perpendicular to the beams, rested on greased aluminum sheets attached to the beams and bore against the edge of the beam flanges by means of 10-mm-thick plates attached to the underside of the planks. This arrangement subjected each

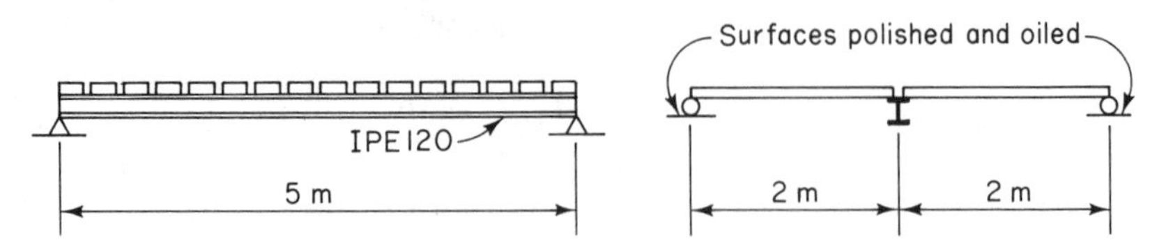

Figure 7–11 Fischer's setup for lateral-buckling tests. (From Ref. 12.)

beam to a lateral force on the top flange in addition to the component of load in the plane of the web. The failure loads for each beam were 353 and 194 kp/m for the 1/12 and 1/9 slopes, respectively, the latter being only slightly larger than the theoretical buckling load.

Fischer concludes that friction between the beam top flange and the structure delivering the load to it, which was virtually eliminated in his tests, is present in practical construction and exerts a stabilizing effect that can impart a stiffness to a system that is much greater than for each beam acting independently. In assessing the applicability of his results to the behavior of rafters in bin and tank roofs, it should be noted that the roof plate would rest on the edge of the flange of a rafter that happens to be tilted and exert a stabilizing force instead of a lateral force as in the tests on beams with inclined webs.

API 650 considers rafters not attached to the roof plate to be laterally supported by friction only if they are 15 in. or less in depth and their slope is 2/12 or less. Unattached open-web joists are not considered to be laterally supported in any case. Fischer's tests suggest that the restriction on solid-webbed rafters is unduly conservative and that these members can be designed as having lateral support. This conclusion is also in agreement with provisions in DIN 4119 for rafters in cone and dome roofs with moment-resistant connections at the compression ring, for which lateral support of the compression flange need be provided only in the negative-moment region (Sec. 7–8). Therefore, simply-supported rafters require only such bracing and circumferential ties as may be needed for erection or other purposes (Sec. 7–11).

Rafters with attached roof plates. If roof plates are welded to the rafters, an effective width of plate can be considered as part of the rafter cross section. An effective-width formula is given in Ref. 6. However, the flat-plate effective-width formulas in Sec. 6–7 give comparable results, and the analysis of service-load maximum bending stresses can be based on a cross section consisting of the rafter and an effective width of roof plate computed by Eq. (6–29), using for b the distance between rafters at the point where the stress is being computed.

7-6. Supported Cone and Umbrella Roofs

Since local buckling of a roof-plate segment bounded by the rafters of a supported cone or umbrella roof has not been investigated, the rafters must be designed for the entire load. Rafter spacing depends on the procedure to be used in erecting the roof plate and on the thickness of the plates. API 650 requires rafter spacing to be not more than 2π ft measured along the circumference of the tank and $5\frac{1}{2}$ ft along inner rings. The limit 2π is the approximate span of a roof plate of the specified minimum thickness, $\frac{3}{16}$ in., that can support the API live load at the allowable bending stress. Thus, with $w = 25 +$

7.7 psf the fixed-end moment $wL^2/12$ gives $L = 6.56$ ft with $F_b = 20$ ksi. DIN 4119 does not specify rafter spacing except to say that it must be such as to make the buckling pressure of the roof at least equal to the dead weight of the plate. The minimum roof-plate thickness is 4 mm (0.16 in.), which is practically the same as the API minimum. Thinner roof plates are used for tanks not required to be designed to API or similar standards. Plates as thin as 16 gage (0.06 in.) are used in standard bolted tanks.

Estimates of the deflection of flat-plate segments of umbrella roofs can be determined by procedures discussed in Sec. 6–17 and 6–19.

7-7. Cone Roofs with Simply Supported Rafters

Rafters that are connected to rings supported on beams or trusses, as in Figs. 7–2a and b, are usually designed and built as simply supported members. The reactions V_1 and V_2 and the moment M at any point are given by (Fig. 7–12)

$$V_1 = \frac{a}{6}(2q_1 + q_2) \tag{7–17a}$$

$$V_2 = \frac{a}{6}(q_1 + 2q_2) \tag{7–17b}$$

$$M = \frac{a^2}{6}\frac{x_1}{a}\left[2q_1 + q_2 - 3q_1\left(\frac{x_1}{a}\right) + (q_1 - q_2)\left(\frac{x_1}{a}\right)^2\right] \tag{7–18}$$

where a = horizontally projected length of rafter
x_1 = horizontal coordinate with origin at V_1
q_1, q_2 = load per unit of horizontally projected length

The moment attains a maximum value at

$$\frac{x_1}{a} = \frac{q_1 - \sqrt{(q_1^2 + q_1 q_2 + q_2^2)/3}}{q_1 - q_2} \tag{7–19}$$

The deflection δ normal to the rafter at any point is given by

$$\delta = \frac{a^4}{36EI_r\cos^2\phi}\left(\frac{x_1}{a}\right)\left[0.8q_1 + 0.7q_2 - (2q_1 + q_2)\left(\frac{x_1}{a}\right)^2 + 1.5q_1\left(\frac{x_1}{a}\right)^3 - 0.3(q_1 - q_2)\left(\frac{x_1}{a}\right)^4\right] \tag{7–20}$$

where I_r = moment of inertia of rafter
ϕ = angle of rafter with horizontal

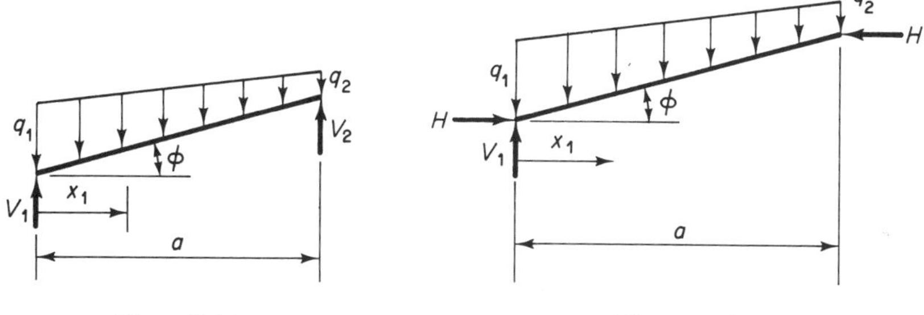

Figure 7-12 Figure 7-13

The reactions V_1 and H and the moment M at any point x_1 for a rafter simply supported at the compression ring (Fig. 7-13) are given by

$$V_1 = \frac{a}{2}(q_1 + q_2) + Q_0 \tag{7-21a}$$

$$H \tan \phi = \frac{a}{6}(q_1 + 2q_2) + Q_0 \tag{7-21b}$$

$$M = M \text{ by Eq. (7-18)} \tag{7-21c}$$

where Q_0 = the load per rafter due to the weight of the roof and appurtenances within the area of the compression ring.

The main rafters of roofs with subrafters support loads in addition to those considered in the preceding equations. Figure 7-14 shows the load on a main rafter with a single ring of subrafters. The reactions V_1 and H for a load varying uniformly from q_3 to q_4 over the length c are given by

$$V_1 = \frac{c}{2}(q_3 + q_4) \tag{7-22a}$$

$$H \tan \phi = \frac{c}{6a}[3a(q_3 + q_4) - c(2q_3 + q_4)] \tag{7-22b}$$

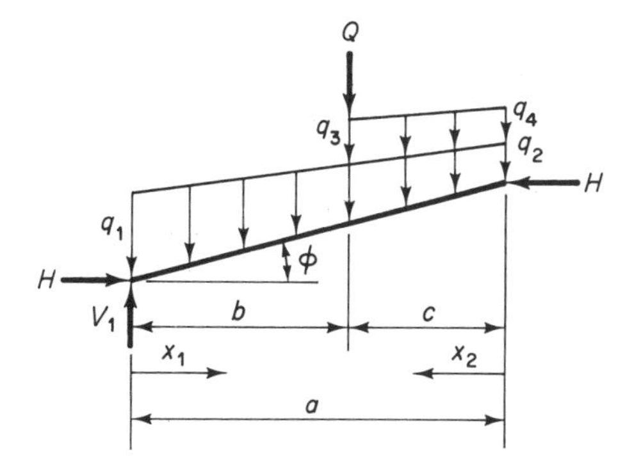

Figure 7-14

The moment at any point $x_2 \leqslant c$ is

$$M = \frac{cx_2}{6}\left[3(q_3 + q_4) - (2q_3 + q_4)\left(\frac{c}{a}\right) - 3q_4\left(\frac{x_2}{c}\right) - (q_3 - q_4)\left(\frac{x_2}{c}\right)^2\right] \quad (7\text{–}23)$$

where x_2 = horizontal coordinate with origin at end 2. The deflection normal to the rafter at $x_2 = c$ is

$$\delta_c = \frac{abc^2}{36EI_r \cos^2 \phi}\left[(2q_3 + q_4)\left(1 - \frac{b^2}{a^2}\right) - 0.3(4q_3 + q_4)\left(\frac{c^2}{a^2}\right)\right] \quad (7\text{–}24)$$

where $b = a - c$ (Fig. 7–14)

Formulas for the deflection at intermediate points are useful for main rafters that support more than one ring of subrafters. The deflection at any point $x_2 \leqslant c$ is given by

$$\delta = \frac{ac^2x_2}{36EI_r \cos^2 \phi}\left[2(2q_3 + q_4) - 1.5(3q_3 + q_4)\left(\frac{c}{a}\right) + 0.3(4q_3 + q_4)\left(\frac{c}{a}\right)^2\right.$$
$$- 3(q_3 + q_4)\left(\frac{c}{a}\right)\left(\frac{x_2}{c}\right)^2 + (2q_3 + q_4)\left(\frac{c}{a}\right)^2\left(\frac{x_2}{c}\right)^2 + 1.5q_4\left(\frac{c}{a}\right)\left(\frac{x_2}{c}\right)^3$$
$$\left. + 0.3(q_3 - q_4)\left(\frac{c}{a}\right)\left(\frac{x_2}{c}\right)^4\right] \quad (7\text{–}25)$$

The deflection at any point $x_1 \leqslant b$ is

$$\delta = \frac{ac^2x_1}{36EI_r \cos^2 \phi}\left[(2q_3 + q_4)\left(1 - \frac{x_1^2}{a^2}\right) - 0.3(4q_3 + q_4)\frac{c^2}{a^2}\right] \quad (7\text{–}26)$$

The reaction H and maximum moment $M_{\max}$ for the concentrated load Q in Fig. 7–14 are

$$H \tan \phi = Q\frac{b}{a} \quad (7\text{–}27)$$

$$M_{\max} = Q\frac{bc}{a} \quad (7\text{–}28)$$

The deflection δ_Q normal to the rafter at the point of application of Q is

$$\delta_Q = \frac{Qb^2c^2}{3EI_ra \cos^2 \phi} \quad (7\text{–}29)$$

The deflection at any point is given by

$$\delta = \frac{Qcx_1}{6aEI_r \cos^2 \phi}(a^2 - c^2 - x_1^2) \qquad 0 < x_1 < b \quad (7\text{–}30)$$

$$\delta = \frac{Qbx_2}{6aEI_r \cos^2 \phi}(a^2 - b^2 - x_2^2) \qquad 0 < x_2 < c \tag{7-31}$$

Intermediate rafters cantilevered to support headers. The following formulas are useful in the design of intermediate rafters supported by beams that frame into main rafters or trusses and are cantilevered to support the end of a header (Fig. 7-4). The rafters are assumed to be simply supported (Fig. 7-15).

$$V_1 = \frac{a}{2}(q_1 + q_2) + Q_h - V_2 \tag{7-32a}$$

$$V_2 = \frac{a}{d}\left[\frac{a}{6}(q_1 + 2q_2) + Q_h\right] \tag{7-32b}$$

$$M_x = x_1\left(V_1 - \frac{q_1 x_1}{2} + \frac{q_1 - q_2}{6a} x_1^2\right) \tag{7-33}$$

where a = horizontally projected length of rafter
d = horizontal distance between supports
Q_h = load from header

The point of maximum moment is at

$$\frac{x_1}{a} = \frac{q_1}{q_1 - q_2}\left(1 - \sqrt{1 - \frac{2(q_1 - q_2)}{q_1^2}\frac{V_1}{a}}\right) \tag{7-34}$$

The deflection normal to the rafter for $0 < x_1 < d$ is

$$\delta = \frac{d^5}{120EI_r \cos^2 \phi}\frac{x_1}{d}\left[20\frac{V_1}{d^2}\left(1 - \frac{x_1^2}{d^2}\right) - 5\frac{q_1}{d}\left(1 - \frac{x_1^3}{d^3}\right) + \frac{q_1 - q_2}{a}\left(1 - \frac{x_1^4}{d^4}\right)\right] \tag{7-35}$$

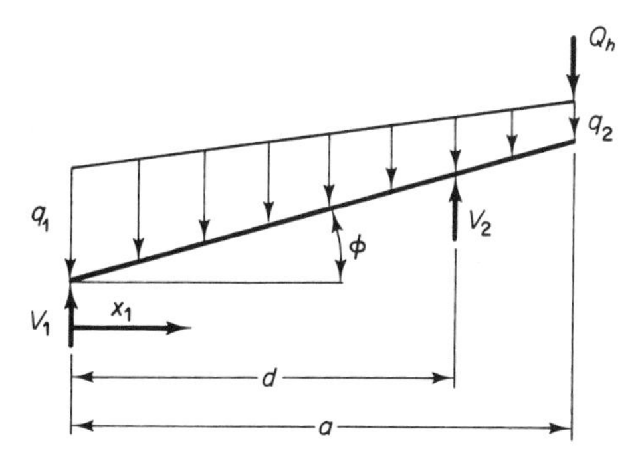

Figure 7-15

Example 7-3

Design a supported conical roof for a bin 17 ft 6 in. in diameter. The roof consists of 12 cold-formed flanged sections joined by bolts through the webs and supported by the bin wall and a center ring 2 ft in diameter. The ring is supported on two cold-formed channels (Fig. 7-16). The loads are 25 lb per square foot of roof with an additional 500 lb on the ring. F_y = 33 ksi, AISI specification.

$$\text{distance between flanges at bin wall} = \frac{17.5\pi}{12} = 4.58 \text{ ft}$$

$$\text{distance between flanges at ring} = \frac{2\pi}{12} = 0.524 \text{ ft}$$

$$\text{beam span} = a = \frac{17.5 - 2}{2} = 7.75 \text{ ft}$$

Allow 5 psf for the weight of the roof, so $w = 25 + 5 = 30$ psf

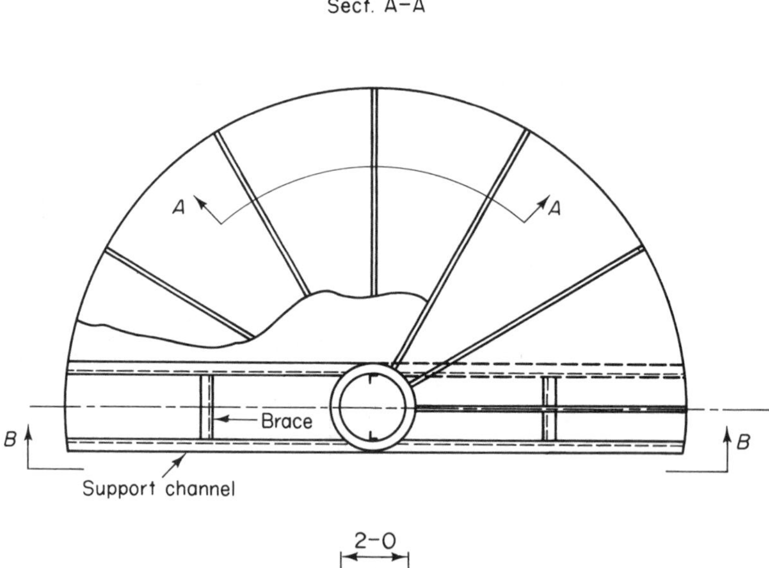

Figure 7-16 Roof plan for Example 7-3.

From Eqs. (7-17)-(7-19),

$$q_1 = 30 \times 4.58 = 137 \text{ psf}$$

$$q_2 = 30 \times 0.524 = 16 \text{ psf}$$

$$V_1 = \frac{7.75}{6}(2 \times 137 + 16) = 375 \text{ lb}$$

$$V_2 = \frac{7.75}{6}(137 + 2 \times 16) = 218 \text{ lb}$$

$$\frac{x_1}{a} = \frac{137 - \sqrt{(137^2 + 137 \times 16 + 16^2)/3}}{137 - 16} = 0.437$$

$$M = \frac{7.75^2 \times 0.437}{6}(290 - 3 \times 137 \times 0.437 + 121 \times 0.437^2) = 584 \text{ ft-lb}$$

Try a 14-gage section ($t = 0.075$ in.) flanged as shown in Fig. 7-17b. Since the roof sheet serves as the tension flange, its effective width is determined by shear lag and

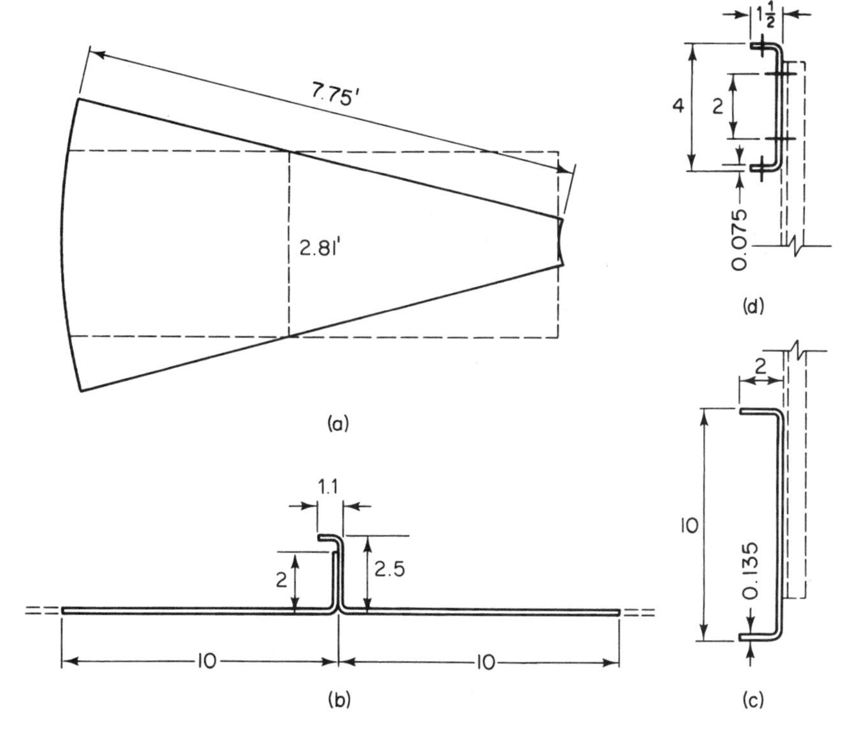

Figure 7-17 Details for roof of Example 7-3.

will be assumed equal to that of a beam having a uniform cross section the same as at the point of maximum moment (Sec. 6–12). The section width at this point is 4.58 - 0.437(4.58 - 0.524) = 2.81 ft, so l/b = 7.75/2.81 = 2.76 (Fig. 7–17a). Since the load is not distributed uniformly, the effective width for a beam with a concentrated load at the quarter point will be used. Therefore, from Table 6–3,

$$w_e = 0.58w = 0.58 \times 2.81 \times 12 = 20 \text{ in.}$$

Cross-sectional properties are determined for a section with square corners, using centerline lengths of elements and a reference axis at mid-thickness of the bottom flange:

Element	l	y	Moment of inertia
Top flange	1.0625	2.425	1.0625×2.425^2
Web	2.425	1.2125	$2.45^3/3$
Web	1.9625	0.98125	$1.9625^3/3$
Bottom flange	20	0	0
	$\Sigma l = 25.45$		$\Sigma I_0/t = 13.52 \text{ in.}^3$

$$\Sigma ly = 7.4426 \qquad y_{cg} = \frac{\Sigma ly}{\Sigma l} = \frac{7.4426}{25.45} = 0.292 \text{ in.}$$

$$\frac{I_x}{t} = \sum \frac{I_0}{t} - y_{cg}^2 \Sigma l = 13.52 - 0.292^2 \times 25.45 = 11.35 \text{ in.}^3$$

$$I_x = 11.35 \times 0.075 = 0.851 \text{ in.}^4$$

$$S_{xt} = \frac{0.851}{0.292 + 0.0375} = 2.58 \text{ in.}^3 \qquad S_{xc} = \frac{0.851}{2.5 - 0.3295} = 0.392 \text{ in.}^3$$

$$f_c = \frac{584 \times 12}{0.392} = 17.9 \text{ ksi}$$

Check for local buckling of the flange (Table 6–1):

$$\text{limiting} \frac{w}{t} = \frac{63.3}{\sqrt{F_y}} = \frac{63.3}{\sqrt{33}} = 11.0$$

$$\text{actual} \frac{w}{t} = \frac{1.1 - 0.0938 - 0.075}{0.075} = 12.4 > 11.0$$

Then, from Eq. (6–32b),

$$F_c = 33(0.767 - 0.00264\sqrt{33} \times 12.4) = 19.1 \text{ ksi} > 17.9 \qquad \text{OK}$$

The theoretically correct value of I, based on $\frac{3}{32}$-in. inside-corner radii, is 0.830 in.^{4}, which differs from the value just computed by only 2.5%. Furthermore, the section modulus is not very sensitive to effective width; thus, $S_{xc} = 0.396$, 0.392, and 0.386 in.^{3} for effective widths of 25, 20, and 15 in., respectively.

Deflection

Since the effective width of the tension flange depends on shear lag, which varies from a maximum at or near the point of maximum moment to zero at each end, the moment of inertia varies from its smallest value at the ring to a maximum at the bin wall. Therefore, only an approximate value of deflection can be obtained from Eq. (7–20). A value based on the moment of inertia already calculated for the cross section of maximum moment will probably be a fair estimate. For the deflection at midspan this gives

$$\delta = \frac{7.75^4 \times 1728}{36 \times 30 \times 10^6 \times 0.851}(109.6 + 11.2 - 290 \times 0.5^2 + 205.5 \times 0.5^3 - 36.3 \times 0.5^4)$$

$$= 0.24 \text{ in.}$$

$$\frac{\delta}{a} = \frac{0.24}{7.75 \times 12} = \frac{1}{388}$$

A conservative estimate of deflection is obtained by using the moment of inertia at the ring support. The tension flange here is $0.524 \times 12 \approx 6$ in. wide. The computations for moment of inertia for the 20-in. width are easily revised for a 6-in. width. Thus,

$$\Sigma l = 11.45 \text{ in.} \qquad y_{cg} = \frac{7.4426}{11.45} = 0.650 \text{ in.}$$

$$\frac{I_x}{t} = 13.52 - 0.650^2 \times 11.45 = 8.68 \text{ in.}^3$$

$$I_x = 8.68 \times 0.075 = 0.651 \text{ in.}^4$$

This value of I_x gives a deflection of 0.31 in.

Center-ring support channels

The concentrated load P at the center of each channel is

$$P = \frac{12V_2 + \pi \times 1^2 \times 30 + 500}{2} = 1605 \text{ lb}$$

$$M = \frac{1605 \times 17.5}{4} = 7022 \text{ ft-lb}$$

Try a $10 \times 2 \times 0.135$ channel (Fig. 7–17c):

$$I_x = \left(2 \times 2 \times 4.9325^2 + \frac{9.733^3}{12}\right) \times 0.135 = 23.50 \text{ in.}^4$$

$$S_x = \frac{23.50}{5} = 4.70 \text{ in.}^3$$

$$f_c = \frac{7022 \times 12}{4.70} = 17.9 \text{ ksi}$$

Check local buckling of the flange:

$$\frac{w}{t} = \frac{2 - 0.135 - 0.1875}{0.135} = 12.4$$

From Eq. (6–32b),

$$F_c = 33(0.767 - 0.00264\sqrt{33} \times 12.4) = 19.1 \text{ ksi} > 17.9 \qquad \text{OK}$$

Check lateral-torsional buckling. Support channels are braced 4.5 ft from the bin wall and at connection to the center ring, so the unbraced length is 4.5 ft:

$$\Sigma l = 2 \times 1.9325 + 9.865 = 13.73 \text{ in.}$$

$$x_{cg} = \frac{2 \times 1.9325 \times 1.9325/2}{13.73} = 0.272 \text{ in.}$$

$$\frac{I_y}{t} = \frac{2 \times 1.9325^3}{3} - 13.73 \times 0.272^2 = 3.80 \text{ in.}^3$$

$$I_y = 3.80 \times 0.135 = 0.512 \text{ in.}^4$$

From Eq. (6–2a), with $C_b = 1$,

$$\frac{L^2 S_{xc}}{dI_{yc}} = \frac{(4.5 \times 12)^2 \times 4.70}{10 \times 0.512/2} = 5354 < 17.77\frac{29{,}500}{33} = 15{,}885$$

$$\frac{F_b}{F_y} = \frac{2}{3} - \frac{0.0188 \times 33 \times 5354}{29{,}500} = 0.554$$

$$F_b = 0.554 \times 33 = 18.3 > 17.9 \text{ ksi} \qquad \text{OK}$$

Center ring (Fig. 7-17d)

$$\text{load on ring} = 12V_2 + \pi \times 1^2 \times 30 + 500 = 3210 \text{ lb}$$

The moment at the support [Eq. (6–67)] is

$$M = 0.159nQR = 0.159 \times 3210 \times 1 = 511 \text{ ft-lb}$$

Try a $4 \times 1\frac{1}{2} \times 0.075$ channel with allowance for the $\frac{17}{32}$-in. bolt holes shown:

$$\frac{I_x}{t} = 2(1.4625 - 0.531) \times 1.9625^2 + \frac{3.925^3}{12} - 2 \times 0.531 \times 1^2 = 11.15 \text{ in.}^3$$

$$I_x = 11.15 \times 0.075 = 0.836 \text{ in.}^4 \qquad S_x = \frac{0.836}{2} = 0.433 \text{ in.}^3$$

$$f_b = \frac{511 \times 12}{0.433} = 14.2 \text{ ksi} \qquad \text{OK}$$

Example 7-4

Design a supported cone roof for a bin 90 ft in diameter. The angle of repose of the stored material is 28°. The live load is 25 psf of horizontal projection, A36 steel.

Assume a 10-ft diameter compression ring and an 18-ft height, which gives a base angle of 24.2°. Assume 20 main rafters with 20 subrafters (Fig. 7-18). This gives a circumferential spacing at the bin wall of $90\pi/40 = 7.06$ ft. The subrafter headers will be put midway between the center of the bin and the wall so that the roof-plate span at this point will not exceed the 7.06-ft span at the wall. The main rafter is assumed to be simply supported at the compression ring.

For a $\frac{3}{16}$-in. roof plate the load is 25 + 7.7 = 32.7 psf = 0.227 psi. Assuming the roof plate to be one-way continuous with a span of 7.06 ft, which is conservative, the span-thickness ratio is $7.06 \times 12/0.1875 = 452$. Using Table 6-6 for the large-deflection theory of plates, compute

$$\sqrt{\frac{p}{E}}\left(\frac{b}{t}\right)^2 = \sqrt{\frac{0.227}{30 \times 10^6}} \times 452^2 = 17.8$$

This gives $(b/t)\sqrt{f/E} = 9$ and $\delta/t = 2.18$, from which

$$f = \left(\frac{9}{452}\right)^2 \times 30 \times 10^6 = 11{,}900 \text{ psi}$$

$$\delta = 2.18 \times 0.1875 = 0.41 \text{ in.} \qquad \frac{\delta}{b} = \frac{0.41}{7.08 \times 12} = \frac{1}{207}$$

For a $\frac{1}{8}$-in. roof plate the stress f is 15,280 psi. The deflection is 0.50 in. The corresponding deflection-span ratio of 1/170 would be acceptable in most cases.

Subrafter

The horizontal span $a = 45 - 22.5 \cos 9° = 22.8$ ft.

$$q_1 = 32.7 \times \frac{90\pi}{40} = 231 + 10 \text{ for rafter} = 241 \text{ plf}$$

$$q_2 = 32.7 \times \frac{44.4\pi}{40} = 114 + 10 \text{ for rafter} = 124 \text{ plf}$$

$$\frac{x_1}{a} = \frac{241 - \sqrt{(241^2 + 241 \times 124 + 124^2)/3}}{241 - 124} = 0.474 \qquad \text{[Eq. (7-19)]}$$

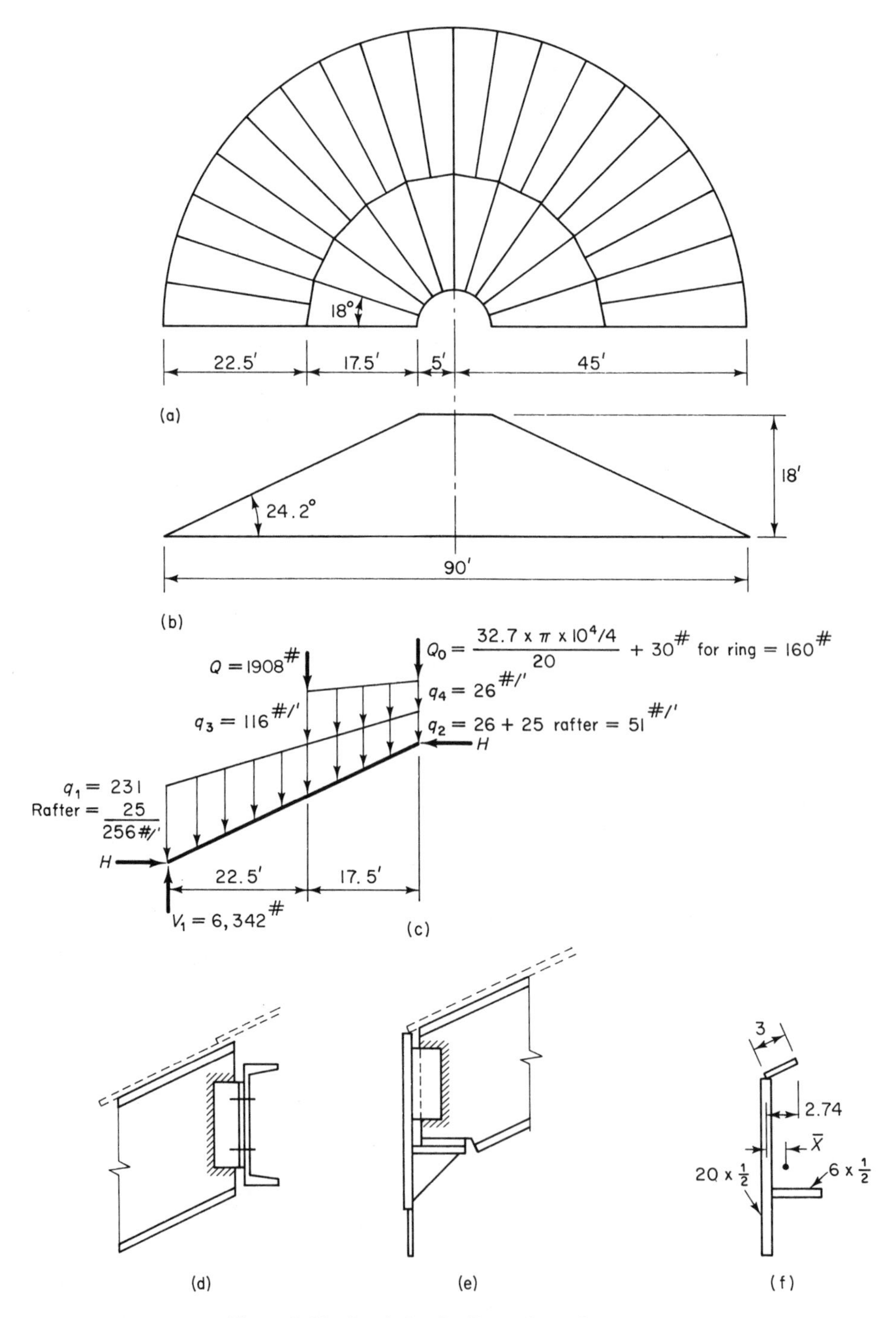

Figure 7-18 Roof plan for Examples 7-4 and 7-5.

$$M = \frac{22.8^2}{6} \times 0.474(482 + 124 - 723 \times 0.474 + 117 \times 0.474^2) = 11.9 \text{ ft-kip}$$

[Eq. (7–18)]

Assume that the subrafter is supported against lateral buckling:

$$S_x = \frac{11.9 \times 12}{24} = 5.95 \text{ in.}^3$$

The lightest W section with $S_x \geqslant 5.95$ in.^{3} is the W8 × 10, for which $S_x = 7.81$ in.^{3}

The AISC limiting span-depth ratio for purlins will be adopted; that is, $L/d = 1000/F_y$ for a fully stressed member, which gives 28 for A36 steel. The corresponding deflection-span ratio is 1/220 for a compact section (Sec. 6–2). However, the W8 × 10 is understressed, so that the recommended limit on L/d can be increased in the ratio of the allowable stress to the actual or, equivalently, the ratio of the section modulus furnished to the section modulus required [Eq. (6–12)]. Therefore,

$$\frac{L}{d} = 28 \times \frac{7.81}{5.95} = 37$$

$$\frac{L}{d} = 22.8 \times \frac{12}{8} = 34 < 37 \qquad \text{OK}$$

If the subrafter is assumed to be laterally unsupported, as would be required by API 650, the W8 × 10 is inadequate according to AISC 1.5.1.4.6a. Thus,

$$r_T = 0.99 \text{ in.} \qquad \frac{L}{r_T} = \frac{300}{0.99} = 303$$

$$\frac{d}{A_f} = 9.77 \qquad \frac{Ld}{A_f} = 300 \times 9.77 = 2931$$

For this case $C_b = 1$.

For $L/r_T \geqslant \sqrt{510{,}000C_b/F_y} = 119$, use the larger of

$$F_b = \frac{170{,}000C_b}{(L/r_T)^2} = \frac{170{,}000}{303^2} = 1.85 \text{ ksi}$$

$$F_b = \frac{12{,}000C_b}{Ld/A_f} = \frac{12{,}000}{2931} = 4.09 \text{ ksi}$$

$$f = 11.9 \times 12/7.81 = 18.3 > 4.09 \text{ ksi}$$

Try W12 × 22, $S_x = 25.4$ in.^{3}:

$$r_T = 1.02 \text{ in.} \qquad \frac{L}{r_T} = \frac{300}{1.02} = 291 \qquad F_b = \frac{170{,}000}{2.91^2} = 2.01 \text{ ksi}$$

$$\frac{d}{A_f} = 7.19 \qquad \frac{Ld}{A_f} = 300 \times 7.19 = 2160 \qquad F_b = \frac{12{,}000}{2160} = 5.56 \text{ ksi}$$

$$f = \frac{11.9 \times 12}{25.3} = 5.64 \text{ ksi} \approx 5.56$$

It will be noted that in this case the section required if the rafter is assumed to have no lateral support is twice as heavy as the one required with lateral support. Considering the results of Fischer's tests (Sec. 7-5), the authors think this is an unnecessary penalty. (The slenderness $L/r_y = 300/0.841 = 357$ of the W8 × 10 is only slightly larger than the 345 of Fischer's beam.)

Header

$$\text{Span} = 2 \times 22.5 \sin 9^\circ = 6.95 \text{ ft}$$

Subrafter $V_2 = (22.8/6)(241 + 2 \times 124) = 1858$ [Eq. (7-17b)]
weight of header = 50
1908 lb.

$$M = 1908 \times \frac{6.95}{4} = 3315 \text{ ft-lb} \qquad S_x = 3.3 \times \frac{12}{22} = 1.8 \text{ in.}^3$$

Use $C4 \times 5.4$, $S_x = 1.93$ in.^{3}, and $L/d = 6.95 \times \frac{12}{4} = 21$.

Main rafter (Fig. 7-18c)

Allow 25 plf for the weight of the rafter.

q_1, q_2, Q_0 load:

$$q_1 = 231 + 25 = 256 \text{ plf} \qquad q_2 = 26 + 25 = 51 \text{ plf} \qquad Q_0 = 160 \text{ lb}$$

$$V_1 = \frac{40}{2}(256 + 51) + 160 = 6300 \text{ lb} \qquad \text{[Eq. (7-21a)]}$$

$$H \tan \phi = \frac{40}{2}(256 + 102) + 160 = 2547 \text{ lb} \qquad \text{[Eq. (7-21b)]}$$

Moment at $x_1 = 22.5$ ft, $x_1/a = 22.5/40 = 0.5625$

$$M = \frac{40^2}{6} \times 0.5625(563 - 768 \times 0.5625 + 205 \times 0.5625^2) = 29{,}380 \text{ ft-lb}$$

[Eq. (7-21c)]

q_3, q_4 load:

$$q_3 = 116 \text{ plf} \qquad q_4 = 26 \text{ plf} \qquad c = 17.5 \text{ ft} \qquad \frac{c}{a} = 0.4375$$

$$V_1 = \frac{17.5}{2}(116 + 26) = 1242 \text{ lb} \qquad \text{[Eq. (7-22a)]}$$

$$H \tan \phi = \frac{0.4375}{6}(3 \times 40 \times 142 - 17.5 \times 258) = 913 \text{ lb} \qquad \text{[Eq. (7-22b)]}$$

The moment at $x_1 = 22.5$ ft is

$$M = 22.5V_1 - (22.5 \tan \phi)H = 22.5(1242 - 913) = 7400 \text{ lb}$$

Q load:

$$Q = \text{load on rafter header} = 1908 \text{ lb}$$

$$V_1 = Q = 1908 \text{ lb} \qquad H = 1908 \times \frac{22.5}{18} = 2385 \text{ lb}$$

The moment at $x_1 = 22.5$ ft is

$$M = 1908 \times 22.5 - 2385 \times 22.5 \tan 24.2^\circ = 18{,}800 \text{ ft-lb}$$

The total moment is

$$M = 29{,}380 + 7400 + 18{,}800 = 55{,}580 \text{ ft-lb}$$

$$V_1 = 6300 + 1242 + 1908 = 9454 \text{ lb}$$

$$H = \frac{2547 + 913}{\tan 24.2^\circ} + 2385 = 10{,}080 \text{ lb}$$

The axial force is

$$P = V_1 \sin 24.2^\circ + H \cos 24.2^\circ = 13{,}070 \text{ lb}$$

The moment at $x_1 = 22.5$ ft is not necessarily the maximum, and, in general, moments at several points should be determined to assure a maximum. In this example, the moment at $x_1 = 22.5$ ft is in fact the maximum.

Assuming the rafter to be supported against lateral buckling,

$$S_x = 55.6 \times \frac{12}{24} = 27.8 \text{ in.}^3 \qquad \text{W14} \times 22 \text{ gives } S_x = 29.0 \text{ in.}^3$$

$$L = 40 \times \frac{12}{\cos 24.2^\circ} = 526 \text{ in.} \qquad \frac{L}{d} = \frac{526}{14} = 37.6$$

The required section modulus to give a deflection equal to that of a fully stressed beam with $L/d = 28$ is $S_x = 27.8 \times 37.6/28 = 37.3$ in.^{3} for a 14-in. section and $37.3 \times 14/16 = 32.6$ in.^{3} for a 16-in. section. Try W16 × 26: $S_x = 38.4$ in.^{3}, $A = 7.68$ in.^{2}, $r_x = 6.26$ in., and $I_x = 301$ in.^{4}

$$f_a = \frac{13.07}{7.68} = 1.70 \text{ ksi} \qquad f_b = \frac{M}{S_x} = 55.6 \times \frac{12}{38.4} = 17.4 \text{ ksi}$$

$$\frac{L}{r_x} = \frac{526}{6.26} = 84 \qquad F_a = 14.9 \text{ ksi}$$

Since $f_a/F_a = 1.70/14.9 = 0.114 < 0.15$, AISC Formula (1.6–2) may be used:

$$\frac{f_a}{F_a} + \frac{f_b}{F_b} = 0.114 + \frac{17.4}{24} = 0.84 < 1$$

Because there were no axial compressive forces in Fischer's tests (Sec. 7–5), there is a question as to the effectiveness of friction in restraining lateral-torsional buckling under combined bending and axial compression. However, it is reasonable to assume that frictional restraint will be effective if bending predominates, and in this case the stress due to bending is over 90% of the total.

Although the span-depth ratio L/d gives a satisfactory check on deflection, deflection will be computed here for illustrative purposes and to show the amplification of beam deflection by the axial force.

Deflection at x_1 = 22.5 ft

q_1, q_2, Q_0 load by Eq. (7–20) with $x_1/a = 22.5/40 = 0.563$:

$$\delta = \frac{40^4 \times 0.563 \times 12^3}{36 \times 30 \times 10^6 \times 301 \cos^2 24.2^\circ}(240.5 - 563 \times 0.563^2$$

$$+ 384 \times 0.563^3 - 61.5 \times 0.563^4) = 1.14 \text{ in.}$$

q_3, q_4 load by Eq. (7–24) with $b/a = 22.5/40 = 0.563$ and $c/a = 0.437$:

$$\delta = \frac{40 \times 22.5 \times 17.5^2 \times 12^3}{36 \times 30 \times 10^6 \times 301 \cos^2 24.2}[258(1 - 0.563^2) - 0.3 \times 490 \times 0.437^2]$$

$$= 0.26 \text{ in.}$$

Q load by Eq. (7–29):

$$\delta = \frac{1908 \times 22.5^2 \times 17.5^2 \times 12^3}{3 \times 30 \times 10^6 \times 301 \times 40 \cos^2 24.2^\circ} = 0.57 \text{ in.}$$

$$\delta = 1.14 + 0.26 + 0.57 = 1.97 \text{ in.}$$

This value of δ is due only to bending from the transverse loads and is increased by the axial compression P. The amplification factor is the same as for the bending stress in the second term of the stress-interaction formula. Therefore,

$$\delta = \frac{1.97}{1 - 1.70/21.2} = 2.14 \text{ in.} \qquad \frac{\delta}{L} = \frac{2.14}{526} = \frac{1}{246} < \frac{1}{220} \qquad \text{OK}$$

The axial-force amplification of deflection can also be taken into account in establishing a limiting L/d by multiplying the value for transverse load by the amplification factor. Thus,

$$\frac{L}{d} = 28\left(1 - \frac{1.70}{21.2}\right) = 25.7$$

Deflection at the point of maximum moment is not necessarily a maximum. In this example, maximum deflection is at $x_1 = 20$ ft, but it is only 2% larger than the deflection at 22.5 ft.

If the rafter is assumed to have no lateral support except at the connection of the header, it must be checked for buckling. The W16 X 26 proves to be inadequate. Try W14 X 43: $A = 12.6$ in.^{2}, $S_x = 62.7$ in.^{3}, $r_x = 5.82$ in., $r_y = 1.89$ in., $r_T = 2.12$ in., and $d/A_f = 3.22$.

$$\frac{Ld}{A_f} = 296 \times 3.22 = 953 \qquad F_b = \frac{12{,}000}{953} = 12.6 \text{ ksi}$$

$$\frac{L}{r_T} = \frac{296}{2.12} = 140 > \sqrt{\frac{510{,}000}{F_y}} \qquad F_b = \frac{170{,}000}{140^2} = 8.6 \text{ ksi}$$

$$\frac{L}{r_x} = \frac{526}{5.82} = 90.4 \qquad \frac{L}{r_y} = \frac{296}{1.89} = 157 \qquad F_a = 6.06 \text{ ksi}$$

$$F'_E = 18.2 \text{ ksi} \qquad C_m = 1$$

$$f_a = \frac{13.1}{12.6} = 1.04 \qquad f_b = \frac{55.6 \times 12}{62.7} = 10.6$$

$$\frac{f_a}{F_a} + \frac{f_b}{F_b}\,\frac{1}{1 - f_a/F'_E} = \frac{1.04}{6.06} + \frac{10.6}{12.6} \times \frac{1}{1 - 1.04/18.2} = 1.06$$

This analysis shows the W14 X 43 to be somewhat overstressed. However, the value of F_a is conservative because it is based on pinned-end bending ($K = 1$) of the 296-in. segment about its weak axis. The adjacent shorter segment, 230 in. long, provides rotational restraint to the 296-in. segment. Therefore, the W14 X 43 is satisfactory. No lighter section is adequate. Thus the API requirement that the rafter be considered laterally unsupported increases the weight of the main rafters by 65%.

Compression ring

Use a bolted rafter connection. Try C12 X 20.7 (Fig. 7-18d). Ring formulas are given in Sec. 6-19, Eqs. (6-71) and (6-72). Then with $Q = H = -10{,}080$ lb we have: At the load points,

$$M = \frac{QR}{2}\left(\cot\frac{180^\circ}{n} - \frac{n}{\pi}\right) = \frac{-10.08 \times 5}{2}\left(\cot\frac{180^\circ}{20} - \frac{20}{\pi}\right) = 1.32 \text{ ft-kip}$$

$$P = \frac{-10.08}{2}\cot\frac{180^\circ}{n} = -\frac{10.08}{2}\cot\frac{180^\circ}{20} = -31.8 \text{ ft-kip}$$

On the outer fiber,

$$f = -\frac{31.8}{6.09} - \frac{1.32 \times 12 \times 0.698}{3.88} = -8.09 \text{ ksi}$$

On the inner fiber,

$$f = -\frac{31.8}{6.09} + \frac{1.32 \times 12 \times 2.244}{3.88} = +3.92 \text{ ksi}$$

Midway between load points,

$$M = \frac{QR}{2}\left(\csc\frac{180^\circ}{n} - \frac{n}{\pi}\right) = \frac{-10.08 \times 5}{2}\left(\csc\frac{180^\circ}{20} - \frac{20}{\pi}\right) = -0.662 \text{ ft-kip}$$

$$P = \frac{-10.08}{2}\csc\frac{180^\circ}{n} = \frac{-10.08}{2}\csc\frac{180^\circ}{20} = -32.2 \text{ kips}$$

On the outer fiber,

$$f = \frac{-32.2}{6.09} + \frac{0.662 \times 0.698}{3.88} = -5.12 \text{ ksi}$$

On the inner fiber,

$$f = \frac{-32.2}{6.09} - \frac{0.662 \times 2.244}{3.88} = -5.62 \text{ ksi}$$

The compression ring need not be checked for buckling because it is supported by the weldment of the center segment of the roof plate to the rafters (Fig. 7–18d).

Tension ring

Since the radial forces on the tension ring are equal to those on the compression ring, the circumferential forces P are numerically the same but of opposite sign, while the moments are $n/R = 45/5 = 9$ times the moments in the compression ring, again of opposite sign. Therefore, at the load points, $P = +32.2$ kips and $M = 9(-1.32) = -11.9$ ft-kip, and midway between, $P = +32.2$ kips and $M = 9 \times 0.662 = 5.96$ ft-kip.

With the detail shown in Fig. 7–18e, the tension ring cross section will be assumed as shown in Fig. 7–18f, with a width $16t$ of the roof plate considered to be effective as part of the ring:

$$\begin{aligned}
3 \times \tfrac{3}{16} &= 0.56 & \times 1.37 &= 0.77\\
6 \times \tfrac{1}{2} &= 3.00 & \times 3.25 &= 9.75\\
20 \times \tfrac{1}{2} &= \underline{10.00} & - & \quad \underline{\;-\;}\\
A &= 13.56 \text{ in.}^2 & & \quad 10.52\\
& & \bar{x} &= 0.78 \text{ in.}
\end{aligned}$$

$$I = \frac{3}{16} \times 3 \times \frac{2.74^2}{12} + 0.56 \times 0.53^2 + \frac{1}{2} \times \frac{6^3}{12} + 3.00 \times 2.47^2 + 10 \times 0.78^2 = 40.6 \text{ in.}^4$$

At the load points,

$$f = \frac{32.2}{13.56} + \frac{11.9 \times 12 \times 1.03}{33.9} = +6.71 \text{ ksi}$$

$$f = \frac{32.2}{13.56} - \frac{11.9 \times 12 \times 5.47}{33.9} = -20.7 \text{ ksi}$$

Between load points,

$$f = \frac{32.2}{13.56} - \frac{5.96 \times 12 \times 1.03}{33.9} = -0.20 \text{ ksi}$$

$$f = \frac{32.2}{13.56} + \frac{5.96 \times 12 \times 5.47}{33.9} = +13.9 \text{ ksi}$$

Tension rings are discussed in Sec. 8-9.

Self-supporting cone roof

Assume $\frac{1}{2}$-in. plate. $w = 20.4 + 25 = 45.4$ psf, and from Eq. (7-14),

$$\frac{t}{D} = \frac{(45.4 \tan 24.2)^{0.4}}{16{,}000 \sin 24.2} = \frac{1}{1963}$$

$$t = \frac{90 \times 12}{1963} = 0.55 \text{ in.}$$

Use $\frac{9}{16}$-in. plate.

The weights of the supported roof (assuming the rafters to have lateral restraint) and the self-supporting roof are as follows:

Supported roof:

$$20 \text{ W8} \times 10 \times 24.67 = 4{,}930$$

$$20 \text{ W16} \times 26 \times 43.85 = 22{,}800$$

$$20 \text{ C4} \times 5.4 \times 6.35 = 750$$

$$1 \text{ C12} \times 20.7 \times 10\pi = 650$$

$$\text{roof plate: } \frac{90^2\pi \times 7.65}{4 \cos 24.2} = \underline{53{,}350}$$

$$82{,}480 \text{ lb}$$

Self-supporting roof:

$$\text{roof plate: } \frac{90^2\pi \times 23.0}{4 \cos 24.2} = 160{,}420 \text{ lb}$$

7-8. Cone Roofs with Rafters Fixed at Ring

Economy in rafter size can be achieved by using a moment-resistant connection of the rafter to the compression ring. Such a rafter is statically indeterminate. The moment M_0 at the connection to the ring is determined by the requirement that the slope γ of the deflected rafter equals the angle of twist of the ring (Fig. 7-19). Using γ for a uniformly distributed torque, Eq. (6-80), simplifies the solution and gives a negligible error. The resulting moment M_0 for the q_1, q_2 loading in Fig. 7-19 is given by

$$M_0 = \frac{a^2(7q_1 + 8q_2)}{120(1 + 3k)} \tag{7-36a}$$

$$k = 0.159n \frac{I_r}{I_0} \frac{R}{L} \tag{7-36b}$$

where I_0 = x-axis moment of inertia of compression ring
I_r = x-axis moment of inertia of rafter
L = length of rafter
R = radius of compression ring
n = number of rafters connected to ring

The rafter-connection moment for the partial loading q_3, q_4 (Fig. 7-14) is

$$M_0 = \frac{c^2}{12(1 + 3k)} \left[2(2q_3 + q_4) - 1.5(3q_3 + q_4)\left(\frac{c}{a}\right) + 0.3(4q_3 + q_4)\left(\frac{c}{a}\right)^2\right] \tag{7-37}$$

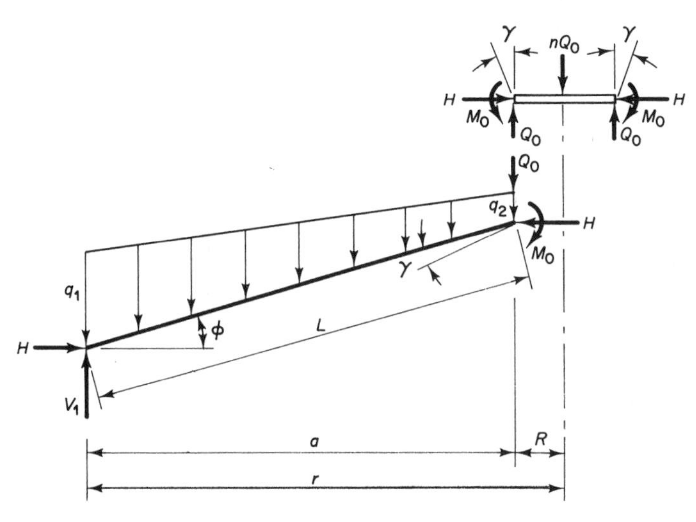

Figure 7-19

For the concentrated load Q in Fig. 7-14 the moment is

$$M_0 = \frac{Qbc(a + b)}{2a^2(1 + 3k)} \tag{7-38}$$

The deflection normal to the rafter at any point is

$$\delta = -\frac{M_0 x_1}{6EI_r a}(a^2 - x_1^2) \tag{7-39}$$

where the negative sign denotes an upward deflection.

Based on recommendations in Ref. 11 which are supported by tests, DIN 4119 requires an analysis of lateral-torsional buckling of the rafters only if the unbraced length of the compression flange in the negative-moment region exceeds $180r_y$. This provision applies to rafters with unattached roof plates as well as to those with attached plates.

Example 7-5

Design the main rafters and compression ring for the cone roof of Example 7-4 using moment-resistant rafter-to-ring connections.

The smallest section will result if the maximum positive moment and the end moment are equal. Since the moment diagram for the end moment alone is triangular, the two moments will be approximately equal if the end moment is two-thirds of the maximum positive moment. From Example 7-4 the moment at the header connection to the main rafter is 55,580 ft-lb. Therefore,

$$M_0 \approx \frac{2}{3} \times 55{,}580 \approx 37{,}000 \text{ ft-lb}$$

The M_0 moments produce bending of the compression ring about the horizontal (x) axis of its cross section. The moment is given by Eq. (6-78):

$$M_x = 0.159nM_0 = 0.159 \times 37 \times 20 = 118 \text{ ft-kip}$$

A welded box section will be used. Stresses due to the ring axial force and bending about the vertical (y) axis will be relatively small for a box section. Allow about 4 ksi, which gives an allowable bending stress for the x axis of 24 - 4 = 20 ksi. Therefore,

$$S_x = \frac{118 \times 12}{20} = 70.8 \text{ in.}^3$$

Try the cross section shown in Fig. 7-20 for which

$$I_x = 2 \times 0.375 \times \frac{14^3}{12} + 2 \times 8 \times 0.500 \left(\frac{12.5}{2}\right)^2 = 484 \text{ in.}^4$$

$$S_x = \frac{484}{7} = 69.1 \text{ in.}^3$$

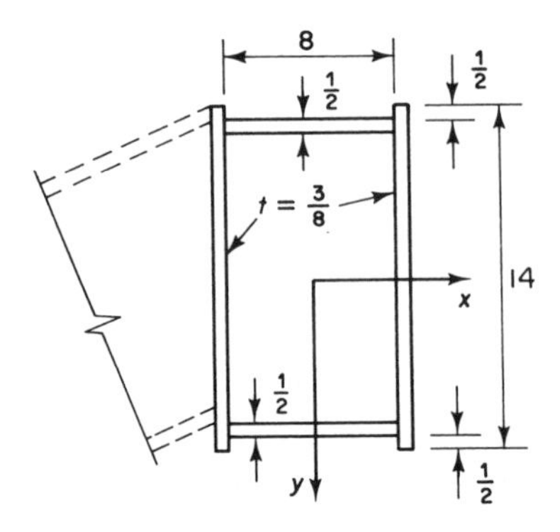

Figure 7-20 Compression ring for roof of Example 7-5.

$$I_y = 2 \times 0.500 \times \frac{8^3}{12} + 2 \times 14 \times 0.375 \left(\frac{8.375}{2}\right)^2 = 227 \text{ in.}^4$$

$$S_y = \frac{227}{4.375} = 51.9 \text{ in.}^3$$

$$A = 2(0.5 \times 8 + 0.375 \times 14) = 18.5 \text{ in.}^2$$

The required rafter section modulus for the estimated value of M_0 is

$$S_x = \frac{37 \times 12}{24} = 18.5 \text{ in.}^3$$

This makes no allowance for the axial-force effect. Since S_x is only 21.3 in.^{3} for a W12 X 19, try a W12 X 22 for which $S_x = 25.4$ in.^{3} and $I_x = 156$ in.^{4} Then from Eq. (7-36b),

$$k = 0.159 \times 20 \times \frac{156}{484} \times \frac{5}{40/\cos 24.2^\circ} = 0.117$$

The moment M_0 for the load $q_1 = 256$ and $q_2 = 51$ plf is given by Eq. (7-36a):

$$M_0 = \frac{40^2(7 \times 256 + 8 \times 51)}{120(1 + 0.351)} = 21{,}710 \text{ ft-lb}$$

For the load $q_3 = 116$ and $q_4 = 26$ plf, Eq. (7-37) gives

$$M_0 = \frac{17.5^2}{12(1 + 0.351)} (516 - 1.5 \times 374 \times 0.4375 + 0.3 \times 490 \times 0.4375^2)$$

$$= 5640 \text{ ft-lb}$$

For the load $Q = 1908$ lb, from Eq. (7-38),

$$M_0 = \frac{1908 \times 22.5 \times 17.5 \times 62.5}{2 \times 40^2(1 + 0.351)} = 10{,}850 \text{ ft-lb}$$

$$M_0 = 21{,}710 + 5640 + 10{,}850 = 38{,}200 \text{ ft-lb}$$

$$M+ = 55{,}580 - \frac{22.5}{40} \times 38{,}200 = 34{,}090 \text{ ft-lb}$$

Check the rafter at the point of maximum positive moment (AISC Formula 1.6–1a).

$$V_1 = 9454 \text{ lb} \qquad \text{(from Example 7–4)}$$

$$H = 10{,}080 \text{ lb} \qquad \text{(from Example 7–4)}$$

$$+\frac{M_0}{h} = 10{,}080 + \frac{38{,}200}{18} = 12{,}200 \text{ lb}$$

$$P = V_1 \sin 24.2^\circ + H \cos 24.2^\circ = 15{,}000 \text{ lb}$$

For a W12 X 22, $A = 6.48$ in.^{2}, $S_x = 25.4$ in.^{3}, and $r_x = 4.91$ in.

$$f_a = \frac{15.0}{6.48} = 2.31 \text{ ksi} \qquad f_b = \frac{34.1 \times 12}{25.4} = 16.1 \text{ ksi}$$

The effective-length coefficient K may be found from an alignment chart.[13] The rafter/ring stiffness ratio is 0.117, which is the value of k computed above to determine M_0. Assume the rafter-bin wall stiffness ratio to be 10, which corresponds to a small amount of rotational restraint. The alignment chart gives $K = 0.73$ for these two values. Therefore,

$$\frac{KL}{r_x} = \frac{0.73 \times 526}{4.91} = 78 \qquad F_a = 15.6 \text{ ksi} \qquad F_E' = 24.5 \text{ ksi} \qquad C_m = 1$$

Web slenderness of the W12 X 22 must be checked by AISC Specification 1.5.1.4.1 to determine if the section qualifies as compact:

$$\frac{f_a}{F_y} = \frac{2.31}{36} = 0.064 \qquad \frac{d}{t} = \frac{640}{\sqrt{36}}(1 - 3.74 \times 0.064) = 81$$

Since $d/t = 47$, the section is compact, and

$$\frac{f_a}{F_a} + \frac{f_b}{F_b}\frac{1}{1 - f_a/F_E'} = \frac{2.31}{15.6} + \frac{16.1}{24}\frac{1}{1 - 2.31/24.5} = 0.89 < 1$$

Check the rafter at the compression-ring end (AISC Formula 1.6-1b):

$$P = H \cos 24.2^\circ = 12{,}200 \cos 24.2^\circ = 11{,}130 \text{ lb} \qquad M_0 = 38{,}200 \text{ ft-lb}$$

$$f_a = \frac{11.13}{6.48} = 1.72 \text{ ksi} \qquad f_b = \frac{38.2 \times 12}{25.4} = 18.0 \text{ ksi}$$

$$\frac{f_a}{0.6F_y} + \frac{f_b}{F_b} = \frac{1.72}{22} + \frac{18.0}{24} = 0.84 < 1$$

The next lighter section (W12 X 19) is inadequate.

The compression flange in the region of negative moment should be braced at intervals not exceeding $180r_y = 180 \times 0.847/12 = 12.7$ ft. However, the point of inflection is only 7.8 ft from the compression ring, so a single ring of bottom-flange bracing will suffice unless additional bracing is needed for erection.

Deflection

At the header-to-rafter connection ($x_1 = 22.5$ ft) the deflection due to M_0, from Eq. (7-39), is

$$\delta = \frac{38{,}200 \times 22.5 \times 12^3}{6 \times 30 \times 10^6 \times 156 \times 40}(40^2 - 22.5^2) = 1.45 \text{ in.}$$

From Example 7-4 the simple-beam deflection at $x_1 = 22.5$ ft for the W16 X 26 ($I_x = 301$ in.^{4}) is 1.68 in. Therefore, the net deflection for the W12 X 22 ($I_x = 156$ in.^{4}) is

$$\delta = 1.68 \times \frac{301}{156} - 1.45 = 1.79 \text{ in.}$$

$$\frac{\delta}{L} = \frac{1.79}{526} = \frac{1}{293} < \frac{1}{220} \qquad \text{OK}$$

It should be noted that the rule $L/d = 28$ which limits the deflection-span ratio to 1/220 for a fully loaded simply supported beam with uniform load is too conservative for equivalent deflection control of a beam with fixed or partially fixed loads.

Compression ring

Ring formulas are given in Sec. 6-19. The moment at each main rafter connection is $M_0 = 38{,}200$ ft-lb, and Eq. (6-78) gives

$$M_x = 0.159nM_0 = 0.159 \times 20 \times 38.2 = 121 \text{ ft-kip}$$

With $Q = H = -12.2$ kips, Eqs. (6-71) give

$$M_y = \frac{QR}{2}\left(\cot\frac{180^\circ}{n} - \frac{n}{\pi}\right) = \frac{-12.2 \times 5}{2}\left(\cot\frac{180^\circ}{20} - \frac{20}{\pi}\right) = +1.60 \text{ ft-kip}$$

$$P = \frac{Q}{2}\cot\frac{180^\circ}{n} = \frac{-12.2}{2}\cot\frac{180^\circ}{20} = -38.5 \text{ kips}$$

$$f = \frac{P}{A} + \frac{M_x}{S_x} + \frac{M_y}{S_y} = -\frac{38.5}{18.5} - \frac{121 \times 12}{69.1} - \frac{1.60 \times 12}{51.9} = -23.5 \text{ ksi}$$

The shearing stress due to torsion of a box section is given by $s_s = M/2At$, where A = the area enclosed by centerlines of sides and t = the thickness at any point. The torsional moment M_t is $M_0/2 = 38{,}200/2 = 19{,}100$ ft-lb. With $A = 8.375 \times 12.675 = 105.7$ in.^{2},

$$f_v = \frac{19{,}100 \times 12}{2 \times 105.7 \times 0.375} = 2890 \text{ psi}$$

Tension ring

Design of the tension ring follows the procedure in Example 7-4.

The moment-resistant rafter-to-ring connection reduces the weight of the main rafters 3510 lb, but the weight of the compression ring is increased 1325 lb. The weight of the roof-supporting structure (rafters, subrafters, headers, and compression ring) is reduced from 30,120 to 27,930 lb (7%). Therefore, in this case the saving in cost of material is not enough to offset the extra cost of fabricating the welded box and the extra field cost of the moment-resistant connections.

7-9. Supported Dome Roofs

Rafters for supported dome roofs can be designed as curved members simply supported at the bin wall and the center ring or, if the compression ring and the rafter connections to it are designed to establish continuity, as a system of intersecting two-hinged arches.

Simply supported rafters. If the center ring is supported on beams or trusses, as in Figs. 7-2a and b, a simply supported curved rafter tends to act as an inclined arch. However, such roofs are so flat that even a slight separation of the ends of the rafter will greatly reduce the arch action. It is suggested in Ref. 13 that the normal analysis for the arch thrust, including the effect of rib shortening, is adequate only if the ratio of arch rise to span equals or exceeds 0.07. For lesser values a more difficult analysis which accounts for deflection due to rib shortening is required. Therefore, the authors suggest that curved subrafters, and curved main rafters supported on center rings that are supported in turn on beams or trusses, be designed as beams if the rise-span ratio is less than 0.07. For these cases Eqs. (7-17)-(7-20) for straight rafters can be used.

If the center ring is not supported on beams or trusses, the reactions V_1 and H for loads distributed as shown in Fig. 7-21 are given by

$$V_1 = \frac{a}{2}(q_1 + q_2) \tag{7-40}$$

$$H = \frac{a^2}{6h_1}(q_1 + 2q_2) \tag{7-41}$$

$$V_1 = \frac{c}{2}(q_3 + q_4) \tag{7-42}$$

$$H = \frac{c}{2h_1}\left[a(q_3 + q_4) - \frac{c}{3}(2q_3 + q_4)\right] \tag{7-43}$$

Design of a dome roof with simply supported rafters is illustrated in Example 7-6.

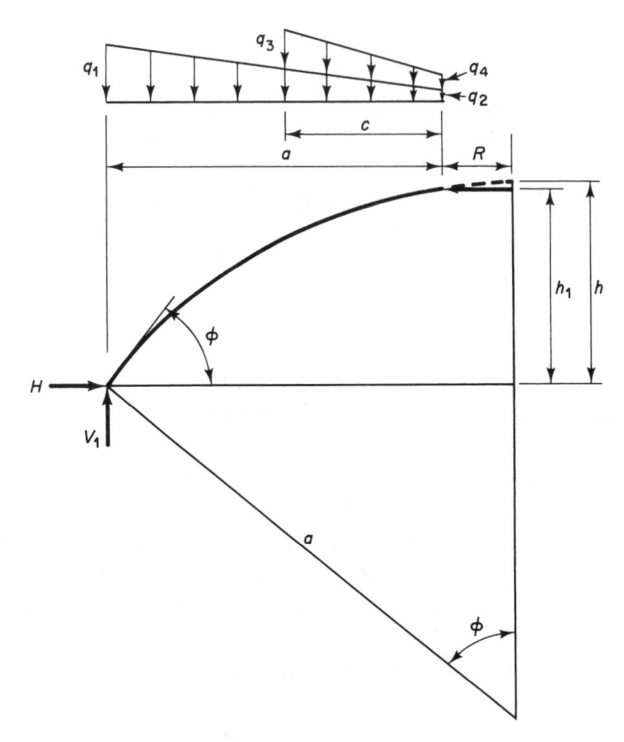

Figure 7-21

Two-hinged arch rafters. The following formulas give the horizontal reaction H for a two-hinged arch consisting of two rafters with moment-resistant connections to the compression ring. The formulas are derived for symmetrical loading, that is, for equal loads on each of the two rafters comprising the arch.

Concentrated load Q at any point (Fig. 7-22a):

$$\Delta \frac{H}{Q} = r^2 - x_Q^2 - 2(a-h)(\phi r - \alpha x_Q - y_Q) \tag{7-44}$$

Uniformly distributed load q_u (Fig. 7-22b):

$$\Delta \frac{H}{q_u r} = \phi\left(\frac{a^2}{2} - r^2\right) \frac{a-h}{r} - \frac{(a-h)^2}{2} + \frac{2r^2}{3} \tag{7-45}$$

Triangularly distributed load q_1 (Fig. 7-22c):

$$\Delta \frac{H}{q_1 r} = \frac{r^2}{4} - \frac{1}{9}(a-h)\left(3\phi r + a - h - \frac{2a^2 h}{r^2}\right) \tag{7-46}$$

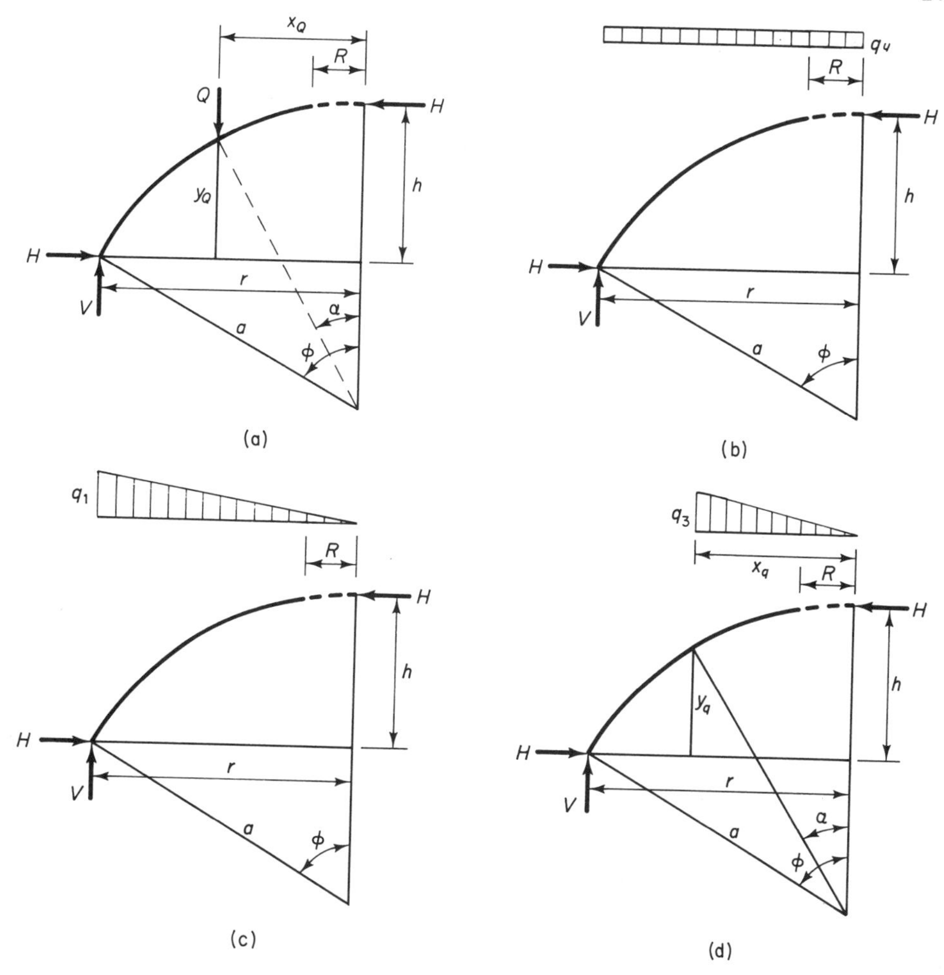

Figure 7-22

Triangularly distributed load q_3 (Fig. 7-22d):

$$\Delta \frac{H}{q_3 x_q} = \frac{r^2}{2} - \frac{x_q^2}{4} - \frac{1}{9}(a-h)\left[9\phi r - 6\alpha x_q + a - h - 8y_q - 2\frac{a^2}{x_q^2}(h - y_q)\right] \tag{7-47}$$

where $\Delta = \phi a^2 - 3r(a-h) + 2\phi(a-h)^2$

The compression ring should be proportioned so that

$$I_o \geqslant \frac{n}{2\pi} I_r \tag{7-48}$$

where I_r = x-axis moment of inertia of rafter
I_0 = moment of inertia about axis in plane of ring
n = number of rafters

The axial-force effect (rib shortening) is neglected in these equations; except for very flat arches (a rise-span ratio of less than about 1/10) the error is negligible.

The stress due to the axial force P and the bending moment M should be checked by Eq. (6-23) using for L the length of the chord connecting the rafter support and the point of inflection in the rafter.

According to DIN 4119, the load to be carried by the rafters can be determined by subtracting the local-buckling pressure p_b of the shell, divided by a factor of safety, from the sum of the live load and weight of the roof plate except that it must not be taken less than half this sum. The local-buckling pressure is given by the following:
For $K \leqslant 1$,

$$p_b = 5.4 \times 10^8 \left(\frac{t}{a}\right)^2 \left(K + \frac{1}{K}\right) \leqslant 21.5 \times 10^8 \left(\frac{t}{a}\right)^2 \text{ psf} \qquad (7\text{-}49)$$

and for $K \geqslant 1$,

$$p_b = 10.8 \times 10^8 \left(\frac{t}{a}\right)^2 \text{ psf} \qquad (7\text{-}50)$$

where $K = l_b^2/58.4at$
t = thickness of shell
a = radius of dome
l_b = diameter of buckle

The three dimensions a, t, and l_b in these formulas must be taken in the same units. The diameter of the buckle is assumed to be the diameter of the circle inscribed in the largest roof-plate panel, taking into account any circumferential members on which the plate is supported.*

Buckling. Lateral-torsional buckling considerations are the same as for cone roofs (Sec. 7-8).

Vertical buckling of a symmetrically loaded two-hinged arch takes the form shown in Fig. 7-23a, which would be restrained by the roof shell. If the individual rafters are simply supported, the buckling mode shown in Fig. 7-23b, which is analogous to the symmetrical buckling mode of a three-hinged arch, must be considered. It involves a downward movement of the compression ring and outward displacements of a portion of each rafter. This mode is also restrained by the roof shell. Finally, there are two buckling modes of a single rafter considered as an inclined two-hinged arch. The first (Fig. 7-23c) is restrained by the roof shell. The second (Fig. 7-23d) is called snap-through buckling. According to a study reported in Ref. 13, snap-

*See footnote on p. 212.

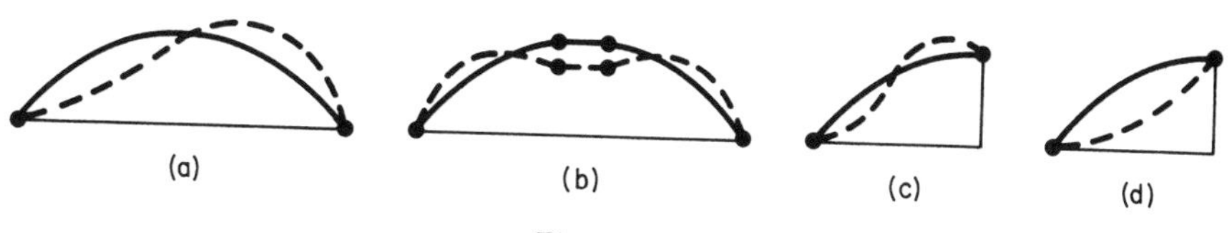

Figure 7-23

through buckling does not occur in arches which satisfy the formula

$$\frac{h}{r_x} \geqslant 4.69 \tag{7-51}$$

where h is the rise of the arch and r_x the cross-sectional radius of gyration for the axis normal to the buckling plane.

Formulas for deflection are too complex for practical purposes. Instead, deflection can be calculated by the formula

$$\delta = \int \frac{Mm\,ds}{EI} \tag{7-52}$$

where M = moment at any point due to applied load
m = moment at any point due to unit load at point where δ is wanted and acting in direction of δ
ds = arc length of element

One procedure for evaluating the integral in Eq. (7-52) is to divide the rafter into a number of elements, compute M and m at the center of each element, and compute $\delta = \Sigma\, Mm\, \Delta s/EI$.

7-10. Unsymmetrically Loaded Dome Roofs

Roofs can be loaded unsymmetrically by snow on part of the roof. DIN 4119 requires that a roof be designed for (1) a full snow load on the projected area and (2) a snow load of about 8 psf on half of the projected area. According to an analysis reported in Ref. 14, the one-sided-load requirement will be satisfied if the rafter moment of inertia satisfies the equation

$$I_x \geqslant \beta \frac{\eta P}{29.4} \left(\frac{r}{100}\right)^2 \tag{7-53}$$

where $\beta = 0.83$ for roofs with attached plates and 1 for unattached plates, η = factor of safety = 1.3, P = largest axial force in rafter (kips), and I_x and r are in inch units.

Example 7-6

Design a supported dome roof for the 90-ft bin of Example 7-4 using the same rafter arrangement (Fig. 7–18a), a radius of curvature of 90 ft, and a shell thickness of $\frac{3}{16}$ in. The rafter-to-ring connections are not to be designed for continuity. Therefore, the reduced load of DIN 4119 (Sec. 7–9) is not applicable, and the design load is 32.7 psf.

Subrafter

The rise of the 22.5-ft subrafter is $90(1 - \cos 7.18°) = 0.71$ ft and the rise-span ratio is $0.71/22.5 = 0.03$. Therefore, arch action is negligible, and the subrafter is designed as a straight member and will be the same as the subrafter of Example 7–4 (W8 X 10).

Header

The header is the same as in Example 7–4: C4 X 5.4.

Main rafter

The main rafter and its loading are shown in Fig. 7–24. The rafter weight is assumed 15 lb/ft. The subrafter reaction on the header is 1908 lb (Example 7–4). Adding the weight of the header (7 X 5.4 ≈ 40 lb) gives 1950 lb. The load Q_r from the ring is given by

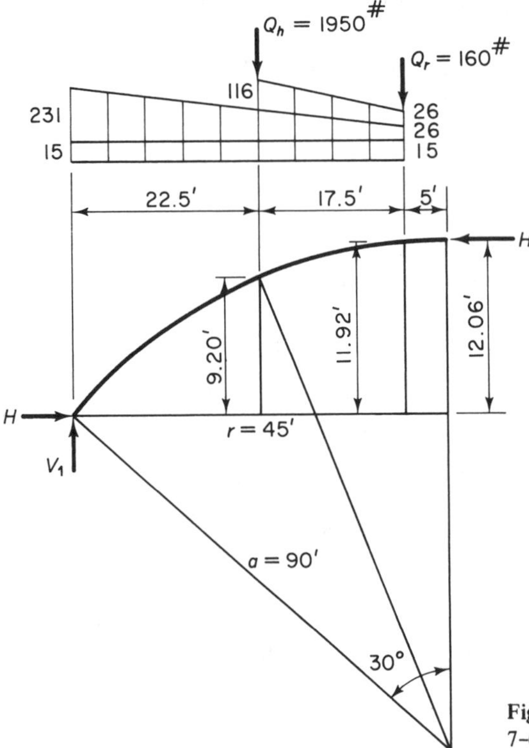

Figure 7–24. Rafter loads for Example 7–6.

$$Q_r = \frac{32.7 \times \pi \times 10^2}{4 \times 20} + \frac{20 \times \pi \times 10}{20} = 160 \text{ lb}$$

where in the second term the weight of the ring is assumed to be 20 lb/ft.

Using Eqs. (7–40)–(7–43) and Q_h and Q_r from Fig. 7–24, we get

$$V_1 = \frac{40}{2}(246 + 41) + \frac{17.5}{2}(116 + 26) + 1950 + 160 = 9092 \text{ lb}$$

$$11.92H = \frac{40^2}{6}(246 + 82) + \frac{17.5}{2}\left[40(116 + 26) - \frac{17.5}{3}(232 + 26)\right]$$

$$+ 1950 \times 22.5 + 160 \times 40$$

$$H = 14{,}620 \text{ lb}$$

The moment at the location of the header load is

$$M = 14{,}620(11.92 - 9.20) - 160 \times 17.5 - 67 \times \frac{17.5}{2} \times \frac{2 \times 17.5}{3}$$

$$- 247 \times \frac{17.5}{2} \times \frac{17.5}{3} = 17{,}519 \text{ ft-lb}$$

$$S_x = \frac{17.5 \times 12}{24} = 8.75 \text{ in.}^3$$

Use Eq. (6–23) to check the combined stress at $x_0 = 22.5$ ft. A W10 X 12, for which $S_x = 10.9$ in.^{3}, turns out to be inadequate. Try a W12 X 14, for which $S_x = 14.9$ in.^{3}, $A = 4.16$ in.^{2}, and $r_x = 4.62$ in.

$$P = \left[9100 - \frac{1}{2} \times 22.5(246 + 131)\right] \sin 14.5^\circ$$

$$+ H \cos 14.5^\circ = 15{,}370 \text{ lb}$$

$$L = \sqrt{40^2 + 11.92^2} = 41.7 \text{ ft}$$

$$\frac{L}{r_x} = \frac{41.7 \times 12}{4.62} = 108 \qquad F'_E = 12.80 \text{ ksi}$$

$$f_a = \frac{15.37}{4.16} = 3.69 \text{ ksi} \qquad f_b = \frac{17.52 \times 12}{14.9} = 14.1 \text{ ksi}$$

$$\frac{f_a}{0.6F_y} + \frac{f_b}{F_b}\frac{1}{1 - f_a/F'_E} = \frac{3.69}{22} + \frac{14.1}{24}\frac{1}{1 - 3.69/12.80} = 0.99$$

The W12 X 14 was assumed to be compact. The flange qualifies, but the web must be checked against AISC 1.5.1.4.1:

$$\frac{d}{t} = \frac{640}{\sqrt{F_y}}\left(1 - 3.74\frac{f_a}{F_y}\right) \qquad \text{if } \frac{f_a}{F_y} \leqslant 0.16$$

$$= \frac{640}{\sqrt{36}}\left(1 - 3.74 \times \frac{3.69}{36}\right) = 66$$

Since $d/t = 60$, the W12 X 14 is compact.

For snow load on one-half of the roof, Eq. (7-57) gives

$$I_x \geqslant \frac{1 \times 1.3 \times 15.77}{29.4}\left(\frac{45 \times 12}{100}\right)^2 \geqslant 19.8 \text{ in.}^4$$

$I_x = 88.6$ in.^{4} for the W12 X 14.

Since there are no negative-moment regions, lateral-torsional buckling need not be considered.

Deflection

To evaluate Eq. (7-56), the rib is divided into four elements 5.69 ft long between the support and header connection and three elements 6.41 ft long between the header and the ring. Values of m are computed for a vertical unit load at the header connection to find the deflection at that point. Computations are given in Table 7-2. The deflection is

$$\delta = \frac{1{,}095{,}000 \times 12^3}{30 \times 10^6 \times 88.6} = 0.71 \text{ in.}$$

Compression ring

The design of the compression ring follows the procedure used in Example 7.4.

Tension ring

The design of the tension ring follows the procedure used in Example 7-4.

TABLE 7-2. DEFLECTION OF RAFTER OF EXAMPLE 7-6

Point	x	y	M	m	ds	$Mm\,ds$
1	2.95	11.71	2,321	0.396	6.41	5,890
2	8.82	10.99	8,392	1.756	6.41	94,460
3	14.63	9.89	14,782	3.833	6.41	363,190
4	20.44	8.38	14,543	4.600	5.69	380,650
5	26.22	6.47	9,577	3.727	5.69	203,100
6	31.86	4.16	5,441	1.566	5.69	48,480
7	37.33	1.48	1,758	-0.126	5.69	-1,260
						1,094,510

Example 7-7

Design the rafters and compression ring for the supported dome roof of Example 7-6 using the DIN 4119 provisions.

Figure 7-25 shows the diameter of the local buckle. Then, from Eq. (7-50),

$$K = \frac{6.52^2}{58.4 \times 90 \times 0.1875/12} = 0.518 < 1$$

From Eq. (7-49),

$$p_b = 5.4 \times 10^8 \left(\frac{0.1875/12}{90}\right)^2 \left(0.518 + \frac{1}{0.518}\right) = 39.8 \text{ psf}$$

Using a factor of safety of 2, we obtain

$$p_b = \frac{39.8}{2} = 19.9 \text{ psf}$$

$$LL + DL - 19.9 = 32.7 - 19.9 = 12.8 \text{ psf}$$

$$0.5(LL + DL) = 0.5 \times 32.7 = 16.4 \text{ psf}$$

Design for 16.4 psf.

Subrafter

The rise of the 22.5-ft subrafter is $90(1 - \cos 7.18^\circ) = 0.71$ ft, and the rise-span ratio is $0.71/22.5 = 0.03$, so that arch action is negligible, and from Fig. 7-12,

$$\text{horizontal span } a = 45 - 22.5 \cos 9^\circ = 22.8 \text{ ft}$$

$$q_1 = 16.4 \times \frac{90\pi}{40} = 116 + 10 \text{ for rafter} = 126 \text{ lb/ft}$$

$$q_2 = 16.4 \times \frac{44.4\pi}{40} = 57 + 10 \text{ for rafter} = 67 \text{ lb/ft}$$

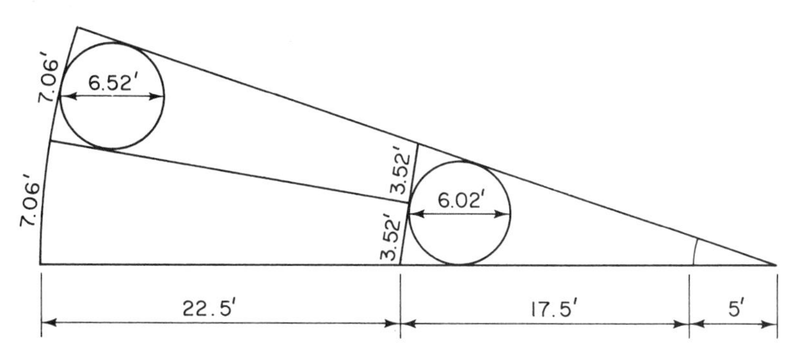

Figure 7-25

From Eq. (7–19),

$$\frac{x_1}{a} = \frac{126 - \sqrt{(126^2 + 126 \times 67 + 67^2)/3}}{126 - 67} = 0.475$$

From Eq. (7–18),

$$M = \frac{22.8^2}{6} \times 0.475(252 + 67 - 378 \times 0.475 + 59 \times 0.475^2)$$

$$= 6287 \text{ ft-lb}$$

$$S_x = \frac{6.29 \times 12}{24} = 3.15 \text{ in.}^3$$

The lightest W section with $S_x \geqslant 3.15$ in.^{3} is the W6 X 9 for which $S_x = 5.56$. However, a 6-in. depth is less than the minimum ($22.8 \times 12/28 = 9.8$ in.) for deflection control of a fully stressed beam, so that the allowable stress must be reduced to $6 \times 24/9.8 = 14.7$ ksi, and the section modulus required is $6.29 \times 12/14.7 = 5.13 < 5.56$ in.^{3}

Header

Span = $2 \times 22.5 \sin 9° = 6.95$ ft. From Eq. (7–17b),

$$\text{subrafter } V_2 = \frac{22.8}{6}(126 + 2 \times 67) = 988$$

$$\text{weight of header} = \underline{\ \ 40\ \ }$$

$$1028 \text{ lb}$$

$$M = 1028 \times \frac{6.95}{4} = 1786 \text{ ft-lb} \qquad S_x = 1.8 \times \frac{12}{22} = 0.98 \text{ in.}^3$$

Use C4 X 5.4, $S_x = 1.93$ in.^{3}, and $L/d = 6.95 \times 12/4 = 21$.

Main rafter

The main rafter and its loading are shown in Fig. 7–26. The rafter weight is assumed 10 lb/ft. The header reaction is $Q_h = 1028$ lb. Equation (7–45) is based on uniform load from the roof rim to the vertex, so the 50-lb weight of the nonexistent 5 ft of rafter within the compression ring provides for a compression-ring weight of $50/1.56 = 32$ lb/ft. Assuming this to be sufficient, no concentrated load need be assumed at the rafter-ring intersection. With these loads and the triangularly-distributed loads q_1 and q_3, Eqs. (7–44)–(7–47) give the following:

$$\Delta = 80.67$$

$$H = 2118 \text{ lb for } Q = 1028 \text{ lb}$$

$$H = 831 \text{ lb for } q_u = 10 \text{ lb/ft}$$

$$H = 3525 \text{ lb for } q_1 = 116 \text{ lb/ft}$$

$$H = 1606 \text{ lb for } q_3 = 58 \text{ lb/ft}$$

The total reaction is $H = 8080$ lb. The corresponding vertical reaction is $V_1 = 4740$ lb. The resulting bending moments are

$$M = +5329 \text{ ft-lb at the header}$$

$$M = -4587 \text{ ft-lb at the connection to the ring}$$

The point of inflection is found to be at $x = 13.9$ ft, or 31.1 feet from the bin wall. The corresponding value of y is 11.0 ft. The resulting chord length L is 33.0 ft at an angle of 19° with the horizontal. The resultant force P at the header parallel to the chord is

$$P = \left(4740 - \frac{126 + 68}{2} \times 22.5\right) \sin 19^\circ + 8080 \cos 19^\circ = 8472 \text{ lb}$$

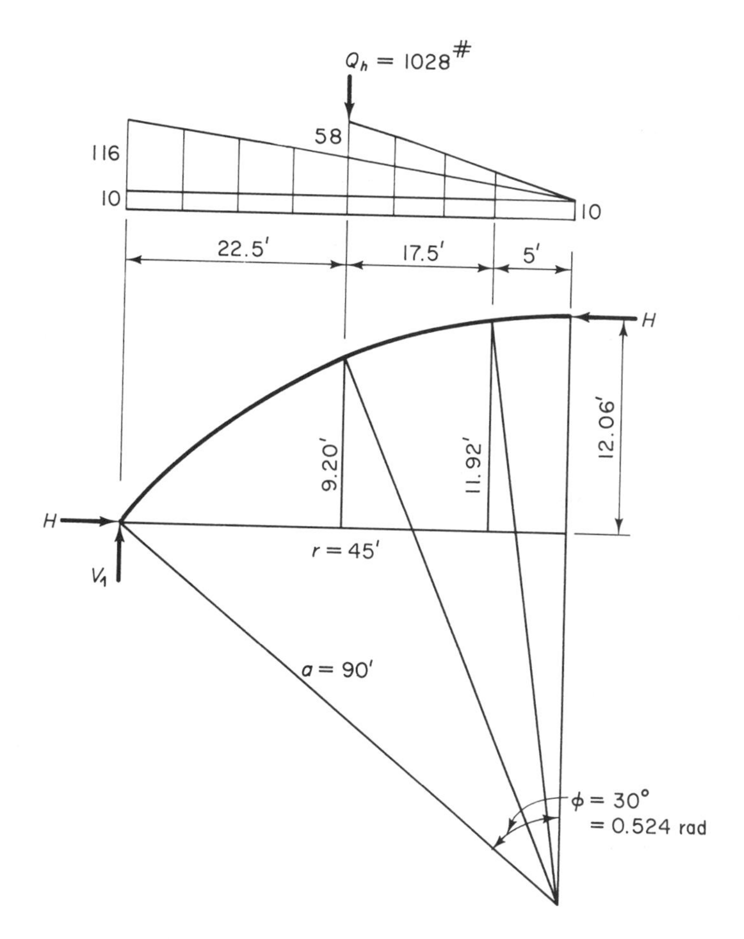

Figure 7-26 Rafter loads for Example 7-7.

For a W8 × 10, $A = 2.96$ in.^{2}, $S_x = 7.81$ in.^{3} and $r_x = 3.22$ in. Then

$$\frac{L}{r_x} = \frac{33.0 \times 12}{3.22} = 123 \qquad F'_E = 9.87 \text{ ksi}$$

$$f_a = \frac{8.47}{2.96} = 2.86 \text{ ksi} \qquad f_b = \frac{5.329 \times 12}{7.81} = 8.19 \text{ ksi}$$

Then from Eq. (6–23)

$$\frac{2.86}{22} + \frac{8.19}{24} \times \frac{1}{1 - 2.86/9.87} = 0.61$$

There is considerable excess capacity, but the next lighter section, W6 × 9, is inadequate. The W8 × 10 could be used to better advantage by decreasing the number of rafters.

The compression flange should be braced laterally in the negative-moment region at intervals not exceeding $180r_y = 180 \times 0.841/12 = 12.6$ ft. Since the point of inflection is at $x = 13.9$ ft, or 8.9 ft from the compression ring, a single ring of bottom-flange bracing will suffice unless additional bracing is needed for erection or other purposes.

Compression ring

In general, open sections such as channels are not suitable as compression rings with moment-resistant connections (Sec. 6–20). Design procedure for a box section is given in Example 7-5. Design requirements for a ring consisting of two wide, parallel ring-plates welded to a channel of the same depth as the rafter are given in Refs. 6 and 11. The required x-axis moment of inertia of the ring is given by Eq. (7–48) as

$$I_0 = \frac{20}{2\pi} \times 30.8 = 98 \text{ in.}^4$$

Tension ring

The design of the tension ring follows the procedure in Example 7–4.

Deflection

The deflection can be computed as in Example 7–6.

7-11. Bracing of Supported Roofs

In general, circumferential bracing of supported roofs is not necessary except as it may be needed for lateral support in the negative-moment regions of rafters with moment-resistant connections to the compression ring or for support during erection. Figure 7–27 shows bracing systems consisting of circumferential ties connected to two adjacent rafters which are trussed together. These ties can be small single-angle members.

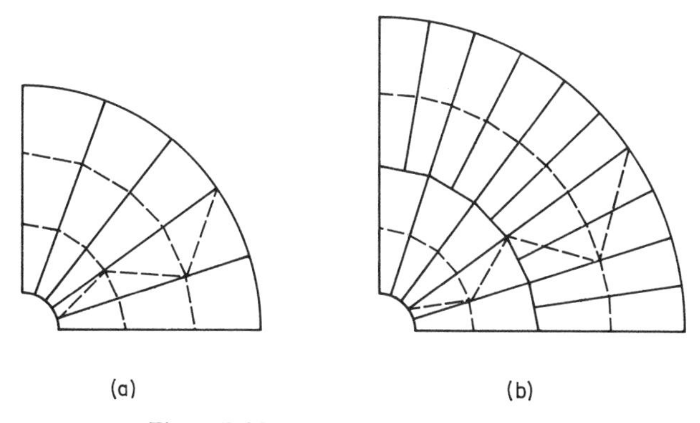

Figure 7-27 Bracing of roof members.

REFERENCES

1. *API Standard 620,* Recommended Rules for Design and Construction of Large, Welded, Low-Pressure Storage Tanks, 5th ed., American Petroleum Institute, Washington, D.C. 1973.
2. *API Standard 650,* Welded Steel Tanks for Oil Storage, 5th ed., American Petroleum Institute, Washington, D.C., 1973.
3. *AWWA Standard D-100,* Steel Tanks, Standpipes, Reservoirs and Elevated Tanks for Water Storage, American Water Works Association, Denver, 1973.
4. S. P. Timoshenko and J. M. Gere, *Theory of Elastic Stability,* 2nd ed., McGraw-Hill, New York, 1961.
5. K. Klöppel and E. Roos, Buckling of Thin-walled Stiffened and Unstiffened Spherical Shells Under Full and One-sided Loading, *Der Stahlbau,* March 1956.
6. *DIN 4119 Deutsche Normen,* Aboveground Cylindrical Steel Tank Construction, Beuth Verlag GmbH, Berlin 30 and Cologne 1, Oct. 1961 (revised June 1979, Feb. 1980).
7. K. Klöppel and O. Jungbluth, Buckling Problems of Thin-walled Spherical Shells, *Der Stahibau,* May 1953.
8. *ASME Boiler and Pressure Vessel Code,* Sect. VIII Rules for Construction of Pressure Vessels, American Society of Mechanical Engineers, New York, 1977.
9. P. Seide, V. I. Weingarten, and E. J. Morgan, *Final Report on the Development of Design Criteria for Elastic Stability of Thin Shell Structures,* Space Technology Laboratories, Inc., Los Angeles, 1960.
10. V. I. Weingarten, E. J. Morgan, and P. Seide, Elastic Stability of Thin-Walled Shells Under Internal Pressure and Axial Compression, *J. Am. Inst. Aero. Astro.,* June 1965.
11. K. Herber, Design of Ribbed Domes for Tank Roofs, *Der Stahlbau,* Sept. 1956.

12. M. Fischer, The Stability of I-Beams Loaded on the Top Flange, *Der Stahlbau,* May 1973.

13. B. J. Johnston (ed.), *Guide to Stability Design Criteria for Metal Structures,* 3rd ed., Wiley, New York, 1976.

14. K. Herber, Bemessung von Tankdächern mit Rippenrostgespärren, *Der Stahlbau,* Sept. 1958.

8

STRUCTURAL DESIGN OF BINS

8-1. Theory of Shells

The membrane theory of shells is based on the assumption that there are no bending moments or transverse shears on a shell element. This leaves only in-plane forces (Fig. 8-1). They are usually called *stress resultants,* which are forces per unit length. Since there are only three stress resultants on an element and three equations of equilibrium for the element, the stress distribution throughout the shell can be determined by consideration of equilibrium.

Membrane theory is a good basis for the design of thin shells except at the boundaries, such as supports, transitions from one shell form to another (as of a cylinder with a conical hopper), etc., where bending stresses and transverse shears will develop in addition to the membrane stress resultants (Fig. 8-2). In the types of shells used for bins and hoppers, such bending is usually localized and is sometimes called an *edge effect.* The bending theory of shells can be used to evaluate these effects. Bending theory is relatively simple for a circular cylindrical shell loaded symmetrically with respect to its axis. Formulas for the hoop-stress resultant, moment, shear, displacement, and slope at any distance from the end are given in Refs. 1 and 2.

8-2. Stresses in Circular Cylinder Wall

An element of a cylindrical bin of radius r acted upon by a normal pressure p_h is shown in Fig. 8-3. The hoop-tension stress resultant N_h is given by

$$N_h = p_h r \tag{8-1}$$

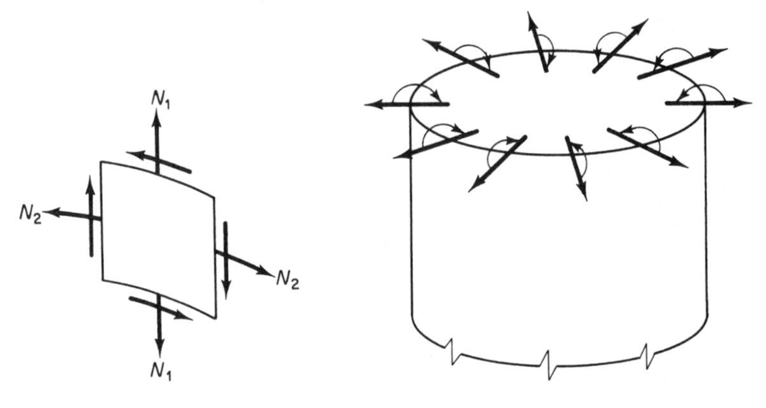

Figure 8-1

Figure 8-2

The vertical stress resultant is determined by the load from the roof and by friction of the contained material on the wall; the latter component is given by (Sec. 5-10)

$$N_y = R(\gamma y - p_v) \tag{8-2}$$

where R = hydraulic radius of cylinder
γ = bulk density of material
y = depth of material above section
p_v = vertical pressure on section

Allowable hoop tension. Hoop tension rarely determines the thickness of the cylindrical shell of a bin, but it may determine the wall thickness of the hopper. The API 650[3] allowable shell tension is 21 ksi, which must be multiplied by the joint-efficiency factor 0.85 of a full-penetration welded butt joint. The steels ordinarily used under this standard are A285 Grade C (yield point 30 ksi), A283 Grades C and D (yield points 30 and 33 ksi, respectively), and A36 (yield point 36 ksi). In terms of these steels the allowable tension, before reduction for joint efficiency, ranges from 58 to 70% of the

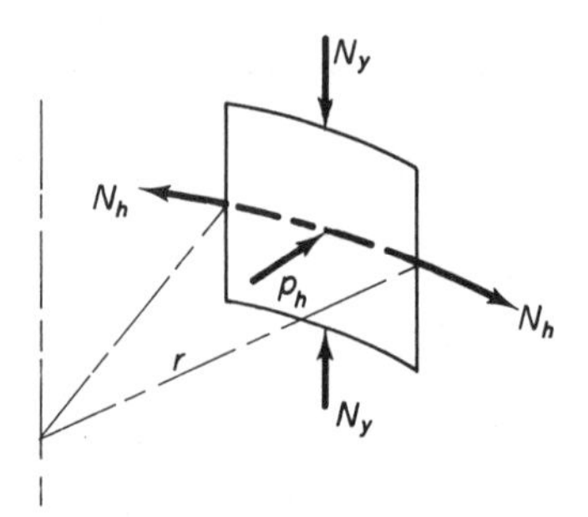

Figure 8-3

yield stress. Other steels permitted by this standard, where severe service conditions justify the use of a premium material, have yield points as high as 60 ksi with no increase in allowable tension unless a more thorough weld-inspection procedure than is required for the 21-ksi allowable is followed. A program of radiographic inspection of complete-penetration joints, particularly the vertical joints, is required for the 21-ksi allowable.

Shell tension permitted by DIN 4119[4] is $\frac{2}{3}F_y$ with a joint-efficiency factor of 0.8. If the vertical welds are fully and the circumferential welds partially (10%) examined by a nondestructive method, the joint-efficiency factor is 0.9. Under this standard the allowable tensions at 0.8 joint efficiency are 16.0, 17.6, and 19.2 ksi for F_y = 30, 33, and 36 ksi, respectively, compared with the API 17.9 ksi at 0.85 joint efficiency. Unless the bin is to be designed to increase the fabricator's options in the steel to be used, the DIN recognition of differing yield stresses is more logical.

Allowable vertical compression. Buckling of axially compressed cylindrical shells and several formulas which have been developed to predict buckling strength are discussed in Sec. 6–13. However, hoop tension reduces the imperfection sensitivity of buckling under axial compression, which increases the buckling strength. Buckling under axial compression and internal pressure is discussed in Sec. 6–16. Compressed air was used to produce the internal pressure in the tests on which Eq. (6–49) of that section is based. Similar results were obtained with hydrostatic pressure and in tests on model silos filled with various granular materials (sand, wheat, etc.).[5] Results of the latter are shown in Fig. 8–4. The tests indicate that granular fill increases the failure load even more than hydrostatic pressure. However, there were significant differences in the wall stresses developed by the different types of contained material. This affects the buckling strength, so that a larger scatter of test results with granular fill could be expected compared with the hydrostatically or air-pressurized cylinders, which was indeed the case.

Since there is no standard specification for the design of bins for granular materials, no one formula has been adopted for general use. The Boardman formula, Eq. (6–42), which, with a factor of safety of 4/3, gives

$$F_a = \frac{3000}{r/t}\left(1 - \frac{100/3}{r/t}\right) \text{ ksi} \tag{8-3}$$

is used.[6] The Donnell formula

$$f_{cr} = E\,\frac{0.6t/r - 10^{-7}r/t}{1 + 0.004E/F_y} \tag{8-4}$$

is also used. It should be noted that this equation gives the buckling stress, which must be divided by a factor of safety.

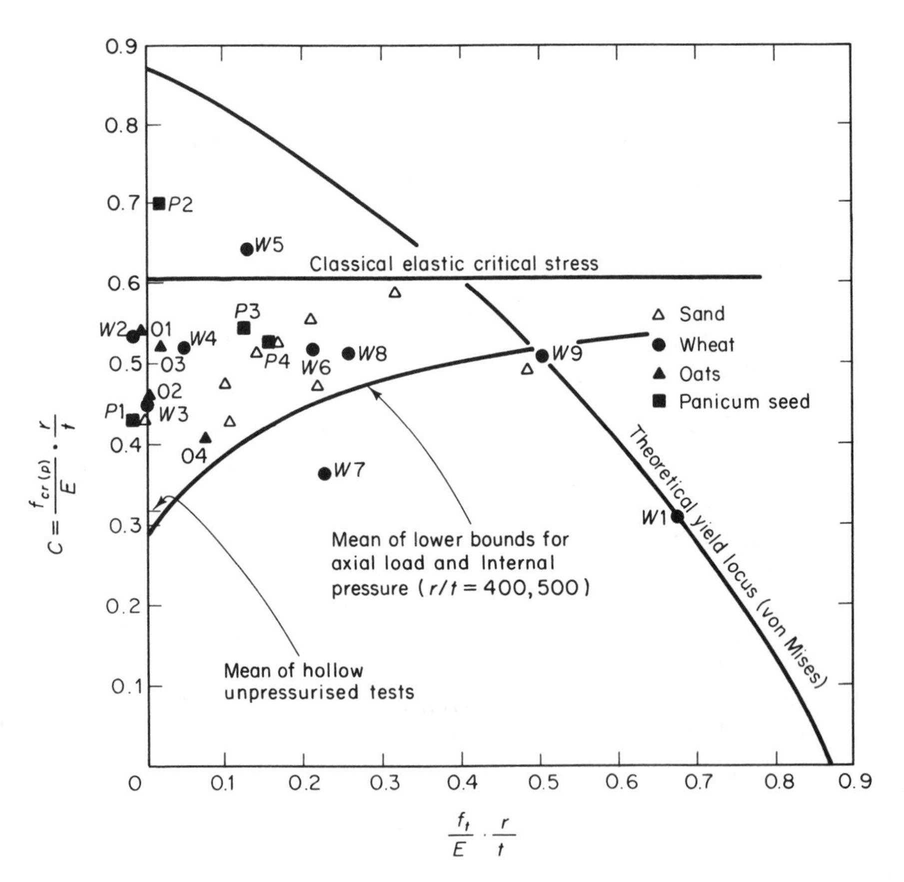

Figure 8-4 Results of tests on axially compressed cylinders containing granular materials (f_t = hoop tension). (From Ref. 5.)

The formulas of the European Convention for Constructional Steelwork (ECCS), which are discussed in Sec. 6-13, give lower-bound curves for hundreds of experimental values and enable the increase in buckling strength due to the lateral pressure of the bin contents to be taken into account (Sec. 6-16). They are repeated here for convenience.

Axial compression with no hoop tension:

$$f_{cr} = \frac{CEt}{r} \tag{8-5a}$$

$$C = \frac{0.374}{\sqrt{1 + 0.01r/t}} \qquad \text{for} \frac{r}{t} \leqslant 212 \tag{8-5b}$$

$$C = \frac{0.315}{\sqrt{0.1 + 0.01r/t}} \qquad \text{for} \frac{r}{t} \geqslant 212 \tag{8-5c}$$

If f_{cr} by Eq. (8–5) exceeds $\frac{3}{8} F_y$, use

$$f_{cr(i)} = F_y \left[1 - 0.347 \left(\frac{F_y}{f_{cr}} \right)^{0.6} \right] \tag{8-6}$$

where f_{cr} is given by Eqs. (8–5).

Axial compression with hoop tension:

$$f_{cr(p)} = \frac{C_p E t}{r} \leqslant \frac{3}{8} F_y \tag{8-7a}$$

$$C_p = C + (0.45 - C) \frac{\rho}{\rho + 0.007} \tag{8-7b}$$

$$\rho = \frac{p}{E} \left(\frac{r}{t} \right)^{3/2} \tag{8-7c}$$

where p = the internal lateral pressure on the cylinder and C is given by Eqs. (8–5b) and (8–5c) If $f_{cr(p)}$ exceeds $\frac{3}{8} F_y$, use Eq. (8–6) with $f_{cr(p)}$ from Eq. (8–7a) substituted for f_{cr}.

The ECCS formulas for elastic buckling, Eqs. (8–5a) and (8–7a), contain a partial factor of safety 4/3 to allow for imperfections in the cylinder. In the inelastic range the partial factor of safety decreases gradually to 1 at $r/t = 0$ (Sec. 6–13). The ECCS suggests that the formulas be used with a factor of safety of 1.5, which, together with the imperfection factor of safety, gives a factor of safety of 2 on the lower bound of test results for elastic buckling, while in the inelastic range it decreases gradually from 2 to 1.5 at $r/t = 0$. However, the ECCS also suggests that a smaller factor of safety may be justified for walls of bins, because of the support from the stored material, but does not specify a value. The authors suggest that the unreduced value 1.5 be used.

It is noted in Ref. 5 that buckling-strength gains from hoop tension may be lost if the wall loses contact with the stored material in the regions of outward buckling. However, inward buckling would still be inhibited, which reduces the imperfection sensitivity of the cylinder. Therefore, some increase in buckling strength can be expected even in this situation. The test results of Fig. 8–4, all of which were on cylinders with $r/t = 400$ and 500, are plotted in Fig. 6–12, which also shows test results for cylinders with no internal pressure. The two lines for cylinders containing granular material are somewhat approximate, since it is not known which of the points in Fig. 8–4 represents $r/t = 400$ and which $r/t = 500$. If tests for other values of r/t were available, it is reasonable to assume that they would plot similarly, with the result that a scatterband of test results for cylinders containing granular material would be established well above the one for cylinders with no internal pressure.

In evaluating bin-wall stresses by the ECCS formulas, the filling value of p_h, which is smaller than the emptying value, should be used to determine p even though the vertical compression is largest during emptying. This is because calculated emptying pressures tend to be envelopes of peak pressures, particularly for mass-flow bins, so that the lateral pressures on some parts of a bin wall may be static while it is under vertical compression from emptying pressures in another part of the bin.

Bin-wall thicknesses are generally such that buckling would be elastic, for which Eqs. (8–5) and (8–7) govern.

Minimum thicknesses. Fabrication and erection and resistance of the shell during erection to local buckling (blow-in) due to wind must also be considered in determining thicknesses of unstiffened shells. API minimum thicknesses are as follows[3]:

D (ft)	Min. t (in.)
$D < 50$	$\frac{3}{16}$
$50 < D < 120$	$\frac{1}{4}$
$120 \leqslant D \leqslant 200$	$\frac{5}{16}$
$200 < D$	$\frac{3}{8}$

Wall courses thinner than the API minimum are used in bins for bulk solids.

8–3. Buckling of Bins with Circumferential Lap Joints

The local-buckling strength of a cylindrical bin with horizontal lap joints may be considerably lower than that of a smooth-walled bin. While being emptied, a 24 × 76 ft lap-jointed grain silo with a central outlet, designed according to DIN 1055 requirements for smooth-walled silos, buckled about 17 ft above the base in a ring in which the wall thickness changed from $\frac{7}{32}$ in. to $\frac{3}{16}$ in. ($r/t = 767$). The silo had been filled with fodder, which produces smaller internal pressures but larger wall-friction forces than grain. It did not collapse, however, and was subsequently filled with grain without having been repaired. This time, the bolts in a vertical joint in the buckled ring failed during emptying, and the silo collapsed toward the side where the bolts had failed.[7]

In the investigation of the collapse, 12 brass cylinders 24 cm long, with $r = 16$ cm and $t = 0.0217$ cm, were tested in axial compression. Six of the cylinders were smooth-walled, while each of the other six had five circumferential lap joints. Three of each type were tested in axial compression alone and three in axial compression together with an internal pressure to simulate the horizontal pressure of the grain. The test cylinders were closely equivalent to the failed silo in terms of the eccentricities of the lapped rings and the ratio r/t. Assuming $t/2$ as the effective eccentricity of each ring at the joint, the ratio of eccentricity to ring height is $0.0217/(2 \times 4) = 0.00271$ for the test cylinder and $4.77/(2 \times 1017) = 0.00234$ for the silo. The values of r/t are $16/0.0217 = 737$ for the test cylinder and $3660/4.77 = 767$ for the silo.

The three unpressurized, smooth-walled test cylinders buckled at an average compressive stress of 8.39 ksi and the corresponding lap-jointed cylinders at an average of 4.28 ksi, a reduction of 49%. The corresponding figures for the pressurized cylinders were 10.75 and 5.86 ksi. Thus, the internal pressure increased the buckling pressure of the smooth-walled cylinder by 28% and of the lap-jointed one by 37%.

The lap-jointed silo can be regarded as a cylinder with axisymmetric imperfections whose amplitude is $t/2$. These imperfections are easy to formulate mathematically, and a buckling analysis was performed by analyzing the cylinder as an assembly of rings. Two extreme cases were considered, one with infinitely stiff joints and the other with hinged joints. The buckling stresses with hinged joints were 4.94 and 5.95 ksi for the unpressurized and

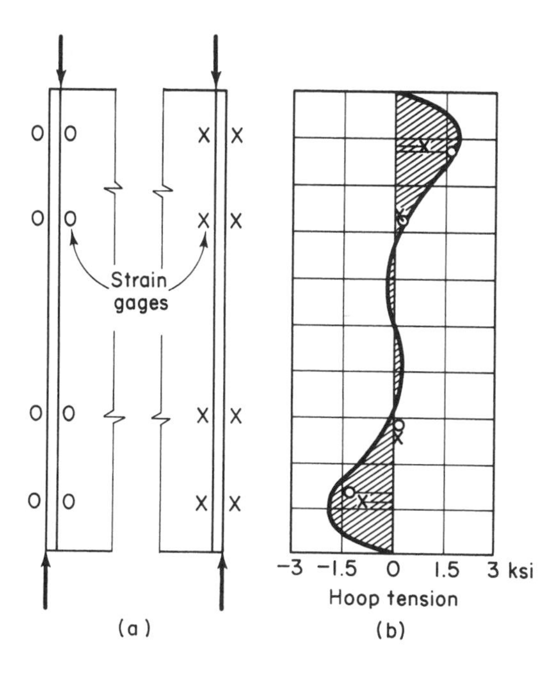

Figure 8-5 Hoop stress in test cylinder: (a) one ring of cylinder (not to scale), (b) curve based on analysis assuming axial compression applied at inner circumference at top and outer circumference at bottom. Test results are shown by x and o. (From Ref. 7.)

pressurized cases, respectively. The corresponding test results were 87 and 98% of these analytical values.

Hoop forces develop in lap-jointed, axially compressed cylindrical shells. Figure 8–5b shows the distribution of these forces in one ring of the test cylinder according to the analysis mentioned above. The numerical values are for an axial compression of 3.06 ksi. These hoop stresses increase at a decreasing rate as the axial stress increases but remain of the same order of magnitude as the axial stress. The hoop tension is significant because it is in addition to the hoop tension caused by the pressure of the bin contents on the wall. In the absence of an analysis to determine the effects of circumferential lap joints in reducing the buckling strength of a cylindrical bin and in increasing the hoop tension in the region of the joint, it would appear advisable to use reduced allowable stresses in checking the bin wall for vertical compression and in determining the number of bolts in the vertical joints of bolted bins, particularly if the stiffening effect of the lateral pressure is taken into account in determining the allowable vertical compression.

8–4. Wall Stresses Due to Unsymmetrical Load

Eccentric drawoff. In addition to membrane forces, bending moments develop in the wall of a bin during flow through an eccentric outlet, because the pressure p_h in the flow channel is smaller than the static (filling) pressure in the nonflowing material (Fig. 8–6). The resulting unsymmetrical loading may cause the bin wall to buckle into the flow channel. Formulas based on analyses of a bin segment subjected to a simplified distribution of the pressures involved, using standard procedures for rings, have been developed.

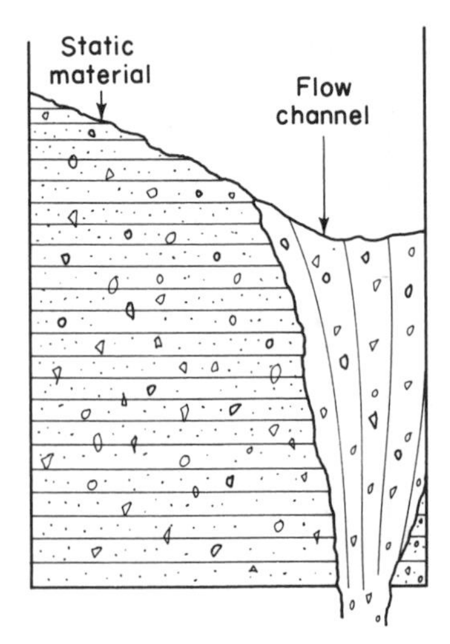

Figure 8–6

Jenike[8] gives the following formula:

$$\frac{r}{t} \leqslant 4.2\sqrt{\frac{F_y}{p_h}} \tag{8-8}$$

Since p_h in this formula is the pressure of the nonflowing material, it should be calculated as a filling pressure. This formula allows the maximum bending moment in the buckled section to reach the yield value. A more liberal value of r/t is obtained by using the plastic moment. Since the shape factor for a rectangular cross section is 1.5, for this case Eq. (8–8) becomes

$$\frac{r}{t} \leqslant 4.2\sqrt{\frac{1.5F_y}{p_h}} \leqslant 5\sqrt{\frac{F_y}{p_h}} \tag{8-9}$$

Jenike's analysis shows that the buckle subtends a central angle of 42°. According to Wozniak[6], the disturbance extends 30° on either side of the outlet. No tests of the validity of Jenike's analysis are reported except that one bin that dented had a value of r/t much larger than would be prescribed by his formula and had dents subtending angles of 46°, 37°, and 40° at three heights. However, his analysis is conservative because it does not account for the interaction between the shell and the contained material as the shell deforms. The DIN 1055 coefficient, Eq. (5–10), for computing an equivalent uniform radial pressure p_h for eccentric withdrawal is claimed to account adequately for the effects of the actual nonuniform pressure.

The interaction between the shell and the contained material can be investigated by considering, as in any statically indeterminate problem, the deformations of the shell and the columns of static material and flowing material. Figure 8–7a shows a vertical section of a 100-ft-diameter self-cleaning coal silo with a seven-drawoff expanded-flow hopper for which one design condition was withdrawal from any two diametrically opposite openings.[9] Pressures were investigated in transverse slices (Fig. 8–7b) using the finite-element mesh shown in Fig. 8–7c. Constant-stress isoparametric quadrilateral elements were used, with the flow-channel boundary approximated by the broken line shown in the figure. Pressures in the static material were computed by Janssen's formula and in the flowing material by Walker's formula (Sec. 5–13). At one section these were 460 and 650 psf, respectively (Fig. 8–7b). Assuming the modulus of elasticity of the coal to be 300 ksf and Poisson's ratio 0.4, the corresponding strains were computed by $\epsilon = (p_h/E)/(1 - \nu)$. Constraints are required to maintain the unbalanced forces due to the stress discontinuities at regional and interelement boundaries, and when they are released, the slice deforms to distribute the unbalanced forces. According to this analysis, the maximum outward radial wall displacement was about 0.15 in. and the corresponding horizontal bending moment about 0.75 ft-kip. The hoop tension varied from 28 kips/ft at the point of maximum

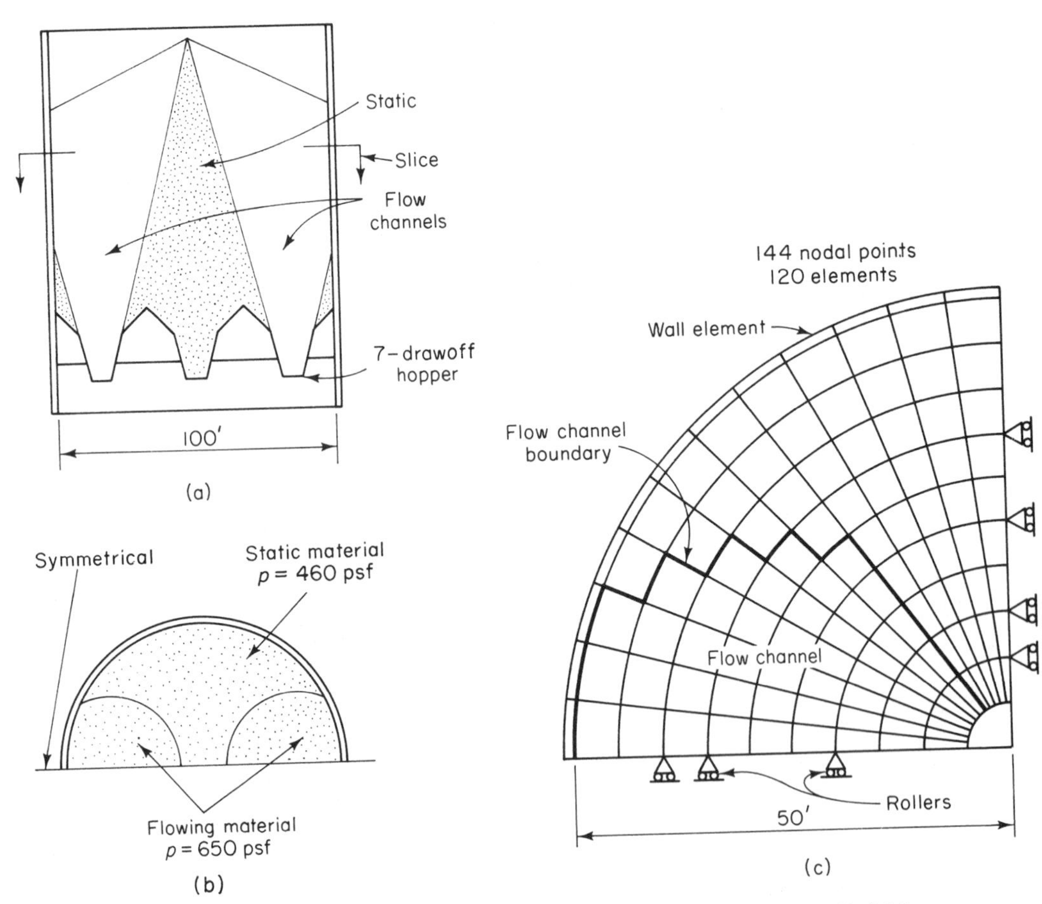

Figure 8-7 Finite-element analysis of eccentric draw-off. (From Ref. 9.)

outward movement to 25 kips/ft at the point of minimum movement. Analysis of the wall for the initial pressures by standard formulas for rings showed wall displacements and moments about 50 times those estimated considering the silo-material interaction.

Nonuniform lateral pressures occur in flat-bottom circular bins which are emptied by trough or tunnel openings, because of the banked-up material that is left after as much of the contents of the bin as the angle of repose allows has been withdrawn (Fig. 8-8). This tends to cause ovalling of the cylinder. Since all the contained material is at rest and since there are major portions of the bin not in contact with it, an analysis of ring sections loaded with the radial pressures from the material in contact with the shell would give some estimate of the wall stresses, and the error in neglecting the interaction between the shell and the contained material would probably be considerably less than in the silo of Fig. 8-7. Formulas for radially loaded rings

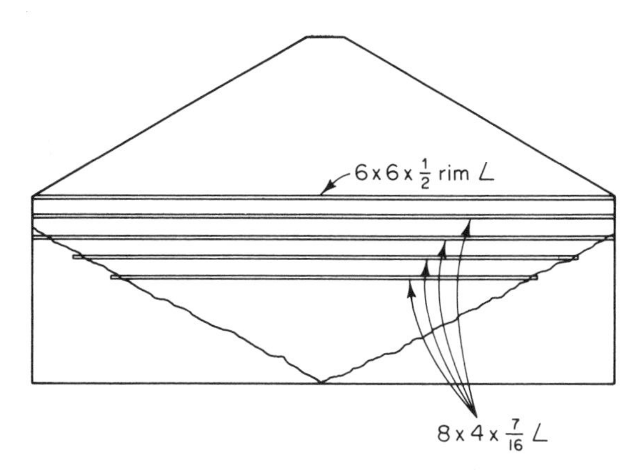

Figure 8-8 Circumferential stiffeners.

in Ref. 10 can be used to make such an analysis. Some designers have used circumferential stiffeners in the upper portion of bins of this type. In one 200 X 64 ft bin four angles were used in the upper 28 ft (Fig. 8-8).

Eccentric filling. The effects of eccentric filling are similar to those of eccentric withdrawal. Off-center filling produces a material surface in the form of a cone with its apex at the point of filling, and this produces bending moments in the shell as well as unsymmetrical membrane forces. The DIN 1055 provision for eccentric withdrawal, Eq. (5-10), can be used to make some allowance for these effects.

Unsymmetrical pressures similar to those from eccentric drawoff can also develop in bins with concentric outlets if segregation occurs during filling. If material with a wide range of particle size is charged into a bin at an angle, the coarse particles tend to flow to one side and the fine particles to the other (Fig. 8-9). An eccentric flow channel is likely to develop within the freer-flowing material, and this will produce differential pressures similar

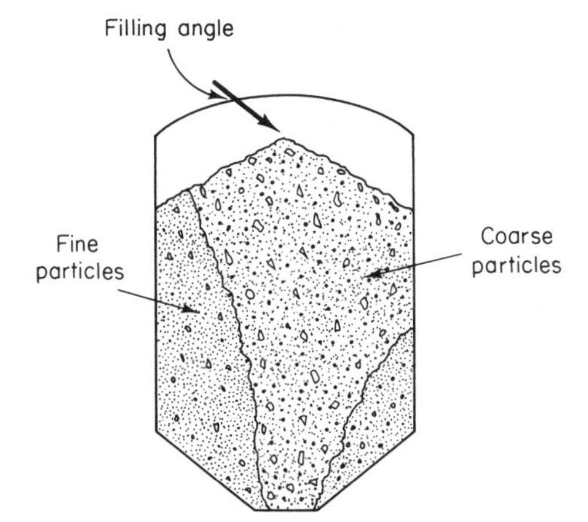

Figure 8-9 Segregation from unsymmetrical filling.

to those from eccentric drawoff. Of course, it is better to avoid this condition by eliminating the cause if for no other reason than to assure uniformity of the discharged material, but if it is likely to exist, it should be anticipated by computing the design pressures as for an eccentric outlet using the DIN 1055 coefficient, Eq. (5–10).

8-5. Wind Buckling of Circular Cylinders

Cylindrical bins are susceptible to buckling under wind load. Vertical compression from the overturning moment may cause buckling near the base, and empty or partially filled bins may buckle on the windward side due to the pressure normal to the surface.

Vertical compression due to overturning moment is usually computed by the elementary beam formula. However, this formula should not be used for bins that are low relative to their height, since they act as deep beams. An analysis of a bin with a height-diameter ratio of 0.5 showed that the overturning effect of wind was taken by the windward 140° of the bin acting as a barrel vault, the maximum compression being about 60° each way from windward, while stresses approached zero on the leeward side.[11] However, overturning compression in low bins is not likely to be large enough to require consideration, particularly if the usual provision in standard specifications for buildings of a one-third increase in allowable stress for wind effects alone, or in combination with stresses from other loads, is followed. Any of the formulas for buckling under uniform compression given in Sec. 6–13 may be used to determine an allowable value. However, if wind alone is being investigated, the ECCS coefficient C_b from Eq. (6–44) may be used instead of the coefficient for uniform axial compression, although some approximation would be involved because of the uniform vertical compression due to the weight of the bin.

The following formula from API Standard 650 gives the ratio of height to thickness of a cylindrical tank that will enable it to resist a wind velocity of 100 mph without buckling on the windward side:

$$\frac{H}{t} = 600 \left(\frac{100t}{D}\right)^{2/3} \tag{8–10}$$

where H = height of cylinder, ft
D = diameter of cylinder, ft
t = average thickness of cylinder, in.

This formula takes into account the suction in open-top tanks or a possible vacuum in closed-top tanks, as well as the external pressure due to the velocity of 100 mph plus a gust factor of 30%. The height for a velocity V may be determined by multiplying H by $(100/V)^2$.

Equation (8–10) is conservative for a closed-top bin because of the differences between the wind-pressure distributions on open-top and closed-top cylinders (Sec. 5–16 and Fig. 5–28). Buckling analyses of a number of cylinders subjected to a closed-top wind-pressure distribution showed the stagnation pressure $q = \rho V^2/2$ at buckling to be at least 1.6 times the stagnation pressure for buckling under a uniformly distributed radial pressure, which is a fair approximation of the resultant pressure distribution on an open-top tank.[12] This result is confirmed by water-tunnel tests, reported in Ref. 13, on closed-top cylinders with a height-diameter ratio of 0.96 and radius-thickness ratios ranging from 420 to 760. The stagnation pressure q at buckling was determined to be

$$q = (0.7 \pm 0.14)p_{cr} \tag{8–11}$$

in which p_{cr} is the uniform radial buckling pressure.

With the ratio 1.6 and the fact that H in Eq. (8–10) is inversely proportional to V^2, the value of H for a closed-top bin can be determined by multiplying H from Eq. (8–10) by 1.6.

If the height of a bin exceeds H by Eq. (8–10), multiplied by the factor 1.6 if the results of Ref. 12 are followed, an intermediate ring stiffener should be used. According to API 650, the required section modulus of an intermediate stiffener is given by

$$S = 0.0001D^2H \tag{8–12}$$

where S = section modulus, in.^{3}, and D and H are as defined in Eq. (8–10). Portions of the shell of $0.6\sqrt{rt}$ above and below the point of attachment to the shell may be included in computing S. The simplest forms of stiffener are (1) a rectangular bar and (2) an angle welded at one toe to the shell.

8–6. Membrane Stresses in Hopper

The meridional tension N_m at a horizontal section of a bin bottom or hopper in the form of a shell of revolution (Fig. 8–10) is found by equating the resultant vertical component to the sum of the resultant of the pressure p_v and the weight W of that part of the material and shell bottom below the section,

$$2\pi r N_m \cos\theta = p_v \pi r^2 + W$$

from which

$$N_m = \frac{p_v r}{2\cos\theta} + \frac{W}{2\pi r\cos\theta} \tag{8–13}$$

where r = radius of horizontal section
θ = angle of N_m with vertical

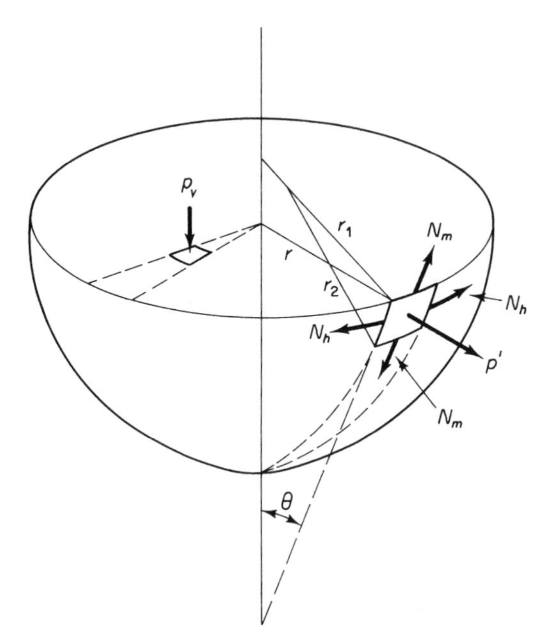

Figure 8-10 Membrane stress resultants in shell of revolution.

Instead of the value of p_v at the section, some designers use γy, where y is the depth of the material above the section.

The membrane forces N_m and N_h (Fig. 8-10) at any point on the surface of a shell of revolution are related by

$$\frac{N_m}{r_1} + \frac{N_h}{r_2} = p' \tag{8-14}$$

where r_1 = radius of curvature of meridian at the point
r_2 = radius of curvature in plane perpendicular to meridian at the point
p' = pressure normal to the surface

Therefore, once N_m is computed from Eq. (8-13), N_h can be found from Eq. (8-14). Formulas for computing p' are given in Sec. 5-12.

Conical bottoms (Fig. 8-11). Since the radius r_1 of a conical bottom or hopper is infinite, the hoop tension by Eq. (8-14) is

$$N_h = p'r_2 = \frac{p'r}{\cos\theta} \tag{8-15}$$

where r is the radius of the horizontal section.

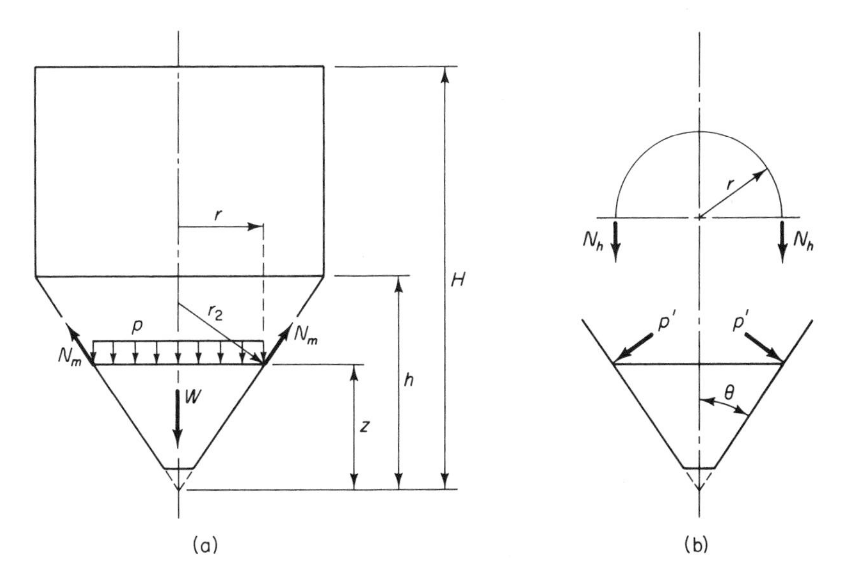

Figure 8-11 Membrane stress resultants in conical hopper.

8-7. Compression Ring

Figure 8-12a shows the meridional stress resultants for the cylinder, hopper, and skirt of a skirt-supported bin with a conical hopper. These forces are not in equilibrium without the shearing forces and bending moments shown in Fig. 8-12b. The accompanying deformation is shown in Fig. 8-12c. It will be noted that the hoop stresses at the juncture are compressive. These and the bending stresses can be reduced by making the shells thicker in the region of bending, or by adding a stiffening (compression) ring at or near the junction, or both. Typical compression rings are shown in Fig. 8-13.

A common procedure for sizing a compression ring is to treat it as a ring loaded radially by the horizontal components of the hopper membrane forces, with effective widths $0.78\sqrt{rt} \leqslant 16t$ of the cylinder, hopper, and skirt considered as parts of the ring. It is recommended in Ref. 6 that it be designed for an allowable stress of 10 ksi. The low value helps to keep the radial displacements small. This reduces the local bending stresses, which could cause fatigue failures in bins subjected to frequent loading and unloading. Higher stresses (15-18 ksi) are suggested if cycling is expected to be infrequent. The ring should also be checked for buckling by Eq. (6-76) of Sec. 6-20.

Compression-ring effectiveness was investigated by Gould et al.[14] and by Wang and Gould[15] in a finite-element analysis of a hydrostatically loaded, column-supported, cylindrical-conical tank 30 ft in diameter by 45 ft high

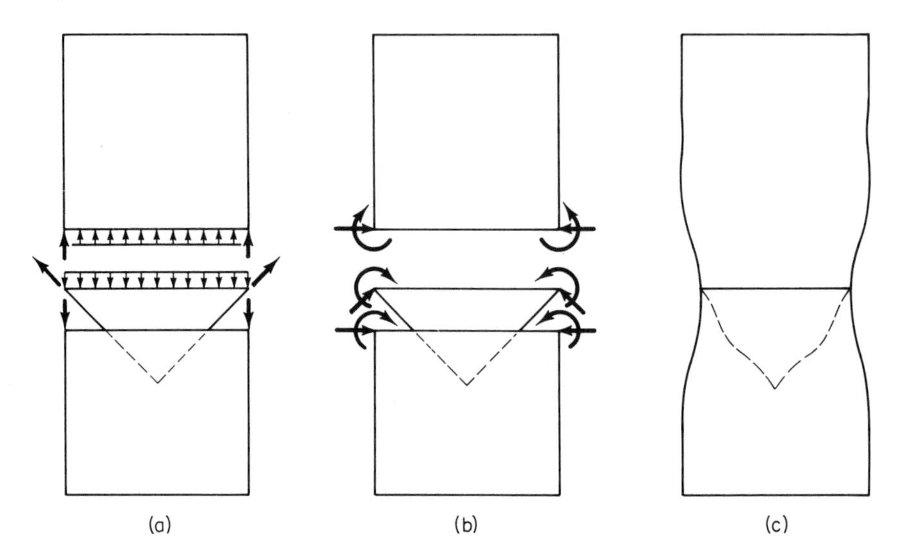

Figure 8-12 Forces and deformation at transition.

overall with (1) no compression ring, (2) a ring proportioned as described above, using an allowable stress of 12 ksi, and (3) an infinitely stiff ring. The radial displacements and hoop-stress resultants at the cylinder-cone junction at the supports were 0.65, 0.12, and 0 in. and 27,000, 7500, and 3000 lb/in. for these three cases, respectively. The stiffener also had a marked effect on the moments, and it can be concluded that a stiffener designed as in method (2) appears to be quite effective in forcing the response toward the ideal rigid-stiffener condition.

Knuckle at junction. The transition from cylinder to hopper can be made gradually by using a knuckle (Fig. 8-13d). This eliminates the need for a compression ring. The hoop-stress resultant in the knuckle is found by Eq.

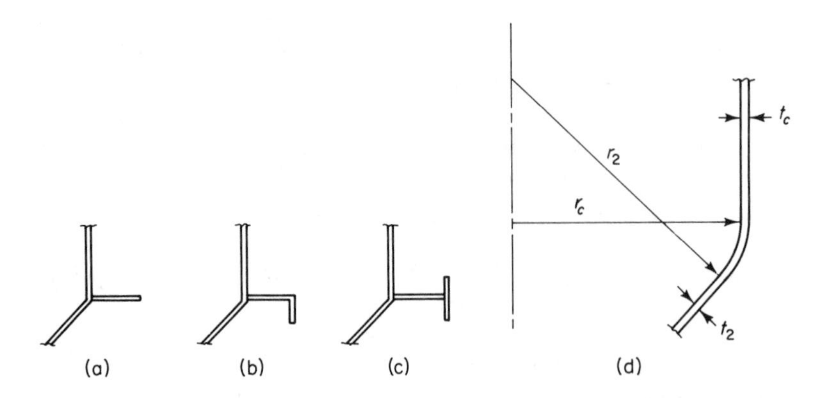

Figure 8-13 (a), (b), and (c) Typical compression rings, (d) transition knuckle.

(8–14) after the meridional resultant is computed from Eq. (8–13). The hoop stress must not exceed the allowable buckling stress. Since there is little information on the buckling of a knuckle connecting a cylinder to a cone, it is customary to evaluate it as for a spherical shell of the same radius as the knuckle. Equations (7–3) and (7–4) may be used for this purpose. However, since the theoretical buckling stress for a uniformly compressed sphere is the same as for an axially compressed cylinder, the latter is also used.[6] The hoop stress should be checked at two points: $0.78\sqrt{r_c t_c}$ from the junction of the knuckle with the shell and $0.78\sqrt{r_2 t_2}$ from its junction with the cone.[6]

8–8. Stresses at Cylinder-Cone Junction

A procedure for computing the secondary bending moments and the hoop stresses at the hopper-cylinder junction of a skirt-supported bin is given in Ref. 16. The analysis is based on the bending theory of shells loaded on an end perimeter by uniformly distributed moments and shears (Fig. 8–2) and takes advantage of the simplification due to the local nature of the bending. The results, which are for a bin with equal thicknesses of cylinder, hopper, and skirt, are given here in a somewhat different and more general form.

The ring stress f_r, which is the same in the ring, cylinder, hopper, and skirt, is given by

$$f_r \left(\frac{A_r}{\sqrt{rt}} + 2.16t \right) = 0.887r \left(p_h + \frac{0.715p'}{\cos\theta} \right) - \sqrt{\frac{r}{t}}\, N_{mh} \sin\theta \qquad (8\text{–}16)$$

where A_r = cross-sectional area of ring, in.^{2}
t = thickness of cylinder, hopper, and skirt, in.
r = radius of cylinder, in.
p_h = horizontal pressure on cylinder at juncture, lb/in.^{2}
p' = normal pressure on hopper wall at juncture, lb/in.^{2}
N_{mh} = meridional force in hopper at juncture, lb/in.
θ = apex half angle of hopper (Fig. 8–11)

The bending stresses are given by

$$f_{bc} = 1.16 \left(2f_r - p_h \frac{r}{t} - 0.5 \frac{p'}{\cos\theta} \frac{r}{t} \right) \qquad (8\text{–}17a)$$

$$f_{bh} = 1.16 \left(f_r - 0.5p_h \frac{r}{t} - \frac{p'}{\cos\theta} \frac{r}{t} \right) \qquad (8\text{–}17b)$$

$$f_{bs} = 1.16 \left[f_r - 0.5 \frac{r}{t} \left(p_h - \frac{p'}{\cos\theta} \right) \right] \qquad (8\text{–}17c)$$

where subscripts c, h, and s of f_b denote cylinder, hopper, and skirt, respectively. These bending stresses must be added to the direct stresses N/t.

The authors have extended Eqs. (8–16) and (8–17) to the case of unequal thicknesses of cylinder, hopper, and skirt. The stress f_r is given by

$$f_r\left(2.63\,\frac{A_r}{\sqrt{r}} + a_1 + a_2 + a_3\right) = a_1 p_h \frac{r}{t_c} + a_2 p' \frac{r}{t_h \cos\theta} - 2.63\sqrt{r}\,N_{mh}\sin\theta \tag{8–18a}$$

The coefficients a in this equation are given by

$$Ta_1 = t_c^{1.5}(2T - t_c^{2.5}) + t_c^2(t_h^2 + t_s^2) \tag{8–18b}$$

$$Ta_2 = t_h^{1.5}(2T - t_h^{2.5}) + t_h^2(t_c^2 - t_s^2) \tag{8–18c}$$

$$Ta_3 = t_s^{1.5}(2T - t_s^{2.5}) + t_s^2(t_c^2 - t_h^2) \tag{8–18d}$$

in which

$$T = t_c^{2.5} + t_h^{2.5} + t_s^{2.5}$$

and subscripts c, h, and s denote cylinder, hopper, and skirt, respectively.

The bending moments are given by

$$M = 0.289tB \tag{8–19a}$$

in which

$$B_c = \frac{f_r}{t_c^{0.5}}\left(a_1 - t_c^{1.5}\right) + p_h r\left(1 - \frac{t_c^{2.5}}{T}\right) + \frac{p'r}{\cos\theta}\,\frac{t_c^{1.5} t_h}{T} \tag{8–19b}$$

$$B_h = \frac{f_r}{t_h^{0.5}}\left(a_2 - t_h^{1.5}\right) - p_h r\,\frac{t_c t_h^{1.5}}{T} - \frac{p'r}{\cos\theta}\left(1 - \frac{t_h^{2.5}}{T}\right) \tag{8–19c}$$

$$B_s = \frac{f_r}{t_s^{0.5}}\left(a_3 - t_s^{1.5}\right) + \frac{t_s^{1.5} r}{T}\left(\frac{p' t_h}{\cos\theta} - p_h t_c\right) \tag{8–19d}$$

Positive moments are clockwise on the cylinder, hopper, and skirt at the junction. With the moment from Eq. (8–19a) and the corresponding stress resultant N, the meridional stress f_m is given by

$$f_m = \frac{N}{t} \pm \frac{6M}{t^2} \tag{8–20}$$

The formulas in this section do not account for compression-ring restraint of rotation of the hopper-cylinder junction. Equation (6–80) shows that twisting of the ring about its circumferential axis is a function of the moment of inertia of the ring about the axis in the plane of the ring. Therefore, a ring of the type shown in Fig. 8–13a has little effect on rotation of the junction, while the other types shown may have a considerable effect.

Allowable stress. Since the stresses due to the local bending at the junction of the cylinder and hopper are not ordinarily investigated, allowable values have not been established. However, guidance is available from experience with similar structures. For example, the ASME Pressure Vessel Code[17] classifies the bending stress at such a junction as a secondary stress, because it is highly localized, and specifies for the maximum combination of primary and secondary stress an allowable value of three times the allowable value for the primary membrane stress acting alone. The allowable primary membrane stress, denoted by S_m, is no larger than two-thirds of the specified minimum yield strength or one-third of the specified minimum tensile strength of the material. Thus, the allowable value of the combined primary and secondary stress can be as large as the material's tensile strength. This is an application of plastic analysis and is analogous to the development of a plastic hinge in a structural frame. It is intended to assure elastic behavior after a few inelastic-stress cycles.

Combinations of primary and secondary stress used in the ASME procedure are found by the maximum shear-stress theory, which, for the two-dimensional state of stress considered here, is expressed by

$$
\begin{aligned}
f_1 &= f_r \\
f_2 &= f_m \\
f_3 &= f_r - f_m
\end{aligned}
\tag{8-21}
$$

where f_r and f_m are the ring and meridional stresses, including the bending-stress components, both positive for tension. The largest of f_1, f_2, and f_3 must not exceed three times the allowable primary membrane stress.

Since the steels ordinarily used for bins are not of pressure-vessel quality, the ASME Pressure Vessel Code allowable primary-secondary stress combination should probably be reduced somewhat for application to bins. Four of the steels listed in Table 2–1 are of pressure-vessel quality, and the code-allowable value of S_m for all four is the maximum, that is, the lesser of $\frac{2}{3}F_y$ or $\frac{1}{3}F_u$. Of the remainder, only A36 is listed in the code, and it is assigned a value of 19.3 ksi, which is only 29% of the minimum tensile strength. Therefore, the authors suggest that S_m be taken as the smaller of $\frac{2}{3}F_y$ or $0.3F_u$ for the steels in Table 2–1 which are not of pressure-vessel quality.

The analysis discussed in this section is used in Examples 8–1 and 8–2.

8-9. Tension Ring

The upper edge of a cylindrical bin is subjected to outward thrusts from a self-supported roof or the rafters of a supported roof. These thrusts must be resisted by the cylinder or, in the usual case, by the cylinder together with a stiffening ring. Such a ring is variously called a tension ring, or, if it is an angle, a rim angle or top angle (Fig. 8–14). In any case, a nominal stiffener,

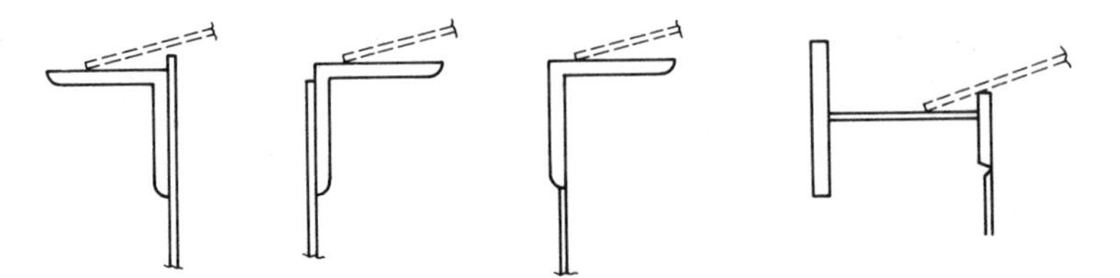

Figure 8-14 Typical tension rings.

such as an angle, is likely to be needed for attachment of a self-supporting roof and to guard against blow-in during erection (Sec. 8-20). In addition, rafters of supported roofs should have vertical stiffeners to distribute the vertical reaction unless the rim angle itself is stiff enough.

Since the tension in a circular ring with a uniformly distributed radial load q is $qD/2$, the required cross-sectional area is given by $A = qD/2F$, where F is an allowable stress. The load q for self-supporting conical and dome roofs is the horizontal component of the membrane stress resultant N_1 at the base (Figs. 7-6 and 7-7). But $N_1 = wD/4 \sin\theta$ for either type of roof, where w is the load per horizontal square foot of area covered and θ is the base angle. Therefore,

$$A = \frac{wD^2 \cot\theta}{8F} \tag{8-22a}$$

Portions of the attached shell and roof plates can be considered as contributing to the area A. The effective widths of such portions are commonly taken as $16t$.

API 650 gives the following formulas for the top-angle area in self-supporting roofs: conical roof:

$$A = \frac{D^2}{3000 \sin\theta} \tag{8-22b}$$

and dome and umbrella roofs:

$$A = \frac{Da}{1500} \tag{8-22c}$$

where A = cross-sectional area, in.^{2}, of angle plus segments of shell and roof plate of effective widths $16t$

D = diameter of bin, ft

a = radius of curvature of roof plate, ft

These formulas are similar to Eq. (8-22a) and are presumably based on the 25-psf live load and 20-psf dead load for which the API formulas for the thickness of these roof shells are based (Secs. 7-2 and 7-3). They can be derived in the same way as Eq. (8-22a) if the membrane stress resultant itself, rather than its horizontal component, is used as the radial load on the ring and F is taken at 16,875 psi. Therefore, these formulas are considerably more conservative than Eq. (8–22a), particularly if F is taken at a more commonly used value, such as $0.6F_y$.

The tension ring for a bin with a supported roof should be analyzed for the moments and circumferential tension given by Eqs. (6–71) and (6–72) of Sec. 6–19, with the horizontal components of the rafter reactions as the radial loads. Effective widths of the attached roof and shell plates can also be considered as parts of the ring in this case. An analysis of this type of ring is given in Example 7–4.

Typical tension rings are shown in Fig. 8–15.

8–10. Column-Supported Bins

Column reactions may be distributed into a cylindrical-bin wall by a ring beam, or girder, or by insert plates at the attachments. A ring girder may consist of a section of the shell as the web, and two flanges one of which also serves as the compression ring at the hopper-cylinder junction (Figs. 8–15a and 8–15b). It may also be located below the junction; an example of this type is shown in Fig. 8–15c.

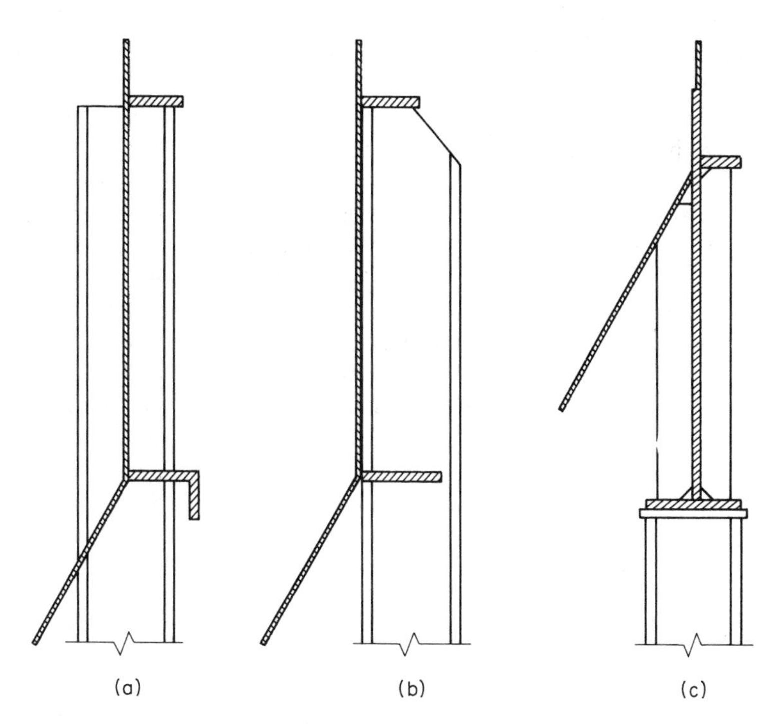

Figure 8–15 Ring girders.

Ring-girder shear. The shear-buckling strength of the web of a straight beam is usually taken to be the buckling strength of a rectangular plate with shears distributed uniformly along the four edges. Similarly, the shear-buckling strength of a curved-beam web can be approximated by the buckling strength of a curved rectangular plate with shear distributed uniformly along its edges, which, for simply supported edges, is given by

$$f_{v(cr)} = 4.39 \frac{E(t/h)^2}{1-\mu^2} \left[\sqrt{1 + 0.0257(1-\mu^2)^{3/4} \left(\frac{h}{\sqrt{rt}}\right)^3} + \frac{0.75 + 0.03h/\sqrt{rt}}{(a/h)^2} \right] \leqslant \frac{F_y}{\sqrt{3}} \qquad \text{(8–23a)}$$

where h = depth of web
t = thickness of web
r = radius of curvature of panel
a = circumferential length of panel

The limiting value $F_y/\sqrt{3}$ is the shear-yield strength. If the second term in the brackets is omitted the equation reduces to Eq. (11–30) of Ref. 18, which is a formula for buckling of a cylindrical shell in torsion; this corresponds to an infinitely long curved plate of depth h. The second term in the brackets was developed by the authors. Results of Eq. (8–23a) are in very good agreement with tabulated values in Ref. 19 for curved panels with $a/h \geqslant 0.5$.

The ultimate strength of a flat-plate panel in shear may be considerably larger than the shear-buckling strength because of the tension field that develops after buckling. Curved panels also develop postbuckling strength, but the postbuckling increment decreases with increasing curvature of the panel and no general method for computing it has been developed.[20] Webs of ring girders of the types shown in Figs. 8–15a and 8–15b are also subjected to hoop tension and vertical compression from the stored solid. Tension in a flat plate increases the shear-buckling strength while compression decreases it.[18] The same effects can be expected in a curved plate. Assuming these effects cancel one another to the extent that they can be neglected, the authors suggest that Eq. (8–23a) be used with a factor of safety of 1.5, which is equivalent to assuming a postbuckling increment of about 10% when compared to the commonly used factor of safety 1.65. Then with E = 30,000 ksi and $\mu = 0.3$, Eq. (8–23a) gives for the allowable shearing stress F_v, in kips per square inch,

$$F_v = \frac{96{,}000}{(h/t)^2} \left[\sqrt{1 + 0.0239\left(\frac{h}{\sqrt{rt}}\right)^3} + \frac{0.75 + 0.03h/\sqrt{rt}}{(a/h)^2} \right] \leqslant 0.38F_y \qquad \text{(8–23b)}$$

Since there is no hoop tension to offset the effects of vertical compression in the web of a ring girder of the type shown in Fig. 8-15c, the authors suggest that the allowable shear by Eq. (8-23b) be reduced 10% for this case. This is equivalent to using Eq. (8-23a) with a factor of safety of 1.65.

The size of transverse stiffeners for a curved web can be determined by formulas for a straight girder.[20] The following is from Ref. 21,

$$I_s = 2.5ht^3\left(\frac{h}{a} - 0.7\frac{a}{h}\right) \tag{8-24}$$

If the web is stiffened on only one side the stiffener moment of inertia I_s is to be computed at the face of the stiffener in contact with the web.

Ring-girder moment. By assuming the girder to be uniformly loaded and continuous but without torsional restraint at the columns, the maximum moment is at the column and is given by

$$M = -Wr\left(\frac{1}{\theta} - \frac{1}{2}\cot\frac{\theta}{2}\right) \tag{8-25}$$

where W is the load on one column and θ is the angle between radii to adjacent columns. This is conservative because the ring girder is restrained torsionally at the supports and along its length by the shell. Formulas which include the effects of torsional restraint at the columns are given in Ref. 10.

The design of a ring girder as an isolated curved beam is conservative because it neglects the participation of the attached shell and hopper. Design of a girder of the type shown in Fig. (8-15a) is illustrated in Example 8-1.

Local reinforcement. The principal difficulty in designing a column-supported bin in which insert plates or similar local reinforcement is used instead of a ring girder has to do with the concentrations of meridional and hoop membrane stress resultants and the meridional and hoop bending moments at the supports. There are no design-code procedures for calculating these stresses. Only rough evaluations can be made, except by finite-element analysis, and design is usually based on tests, experience, and in-house empirical rules.

A detailed finite-element analysis of the stress concentrations at the columns of a hydrostatically loaded cylindrical tank is reported in Refs. 14 and 15. The tank is 30 ft in diameter by 30 ft high and has a conical bottom with an apex half angle of 45°. The bottom ring is 0.312 in. thick, and each of the four columns is attached to a 24 × 90 in. insert plate 0.469 in. thick (Fig. 8-16). The design was provided by a tank manufacturer. The compression ring at the cylinder-hopper junction was designed for an allowable axial stress of 12 ksi.

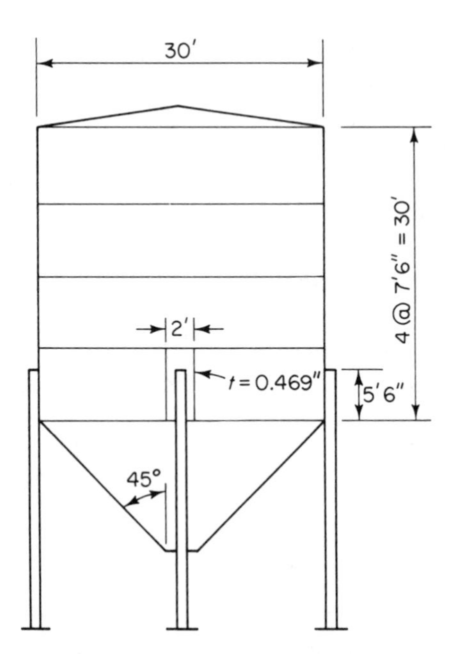

Figure 8-16 Tank analyzed in Ref. 15.

Meridional membrane stress resultants at four positions are shown in Fig. 8-17. The ASME Pressure Vessel Code, discussed in Sec. 8-8, may be used to determine allowable stresses. This code classifies membrane stresses at supports as local primary membrane stresses, for which the allowable value is $1.5S_m$. Assuming the steel to be A283 Grade C, F_y is 30 ksi and F_u 55 ksi so $S_m = 0.3 \times 55 = 16.5$ ksi (the reason for using $0.3F_u$ instead of the

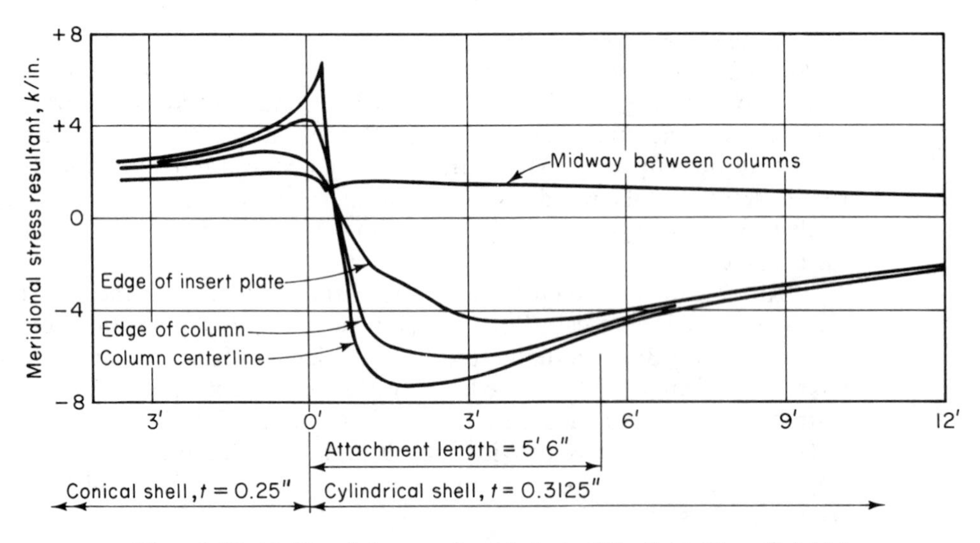

Figure 8-17 Meridional stress resultants in tank of Fig. 8-16. (From Ref. 15.)

code value $\frac{1}{3}F_u$ is discussed in Sec. 8–8). The figure shows N_y to be about 4000 lb/in. at the edge of the insert plate and immediately above it, which gives a compressive stress of 12,800 psi in the $\frac{5}{16}$ - in. shell. The hydrostatic pressure at midheight of the connection is 62.5 × 27.25 = 1703 psf, which gives a hoop tension of 1703 × 15 = 25,545 lb and a hoop stress of 25,545/ 12 × 0.3125 = 6800 psi. Then by Eqs. (8–21),

$$f_1 = +6800 \text{ psi}$$

$$f_2 = -12{,}800 \text{ psi}$$

$$f_3 = 6800 - (-12{,}800) = 19{,}600 \text{ psi}$$

$$1.5S_m = 1.5 \times 16{,}500 = 24{,}750 > 19{,}600 \text{ psi}$$

There are also meridional and circumferential bending moments in this region of the attachment. They are relatively small, however, and if they are included in computing f_1, f_2, and f_3, the allowable stress is $3S_m$ because the bending stresses are secondary.

The cylinder must also be checked for buckling at the attachment. According to the ECCS lower-bound formulas, Eqs. (8–7), the buckling stress for this cylinder is 13,600 psi using for p in Eq. (8–7c) the hydrostatic pressure 62.5 × 24.5 = 1530 psf = 10.6 psi at the top of the attachment. This is only slightly larger than the meridional stress at the attachment. However, the ECCS formula contains a factor of safety of 4/3, and since the compressive stress due to the column reaction is largely local, it is acceptable.

The distribution of the hoop stresses and the meridional moments at the cylinder-hopper junction were found to be similar to those in a skirt-supported bin, which suggests that they are for the most part a result of the geometric discontinuity rather than the concentrated column reactions. Although there is a considerable amplification of the hoop stress in the immediate vicinity of the column compared to values between columns, the meridional and circumferential moments are not much larger at the attachment than they are between attachments. This suggests that the stress concentrations at the cylinder-hopper junction can probably be calculated with sufficient accuracy by the analysis of Sec. 8–8.

Since the local compression in the cylinder at the attachment is in addition to compression from the friction on the wall of a bin containing granular material, the bottom ring in column-supported bins must be thicker than would be needed for the vertical compression due to friction alone.

Two longitudinal distributions of the column reaction on the shell were investigated in Ref. 14. The resulting distributions of N_y differed considerably over the length of the attachment at the column centerline but were practically the same in the region above the attachment.

The procedure described suggests that a column attachment can be designed satisfactorily by using a finite-element analysis to determine the meridional stress resultants, with allowable values from the ASME pressure-vessel code[17] and the ECCS buckling formulas. Stress resultants at the cylinder-hopper junction can be determined by the finite-element analysis or the procedure of Sec. 8–8. Since the effects of the column are localized, the finite-element analysis can be restricted to the immediate region of the attachment.

Column eccentricity. If the column attachment is not concentric with the cylinder wall, there is a bending moment that is resisted by a radial couple (Fig. 8–18). An analysis in Ref. 22 shows that the local stresses in the shell arising from the eccentricity are relatively minor, so that the column takes practically the entire moment. Therefore, this should be taken into account in designing the column.

Bracing. Lateral-force bracing of column-supported bins usually consists of round or square rods with double clevis plates at the ends for attachment to gusset plates welded to the columns. Turnbuckles are provided for adjustment. The maximum bracing rod tension is given by

$$T = \frac{2H}{n \cos \theta} \tag{8-26}$$

where H = horizontal load from wind or seismic forces
θ = angle of diagonal brace with horizontal
n = number of brace panels

If the bracing system is one in which the diagonal bracing is capable of taking compression as well as tension in each panel, n in Eq. (8–26) should be taken as twice the number of brace panels.

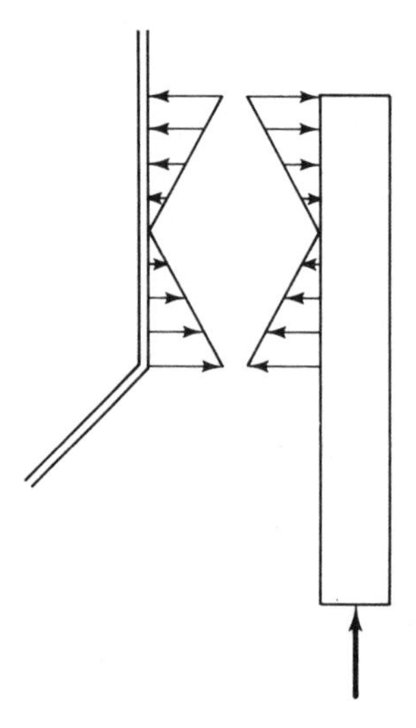

Figure 8–18

8-11. Membrane Forces in Rectangular-Bin Hoppers

Rectangular bins and bunkers usually have pyramidal bottoms with one or more openings (Figs. 1-1 and 1-2). Wedge hoppers are also used (Fig. 4-2b). Rectangular bins may also be flat-bottomed, with multiple wedge openings (Fig. 4-3c).

The membrane forces in a symmetrical pyramidal hopper are shown in Fig. 8-19. If the vertical components of the meridional forces are assumed to be distributed uniformly on the perimeter of a horizontal section, N_m is given by

$$N_m = \frac{W + (p_{va} + p_{vb})ab/2}{2(a + b)\cos\theta} \tag{8-27}$$

where a is the length and b the width of the cross section, p_{va} and p_{vb} are the vertical pressures corresponding to sides a and b, respectively, W is the weight of that portion of the hopper and hopper contents below the section, and θ is the angle of the hopper wall with the vertical.

If $a > b$, $p_{va} < p_{vb}$ because the hydraulic radius is smaller for the longer side. This suggests that the vertical component of the meridional force on wall b is larger than on wall a. Assuming $W/4$ and the resultant vertical pressure on the triangular area adjacent to each wall to be carried by that wall, the following formulas result:

$$N_{ma} = \left(\frac{W}{a} + p_{va}b\right)\frac{1}{4\cos\theta_a} \tag{8-28a}$$

$$N_{mb} = \left(\frac{W}{b} + p_{vb}a\right)\frac{1}{4\cos\theta_b} \tag{8-28b}$$

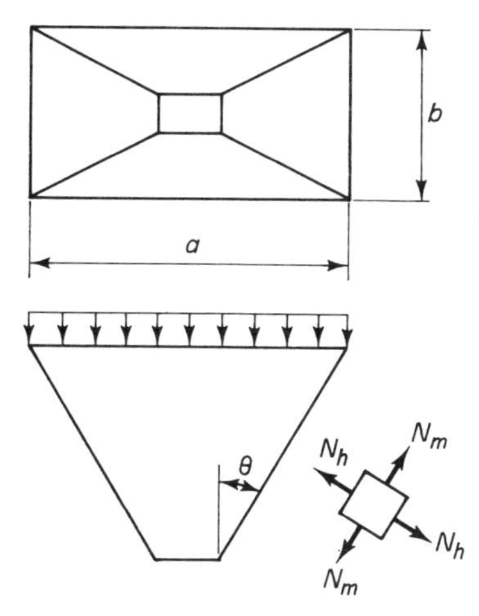

Figure 8-19 Stress resultants in pyramidal hopper.

The horizontal membrane force N_h is given by

$$N_{ha} = \tfrac{1}{2}(p_{vb} + \gamma_h \sin\theta_b)\, b \cos\theta_a \tag{8-29a}$$

$$N_{hb} = \tfrac{1}{2}(p_{va} + \gamma_h \sin\theta_a)\, a \cos\theta_b \tag{8-29b}$$

where γ_h = the weight of the hopper wall per unit of area.

Unsymmetrical pyramidal hoppers. Figure 8-20 shows the assumed distribution of the vertical components of the wall forces at a horizontal cross section of an unsymmetrical pyramidal hopper due to the weight W of the portion of the hopper and its contents below the section. The magnitudes of the corner forces are given by[23]

$$N_{1v} = \frac{W}{2(a+b)} t_x t_y \tag{8-30a}$$

$$N_{2v} = \frac{W}{2(a+b)} (2 - t_x) t_y \tag{8-30b}$$

$$N_{3v} = \frac{W}{2(a+b)} (2 - t_x)(2 - t_y) \tag{8-30c}$$

$$N_{4v} = \frac{W}{2(a+b)} t_x (2 - t_y) \tag{8-30d}$$

where t_x and t_y are coefficients given in Table 8-1. These coefficients are functions of the coordinates x_g and y_g of the center of gravity of the stored

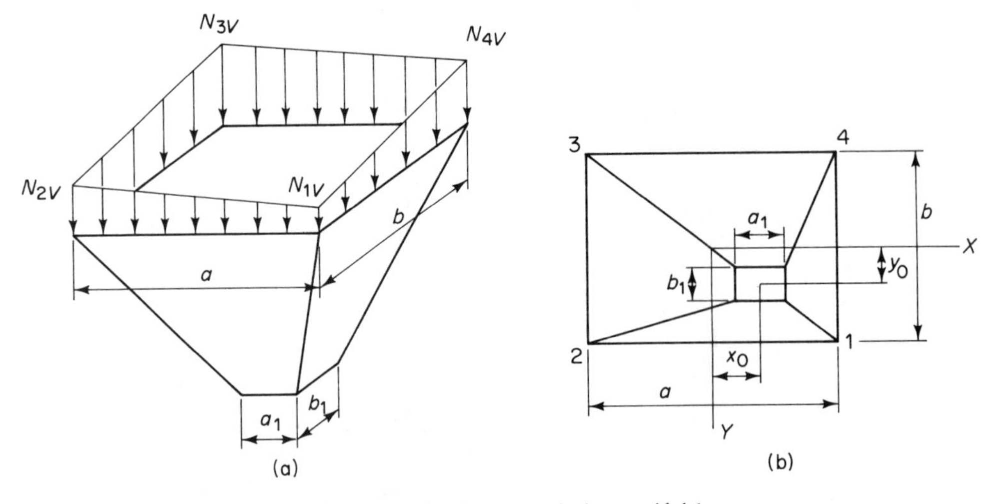

Figure 8-20 Unsymmetrical pyramidal hopper.

TABLE 8-1. COEFFICIENTS t_x AND t_y FOR EQS. (8-30)[a]

	t_x				t_y		
$\frac{x_g}{a}$	$\frac{a}{b} = 1$	1.5	2	$\frac{y_g}{b}$	$\frac{a}{b} = 1$	1.5	2
0.30	1.60	1.67	1.72	0.30	1.60	1.55	1.51
0.35	1.45	1.50	1.54	0.35	1.45	1.41	1.39
0.40	1.30	1.33	1.36	0.40	1.30	1.27	1.26
0.45	1.15	1.17	1.18	0.45	1.15	1.14	1.13
0.50	1.00	1.00	1.00	0.50	1.00	1.00	1.00
0.55	0.85	0.83	0.82	0.55	0.85	0.86	0.87
0.60	0.70	0.67	0.64	0.60	0.70	0.73	0.74
0.65	0.55	0.50	0.46	0.65	0.55	0.59	0.61
0.70	0.40	0.33	0.28	0.70	0.40	0.46	0.48

[a]From Ref. 23.

material, measured from corner 1 (Fig. 8-20), which are given by

$$x_g = x_0 h \frac{(a + a_1)(b + b_1) + 2ab}{12V} \tag{8-31a}$$

$$y_g = y_0 h \frac{(a + a_1)(b + b_1) + 2ab}{12V} \tag{8-31b}$$

where V is the volume of the hopper below the cross section being considered. V is given by

$$V = \frac{h}{6} [(2a + a_1)b + (2a_1 + a)b_1] \tag{8-32}$$

where a_1 and b_1 are the dimensions of the hopper outlet parallel, respectively, to the hopper dimensions a and b and h is the distance from the outlet to the section being investigated.

The hopper-wall forces due to the pressure p_v of the contents above the hopper cross section being investigated are difficult to determine. They tend to concentrate in the stiffer regions of the hopper support structure and are therefore dependent on the construction and the aspect ratio a/b of the cross section. In the absence of a definitive analysis the vertical components can be assumed to be distributed uniformly on the perimeter.

If the hopper aspect ratio a/b is large, the effect of the end walls in supporting the hopper contents is small and the forces on the long walls can be determined by considering equilibrium of a unit length of hopper. Figure 8-21a shows a cross section of a hopper with the weights W_1, W_2, and W_3 of three segments of the hopper contents. The forces acting on the two walls

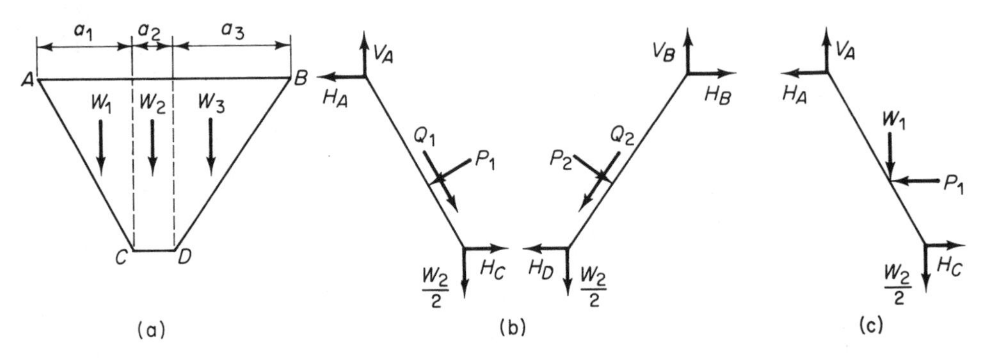

Figure 8-21 Forces in long wedge-type hopper.

are shown in Fig. 8-21b, where P and Q are the normal and frictional components of the pressures on the wall. The three reactive forces on each wall can be determined with P and Q known. P and Q can be determined by any of the theories of granular pressure discussed in Chapter 5 by integrating the pressures p' and q' given by Eqs. (5-22) and (5-23). The simpler Eqs. (5-24) and (5-25) correspond to the Rankine pressure field, for which the horizontal resultant pressure P and the weight W can be used instead (Fig. 8-21c). This is the procedure commonly used. If the hopper is part of a bin, the additional loads $W_4 = \gamma y(a_1 + a_2/2)$ and $W_5 = \gamma y(a_3 + a_2/2)$, where y is the height of contents above the hopper, must be included in calculating the reactions.

If the hopper is unsymmetrical, H_C and H_D are not equal, and the hopper floor must equilibrate the difference. It does this as a horizontal beam supported by the end walls.

Wall bending due to normal pressures. The thickness required to resist normal-pressure bending of the walls of pyramidal hoppers can be determined by using an equivalent rectangle.[24] The equivalent rectangle for a triangular wall is shown in Fig. 8-22a. The equivalent rectangle for a trapezoidal wall (Fig. 8-22b) for which $a_2/a_1 < 4$ is given by

$$a_{eq} = \frac{2a_2(2a_1 + a_2)}{3(a_1 + a_2)} \tag{8-33a}$$

$$b_{eq} = h - \frac{a_2(a_2 - a_1)}{6(a_1 + a_2)} \tag{8-33b}$$

If $a_2/a_1 \geqslant 4$, the equivalent rectangle is determined as for the triangle formed by extending the sloping sides to their intersection.

With the equivalent rectangle known, the plate thickness is determined by Eqs. (6-55) with k_y from Eq. (6-51).

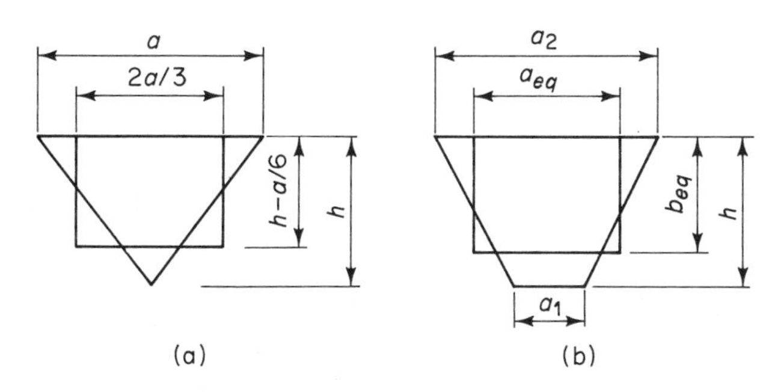

Figure 8-22

8-12. Rectangular-Bin Walls

Wall plate is usually supported on stiffeners, and may be designed for one-way or two-way bending, depending on the arrangement and spacing of the stiffeners. Either the small-deflection theory (Sec. 6-17) or the large-deflection theory (Sec. 6-19) may be used. The difference is not large for relatively thick plates but may be considerable for relatively thin plates. Bin walls are also subjected to direct tension from the pressures normal to the walls framing into them and to compression from friction of the contents on the wall.

Stiffeners are designed for the load from the tributary area of plate using an effective width of the plate as part of the cross section. Stiffeners may also be subjected to direct tension.

Walls of rectangular bins which are supported on columns transfer the load to the columns in a manner which depends on the depth of the bin relative to the column spacing. Bins and bunkers of the type shown in Fig. 1-1a may have walls much higher than they are wide, but they are also built with steep-walled hoppers attached to shallow vertical walls. Bunkers with wedge hoppers may also have vertical walls covering a wide range of span-depth ratios. Therefore, there are many cases for which the stresses are not readily calculable, because the beam formula $f = Mc/I$ does not apply for span-depth ratios less than about 1.5.

Bins with shallow walls. Analysis of the vertical girder of a shallow-walled bin requires simplifying assumptions since the hopper plate participates in the beam action. However, design is often based on the assumption that the girder acts independently of the hopper, loaded with the hopper membrane forces and the stored-solid horizontal pressures above the hopper.

Composite action of a hopper bunker with shallow vertical walls and a deep wedge-type hopper was investigated by strain-gage measurements on a $\frac{1}{16}$ scale model.[25] The bunker was 25 ft wide by 59 ft long with an 8-ft vertical wall and a hopper 27 ft deep. The model was loaded with the hopper wall attached and also with it removed. Stresses were calculated for a cross section consisting of only the girder and for one which included 400 thicknesses of hopper plate, using the simple beam formula with the principal axes of the cross sections. The calculated values were in reasonably good agreement with the experimental ones in both cases. Girder stresses were considerably smaller in the composite section. For economy in design this composite action should be taken into account. The result is a beam of uniform depth if the hopper is of the wedge type and one of variable depth if it is pyramidal. There will also be composite action in beam-supported bunkers without vertical walls. Since there is only the one experimental verification of the effective width, a more conservative value than $400t$ should be used. The authors suggest $200t$ but not more than half the depth of the hopper wall.

Bins with deep walls. Finite-element analyses of deep-walled steel coal bunkers show that the principal vertical-load transfer to the columns is just above the bend line.[26] This is in agreement with experimental investigations of the beam behavior of reinforced-concrete walls. Deep-walled bins and bunkers are usually designed without an analysis of the bending behavior unless a finite-element analysis is made. Instead, a horizontal beam at the bend line, usually a wide-flange or T section with its web perpendicular to the vertical wall or the hopper wall, depending on its position relative to the bend line, is designed for the reactive forces from the vertical-wall stiffeners and the hopper-wall stiffeners. Only reactions normal to the major axis of the beam are considered.

Folded-plate analysis. The combined action of adjoining walls of intermediate and large span-depth ratios can be taken into account by a folded-plate analysis. Since the bin plates are relatively thin they are stiff in their own planes but flexible perpendicular to them, and under the normal pressures from the stored solid they transfer load in the manner of a folded-plate roof. The plates are assumed to resist shear, but not in-plane bending, so there must be a member at each fold to act as a beam flange. The in-plane loads are found by computing the fold-line reactions from load perpendicular to the plates, and resolving them into components parallel to the planes of the plates intersecting at each fold.[27,28]

Allowable plate stress. Since there are several sources of stress in the wall plates, combinations which it is reasonable to consider, and the allowable values of such combinations, must be established. The most likely maximum is a combination of stress due to bending from the pressures of the solid plus stress due to load transfer to the columns (girder action) plus tension from pressure on the walls perpendicular to the wall under consideration. Maximum values of these three do not occur at the same point, and even if the maximum sum can be determined, it is at a point on the surface of the plate. It is extravagant to require that such a maximum be less than an allowable stress prescribed by a standard specification, because, in general, such allowables are intended for situations where yielding may initiate over an entire area, as in a tension member, or along a line, as in the flange of a beam at the section of maximum moment, rather than at a point on a surface.

The following allowable stresses for bending of a transversely loaded plate supported on all four edges were suggested in Secs. 6–17 and 6–19: $0.75F_y$ and $0.9F_y$ for hinged and fixed edges, respectively, using the small-deflection theory, and $0.65F_y$ and $0.9F_y$ for hinged and fixed edges, respectively, using the large-deflection theory. These allowable stresses do not provide for stresses from other sources. The authors suggest that such provision be made by reducing them and that the plate be designed only for the bending due to the normal pressure p, using the large-deflection theory as a more realistic representation of plate behavior in a bin. The following allowable values are suggested:

Fixed-edge plate in column-supported bin:	$0.75F_y$
Fixed-edge plate in skirt-supported bin:	$0.90F_y$
Hinged-edge plate in column-supported bin:	$0.50F_y$
Hinged-edge plate in skirt-supported bin:	$0.65F_y$

It will be noted that no reduction is made for a skirt-supported bin. This is because (1) there is no in-plane bending of the wall, and (2) the maximum stress by the large-deflection theory includes a uniform tension f_a which is adequate to equilibrate the pressure on the walls perpendicular to the wall under consideration. The allowance for column-supported bins may appear small, but it should be noted that the maximum values of the two bending stresses involved are not likely to occur at the same point, while the tension due to pressure on the perpendicular walls is provided for by the tension f_a, as noted.

A finite-element analysis will give the membrane stresses due to in-plane bending of the bin walls, but the local bending due to the wall pressure p is not determined because this pressure is converted into equivalent loads at the nodes. The local-bending stress is readily determined by using Tables 6–5 and 6–6. However, maximum values of these two stresses are not generally at the same point. The local-bending stress is maximum near the midpoint of the panel if the plate is simply supported and near the midpoint of the stiffeners if it is fixed-edge-supported. It is suggested that the local-bending stress be added to the finite-element stress midway between horizontal stiffeners. Also, as noted, the local-bending stress includes a uniform tension f_a, so that tension due to pressure on the walls perpendicular to the wall being investigated need not be considered. It is also suggested that allowable values be taken at $0.9F_y$ for the combined stress and $0.65F_y$ for the in-plane bending stress at the horizontal stiffeners.

Multicell bins. Elastic analysis shows that for multicell bins with continuous plating and horizontal stiffeners moments at the centers of the walls and stiffeners are larger if fully loaded cells alternate with empty ones than if all cells are fully loaded. Also, bending moments at the junctions of adjoining walls and horizontal stiffeners are greater if the cell on each side is loaded and the remaining cells are alternately full and empty. This is the same as the effect on beam moments of the so-called checkerboard loading in buildings. While this can be an important consideration in proportioning reinforcement in the walls of reinforced-concrete bins, it need not be investigated in multicell bins of steel, because the ultimate strength of a cell wall or stringer is not attained until plastic hinges develop at midpanel and the supports. This mechanism develops independently of the manner in which adjacent walls may be loaded. Therefore, elastic-analysis bending moments of a uniformly loaded fixed-end beam can be used to design the walls and horizontal stringers of a multicell bin, taking into account the plastic strength by using a larger allowable stress as described in Sec. 6–17 and 6–19, without any consideration of the manner in which other cells may or may not be loaded.

8-13. Temperature Stresses in Cylindrical-Bin Walls

Except near the bottom of ground-supported flat-bottom bins, the shell of an empty cylindrical bin develops little or no hoop stress from temperature changes, because the bin diameter can change freely to accommodate expansion or contraction of the wall. However, if the bin is not empty, the stored solid resists contraction of the shell, and a temperature drop will produce hoop tension. If the stored solid were incompressible, the hoop tensile stress would be αET, where α = the coefficient of expansion, E = the modulus of elasticity, and T = the temperature drop. The reduction in this value due to radial compaction of the solid can be determined if the stress-strain properties of the solid are known. The pressure p_h on a cylindrical shell due to a temperature drop T is given by [29]

$$p_h = \frac{\alpha ET}{r/t + (1 - \mu)E/E_1} \tag{8-34}$$

where r = radius of cylinder
t = thickness of cylinder wall
μ = Poisson's ratio of solid
E_1 = modulus of elasticity of solid

The resulting hoop tensile stress $f = p_h r/t$ is

$$f = \frac{\alpha ET}{1 + (1 - \mu)(E/E_1)(t/r)} \tag{8-35}$$

Increases in the hoop tensions in two steel bins 86 ft in diameter by 64 ft high, one empty and one filled with wheat, were measured on 4 days during overnight temperature drops of about 50°F on each of four days.[30] The lower plate was $\frac{1}{2}$ in. thick, and strain gages were installed 6 ft above the base ring. The temperatures ranged from 102° to 118° at about 4 p.m. and from 47° to 59° at about 7 a.m. Stresses were -145, +174, +232, and +116 psi in the empty bin for temperature drops of 52°, 49°, 54°, and 47°, respectively, and +3959, +2792, +3256, and +2673 psi in the full bin for drops of 59°, 53°, 59°, and 50°, respectively.

The modulus of elasticity of the wheat was determined using procedures essentially as outlined in ASTM Designation D1883-51T, Tentative Method of Test for Bearing Ratio of Laboratory Compacted Soils. The wheat was placed in a steel cylinder 6 in. in diameter by 7 in. high and compacted to maximum density by vibration under surcharge after which a series of axial loads was applied to the entire surface area by a piston and plunger arrangement in a testing machine. The tests were carried to a compressive stress of 130 psi (18,700 psf). The stress-strain plot was a straight line, except that initially there was a slight decrease in the modulus during the first 20 psi of load. The wheat showed appreciable rebound upon unloading, and the modulus of elasticity remained constant at about 11,400 psi during seven applications of load. The modulus was also determined in a triaxial compression test and found to be 10,400 psi. There was no crushing or damaging of the wheat grains during these tests.

With $E_1 = 11{,}000$ psi, $\mu = 0.4$ (this value is assumed; it was not determined in the tests), $\alpha = 6.5 \times 10^{-6}$, and $E = 30 \times 10^6$ psi, Eq. (8–34) gives $f = 4147$ psi for $T = 55°$. This is in fair agreement with the four strain-gage values quoted above, which averaged 3170 psi for an average temperature drop of 55° (the formula gives $f = 3760$ and 2940 psi for $\mu = 0.3$ and 0, respectively).

It will be noted that these temperature stresses are not particularly large and can ordinarily be covered by the factor of safety, since standard specifications usually permit an increase in the allowable stress if temperature effects are combined with other service stresses. However, if the same temperature drop were to occur at low temperatures, say a fall from 20° to -30°, a brittle-fracture crack might be initiated, which could result in a failure of a vertical joint following a succession of low-temperature cycles unless the steel has adequate notch toughness at these temperatures and the welded joints, particularly the vertical ones, are free of flaws that will result in stress concentrations. Therefore, provision for temperature stresses in bin walls is not necessarily assured by keeping the sum of the temperature stress and the stress caused by pressures from the solid at or below an allowable value but rather by protecting the bin against brittle fracture by choosing a steel that has adequate strength and notch toughness at the expected minimum service temperatures (Sec. 2–1) and by thorough inspection of the weldments.

8-14. Flat-Bottom Bins

The steel bottom of a grade-supported flat-bottom bin should be attached to the shell by continuous fillet welds on both sides. The bottom plate should project at least 1 in. If there is no ringwall, a wider projection, up to about 6 in., can be used to reduce the bearing pressure on the soil. If the bin has a full-area concrete-slab foundation at grade, a steel bottom may or may not be used. If one is not used, a steel ring to which the shell can be welded must be provided. The ring must be embedded in the slab and/or anchored to it to seal the bin against contamination of the contents by outside moisture. The anchorage must be secure enough to resist the tendency of the ring to lift off the slab under wind load when it is empty and from out-of-round deformation of the shell because of nonuniform radial pressures due to banked-up solid after emptying by discharge into a tunnel, or through eccentrically located point openings, or from unevenly banked-up solid following removal by other methods of the solid left after discharge through a central point outlet.

8-15. Design for Earthquake Resistance

Earthquake effects on bins are determined by computing the bending stresses at a section due to the moment of the lateral forces above the section. The elementary beam formula can be used except for bins which are low relative to their diameters. Simplified lateral-force formulas, which do not require an evaluation of the fundamental period of vibration, are usually used. They are discussed in Sec. 5-18.

Allowable stresses may be increased by one-third for seismic effects either alone or in combination with other forces.

8-16. Openings in Bin Walls

Bin walls must be reinforced at openings. The cross-sectional area of the reinforcement should be at least equal to the cross-sectional area of shell lost because of the opening. The concentration of hoop tension immediately above and below an opening tends to flatten a cylindrical wall at these points, while the concentration of vertical compression tends to buckle it adjacent to the other two edges. Therefore, the reinforcement should be shaped to stiffen the edges in the radial direction as well as in the vertical direction. An angle is good for this purpose.

A stress analysis of a framed opening will give a rough approximation at best unless it is by a finite-element procedure. One possibility is to analyze the reinforcement as a rigid frame loaded with the vertical compression and, except where the opening is in a skirt support, the hoop tension. An effective width of shell can be considered as part of the cross section in computing the required area and the section properties for a bending analysis. Since both the hoop tension and the vertical compression concentrate in the regions adjacent to an opening, the calculated stress in such an analysis can be considered acceptable at a small factor of safety relative to the yield point.

Openings at the bottom of a skirt or a flat-bottom bin require careful consideration, particularly if the floor is a concrete slab with an embedded ring plate to which the shell is attached. Such a ring must be anchored to the slab, and if it is interrupted at the opening there must be anchor bolts at the ends.

8-17. Concentrated Loads on Bin Shell

Concentrated loads at the top of a bin shell must be distributed into the shell sufficiently to prevent buckling. This is usually accomplished by using a stiffener. Figures 7-18e and 7-18f show a rafter seat consisting of a $6 \times \frac{1}{2}$ in. ring, which is part of the tension ring, with triangular stiffeners at each rafter. The required length of stiffener is usually determined by assuming the load to radiate into the shell from the point of support; an angle of spread of 30° on each side of the stiffener is suggested in Ref. 6, but 45° is also used. The length l of the stiffener must be such that the compressive stress in the shell, assumed uniformly distributed on the area $2tl \tan 30°$ (or 45°), does not exceed the allowable value according to the buckling formula used in the design of the shell. Rafters for bins of large diameter may require support stiffeners several feet long.

8-18. Corrugated-Sheet Bins

Bins with corrugated steel sheet placed with the corrugations in the circumferential direction usually require vertical stiffeners because the extensional stiffness of the sheet perpendicular to the corrugations is much smaller than the stiffness in the direction of the corrugations. Therefore, hoop tension due to pressure of the stored solid determines the sheet thickness while vertical compression can be considered to be taken entirely by the stiffeners. However, despite the low stiffness normal to the corrugations, the sheet can take some vertical compression. This can be taken into account by using an effective width of sheet as part of the stiffener cross section. An effective-width chart is given in Ref. 31.

Although stiffener buckling is resisted by elastic support from the shell, stiffeners are often designed as unsupported. The stiffened corrugated shell can also be analyzed for buckling under vertical compression by procedures for orthotropic cylindrical shells.[32] Formulas for the extensional and bending stiffnesses involved are given in Ref. 31 and other sources. Analyses of the stress resultants due to nonsymmetric vertical load resulting from wind, eccentric discharge and other nonaxially symmetric loads can also be made as for orthotropic cylinders.

8-19. Recommendations for Design

Many designers and manufacturers do not distinguish between filling and emptying pressures in their design of bins and use the Janssen or Reimbert formulas with values of K and μ' based on their experience, or on laboratory tests of the material, or which they consider to be conservative. However, it should be noted that values that may be considered conservative do not necessarily lead to consistent results. For example, the lateral pressure p_h on a bin wall increases with a decrease in the coefficient of friction μ', which is conservative in computing hoop tension, but the corresponding vertical compression in the wall decreases, which is unconservative with respect to checking the wall for vertical buckling.

The Janssen formulas and the Reimbert formulas are generally used. The value $K = 0.4$ appears to be common, with μ' determined by experience or experiment or chosen from tables of material properties.

Funnel-flow bins. Both DIN 1055 and ACI313-77 specify filling and emptying pressures, DIN by the Janssen formulas with one set of values of K and μ' for filling and another for emptying, and ACI by either the Janssen or the Reimbert formulas with emptying pressures to be determined by multiplying filling pressures by overpressure coefficients discussed in Sec. 5-5.

Comparisons with some test results in Fig. 5-14 and Table 5-6 suggest that DIN and ACI filling pressures, p_h in particular, do not differ significantly. Comparisons of emptying pressures in Fig. 5-15 and Table 5-7 show the vertical pressure p_v to be larger by the ACI, which according to Eq. (5-20) should give a smaller vertical compression F_c in the wall, but F_c by the ACI is larger. The reason for this discrepancy is explained in Example 5-2. It should be noted that vertical reinforcement in reinforced-concrete bins is usually established by minimum requirements rather than by vertical compression in the wall, while wall thickness in steel bins is usually determined by vertical buckling of the wall rather than by the hoop tension. Therefore, differences in predicted values of vertical compression are not as significant for reinforced-concrete bins as for steel bins, and since experience with DIN 1055 covers a number of years, the authors recommend it for steel funnel-flow bins.

Shallow bins. Shallow bins may be designed for Janssen, Reimbert, DIN filling pressures, or Rankine active pressures without regard to differences between filling and emptying pressures (Sec. 5-9).

Mass-flow bins. Tests discussed in Sec. 5–15, results of which are shown in Figs. 5–24 and 5–25, suggest that good estimates of filling pressures p_h in the cylinder are given by the Janssen formula with $K = 0.4$ or the DIN values of K and μ' and, somewhat more conservatively, by the Walker formulas, Eqs. (5–30). Emptying pressures are underestimated by Walker's formula but are in reasonably good agreement with Janssen's formula using the DIN values of K and μ' for emptying or the ACI overpressure coefficients. The Jenike-Johanson theory (Sec. 5–14) gives the largest values, and Fig. 5–24 suggests that they may be overly conservative. The authors believe the results of Fig. 5–24 to be more reliable than those of Fig. 5–25, because the latter are reduced values of the experimental pressures. The reason for the reduction is explained in Sec. 5–15. Therefore, it appears that both DIN and ACI give satisfactory estimates of the lateral flow pressures in mass-flow cylinders, while the Jenike-Johanson formula may be used for more conservative results.

Calculated values of the vertical compression F_c in the cylinder wall in the tests referred to are given in Tables 5–8 and 5–15 for the Walker and the Jenike-Johanson theories. They are, respectively, 441 and 603 plf at the bottom of the cylinder. The ACI procedure (Sec. 5–10) gives 1103 plf, using the Jenike-Johanson value of the Janssen pressure p_v in Table 5–14. This is certainly excessive, particularly since the Jenike-Johanson value is a theoretical upper bound (Sec. 5–14). Therefore, the authors recommend that the Jenike-Johanson formulas be used for the cylinder of steel mass-flow bins. Vertical buckling usually determines the wall thickness, so overestimated values of p_h are of little consequence.

Mass-flow hoppers. Jenike's formula for hopper filling pressures is in good agreement with test results, especially those of Fig. 5–24. Jenike's emptying-pressure formula is also in good agreement, as is Walker's formula. Jenike's formula appears to be somewhat more conservative and requires a coefficient to be read from a chart. Therefore, the authors recommend Walker's formula for the emptying pressures and Jenike's for the filling pressures. It should be noted that emptying pressures are larger at the transition than filling pressures, while the reverse is true in the lower part of the hopper.

The high pressures at the transition of a mass-flow hopper are especially important in hoppers with bolted seams because of the possibility of an unbuttoning failure precipitated by failure of one or two bolts in the region of highest stress. Therefore, bolt spacing at the transition should be carefully checked.

Vertical buckling. Allowable vertical-buckling stresses are usually determined by formulas for buckling under uniform axial compression, with no allowance for the increase in buckling strength due to lateral pressure from the stored solid. However, Fig. 8–4 and the discussion in Sec. 8–2 suggest that recognition of this effect is justified, and the ECCS Eq. (8–7) is recommended for this purpose. However, this formula, or any formula for the increase in buckling strength due to pressure of the stored solid, should not be used unless the properties of the solid are sufficiently well established to enable the hopper outlet to be proportioned to prevent arching and the wall pressures to be predicted with some confidence. Lateral support of the bin wall below an arch may be lost during discharge and vertical compression in the wall increased because of the loss of direct support of the solid above the arch. Lateral support is also questionable in bins with eccentric outlets. Finally, lateral support should not be taken into account if the ECCS tolerances shown in Fig. 6–13 are not likely to be satisfied.

Corrosion allowance (see also Sec. 3–8). Bin walls may experience a gradual loss in thickness due to corrosion and wear. Corrosion depends on the chemical characteristics of the material to be stored and its moisture content; it may be nonexistent for some materials but with others may even require protective coatings of the walls. Wear results from sliding of the material along the walls and, of course, depends on the abrasive characteristics of the material and the frequency with which the bin is charged and discharged. In general, an allowance for these effects should be made except where they are known from experience to be negligible, or where the material is known to be nonabrasive, and inert in respect to producing corrosion. An allowance of $\frac{1}{16}$ in. is commonly used. Smaller allowances can be made because plates to $\frac{1}{2}$ in. thick are available in increments of $\frac{1}{32}$ in. and sheet metal in increments of about $\frac{1}{64}$ in. However, some of these thicknesses may not be readily available.

8–20. Fabrication and Erection*

Plans and specifications should be prepared so as to maximize options in fabrication and erection. This will enable the fabricator and erector to choose options that are best suited to their methods, equipment, experience, and talents and is likely to result in more bids and lower cost to the owner. For example, a specification may give the properties of the bulk solid to be handled, storage capacity, type of flow, ranges of diameters, heights and other dimensions, environmental loads, soil information, zoning requirements, etc., and leave it to the fabricator to carry out the design so as to best accommodate its and the erector's methods. This is an example of a *per-*

*Material for this section was supplied by Reece Stuart III of the Pittsburgh-Des Moines Corporation.

formance specification, which generally maximizes the fabricator-erector options.*

If bids from fabricators who are not staffed to perform structural design are desired, a more or less specific set of plans and specifications must be provided. Even in this case, however, cost-saving options can be provided. For example, some requirements may be mandatory, while others may be satisfied by "approved equal" or "manufacturer's or fabricator's standard." Also, if shell-plate thicknesses are shown on the plans, they can be specified as "minimum," which allows the fabricator to "round up" odd plate thicknesses to the next even gage. Depending on how near the specified thickness is to an even gage as defined by mill pricing practice and depending on the fabricator's inventory, it may be more economical to round up the thickness to the next even gage. As another example, one fabricator may be well set up to produce tubular columns, while another, although able to produce them, may find it more economical, even at a weight penalty, to provide rolled wide-flange columns. If either type will satisfy the purchaser's requirements, the plans and specifications should leave the choice to the fabricator.

A third approach, which provides less freedom than a pure performance specification but more than a fully detailed design with "or equal" options, involves fully detailed plans and specifications accompanied by an invitation to the fabricator to suggest changes to improve economy of fabrication and erection subject to the designer's approval.

A designer who is not familiar with fabrication and erection practices, plant or erection-equipment limitations, costs of alternative design details, preferences as to plate layouts, and the like should consult with fabricators and erectors, who will usually welcome the opportunity to help with plans and specifications that will result in an economical structure.

Cylindrical shells. Heights of shell rings and the lengths of the individual plates in the rings should be selected for maximum economy of fabrication and erection. Wide plates mean fewer pieces to fabricate and erect and fewer seams to weld and inspect, while narrow plates tend to reduce the total weight of the shell. Furthermore, the economical ordering width (mill cost) usually depends on the thickness and might be, for example, over 72 in. through 90 in. for a particular thickness. There is no formula to balance out the economic factors that affect ring-width economy. For plates $\frac{1}{4}$ in. and up, many fabricators have standardized on shell rings from 90 to 100 in. high, with 96 in. a width likely to be found in the stock of many suppliers.

*Author's note: Bids solicited under this type of specification can result in competitive designs that are not equal in performance. Differences in factors of safety in respect to one or more failure modes may also be unacceptably large unless methods of calculating bin loads, shell buckling, and other items not otherwise covered by reference to standard specifications are stated in the invitation to bid.

Economical widths of thinner plates tend to be narrower. The length of the individual plates making up a ring is of little importance to the designer but may be quite important to the fabricator or erector, depending on shop equipment, field-handling capacities, etc.

In judging the relative economy of stiffened and unstiffened shells, the sizes of the stiffeners, their locations relative to plate seams, and their effect on costs of fabrication must be considered. The size and location of the stiffeners may, for instance, materially affect the practicability of the use of automatic welding equipment on the shell seams.

Shell-rim stiffening. Open-top cylindrical bins or hoppers must have rim stiffeners for wind resistance. The AWWA D100 and API 650 rules can be used for proportioning such stiffeners. These requirements can be modified if the structure is protected from wind, as when it is enclosed in a process building. Even in the more usual case where the bin is outdoors and has a roof, the shell must be protected from wind damage when it has been erected but the roof is not yet attached. This responsibility may fall mainly on the erector, but most designs provide for a rim angle or stiffener of some sort, often erected at the time the top ring is in place, to help make the shell round and provide some stiffening while the roof is being erected. If the structure must be left without a roof during construction, overnight, or over a weekend, the erector may employ additional shell-top stiffening elements, such as temporary stiffening rings, cables to ground anchors, or the like.

Roof structures. Some producers of open-web steel joists have computer programs that enable them to provide designs for joist-supported conical roofs for several standard slopes and several live loads and for almost any bin diameter up to 200 ft or more. The bin fabricator supplies the compression ring, connections to the shell, and the roof sheeting or plates, with the joist manufacturer supplying the joists and bridging. This roof structure is easy to erect and is very cost effective.

Connections of the rafters to the center ring of self-supporting conical roofs must be capable of preventing rotation of the compression ring when the ring's erection support is removed. Such structures have been known to collapse upon removal of the temporary support because of rotation of the ring, which allows the roof structure, not yet stabilized by roof sheeting or other attachments, to turn inside out. Prevention of rotation is especially important if the ring supports a heavy load such as an overhead conveyor gallery.

The most critical time in the life of a spherical-dome roof may well be during the erection of stiffeners or temporary supports, before the roof plates or sheets are erected. In the case of the stiffened dome the lateral stability of the radial stiffeners may depend on their being attached to the roof plates or sheets. If, in addition, the plates or sheets are designed to act compositely with the radial stiffeners, there is little vertical strength or stiffness until the stiffeners and sheets are connected to each other. Since economical plate layouts may result in plate seams that have no relationship to the stiffener locations, these connections must often be made "in the air" after the stiffeners have been erected on temporary supports and connected to one another so that the roof sheets or plates can be laid on top.

An erection technique that has been employed successfully in the erection of thin-plate dome roofs is the construction of the entire roof inside the shell, near the bottom, after which it is winched or "blown" to the top of the shell, where it is attached. In the blowing technique, the roof edge is fitted with a flexible seal that wipes the inside of the shell while a high-volume, low-pressure fan-type blower near the bottom of the shell provides the pressure (a few inches of water) to raise the roof. A system of cables and pulleys is used to keep the roof level during this operation. The winching or blowing techniques allow the light structural members and roof plates to be erected near the bottom of the bin, where it is much easier to provide shoring. Furthermore, the cranes or other machines used to erect the roof can be smaller, and the under side of the dome is more accessible for work to be done on it. This type of erection often results in a better-shaped, more economical roof.

Example 8-1

The bin of Fig. 8-23 is designed to hold 1800 tons of stoker coal. The $f_c p_1$ and γp_1 curves are given in Fig. 8-24. The effective angle of friction δ is 50° and is independent of p_1. The angles of internal friction and of friction on steel are $\phi = 42°$ and $\phi' = 35°$.

Roof live load 25 psf, wind load 30 psf, shape factor 0.6 (Table 5-20). Use DIN 1055 and, where applicable, the AISC specification. A283 Grade C steel for shell.

Roof

The roof is self-supporting conical with a base angle $\tan^{-1} 12/18 = 33.7°$. Assuming plate $\frac{1}{4}$ in. thick, the roof load is 25 + 10.2 = 35.2 psf. From Eq. (7-14),

$$\frac{t}{D} = \frac{(35.2 \tan 33.7°)^{0.4}}{16{,}000 \sin 33.7°} = \frac{1}{2512}$$

$$t = \frac{36 \times 12}{2512} = 0.172$$

Use $\frac{3}{16}$-in. plate (7.65 psf).

Rim angle

From Eq. (8-23),

$$A = \frac{D^2}{3000 \sin \theta} = \frac{36^2}{3000 \sin 33.7°} = 0.78 \text{ in.}^2$$

Use a 3 × 3 × $\frac{1}{4}$ angle, $A = 1.34$ in.^{2}

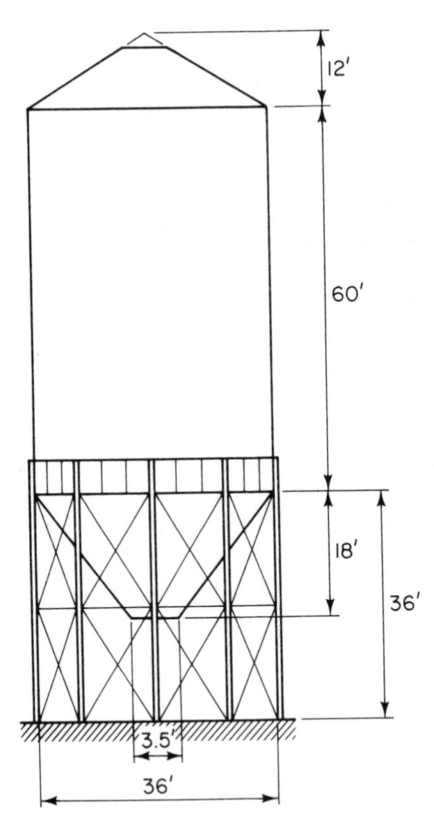

Figure 8-23 Bin for Example 8-1.

Hopper slope

The hopper slope $\theta = \tan^{-1}$ 16.25/18 = 42°. With $\phi' = 35°$ this gives funnel flow (Fig. 4-20).

Outlet (Sec. 4-13)

From Fig. 4-17, $ff = 3.0$ for $\phi = 42°$ and $\delta = 50°$. In Fig. 8-24 the plot of $f_c = p_1/ff$ intersects the entended f_c curve at $f_c = 30$ psf. The corresponding value on the extended γ curve is 40 psf.

From Fig. 4-16, $G(\phi) = 3.6$. Then from Eq. (4-6), $d = G(\phi)f_c/\gamma = 3.6 \times 30/40 = 2.7$ ft. Use 3.5 ft. This is the diameter needed to prevent piping, and it is also large enough to prevent arching (Sec. 4-9).

The 3.5-ft opening should be checked for rate of discharge if this is an important consideration (Sec. 4-15).

Cylinder wall (Sec. 5-5 and Table 5-3)

The formulas needed are Eqs. (5-8a) and (5-8b), (5-10), and (5-20):

$$p_v = \frac{\gamma R}{\mu' K} f\left(\frac{\mu' K y}{R}\right)$$

$$p_h = \frac{\gamma R}{\mu'} f\left(\frac{\mu' K y}{R}\right) = K p_v$$

$$c_1 = 1 + 0.2\left(c_2 + \frac{eC}{1.5A}\right)$$

$$F_c = R(\gamma y - p_v)$$

where R = hydraulic radius = $D/4$. Values of $f(\mu' Ky/R)$ are given in Table 5-1.

For filling pressures, $\phi' = 0.75\phi = 0.75 \times 42° = 31.5°$, $\mu' = \tan\phi' = 0.613$, $K = 0.5$ (Table 5-3).

For emptying pressures, $\phi' = 0.60\phi = 0.60 \times 42° = 25.2°$, $\mu' = \tan\phi' = 0.471$, $K = 1$ (Table 5.3).

In evaluating c_1, $c_2 = 0$ for coal and since the outlet is concentric, e = 0. Therefore, $c_1 = 1$. With the weight of coal at 50 pcf (Fig. 8-24) and $\mu' K/R = 0.613 \times 0.5/(36/4) = 0.0339$, the filling pressure $p_{hf} = 738f\ (0.0339y)$. This is less than the emptying value, but it is needed to determine the allowable vertical compression by the ECCS formulas which take the beneficial effect of hoop tension into account. Values are given in Table 8-2.

For emptying pressures $\mu' K/R = 0.471 \times 1/\frac{36}{4} = 0.0523$, which with $\gamma = 50$ pcf gives $p_{ve} = 956f(0.0523y)$. With $K = 1$, $p_{he} = p_{ve}$. Values are given in Table 8-2.

The membrane stress resultants $N_y = F_c$ + roof load + shell weight are given in Table 8-3. The corresponding stresses $N_y/12t$ and the allowable values $f_{cr}/1.5$ are

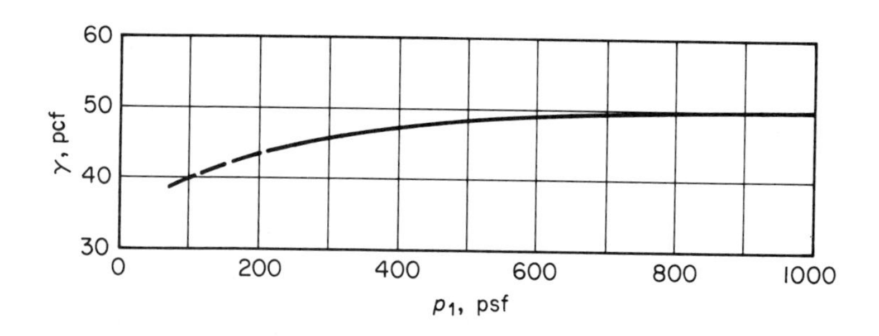

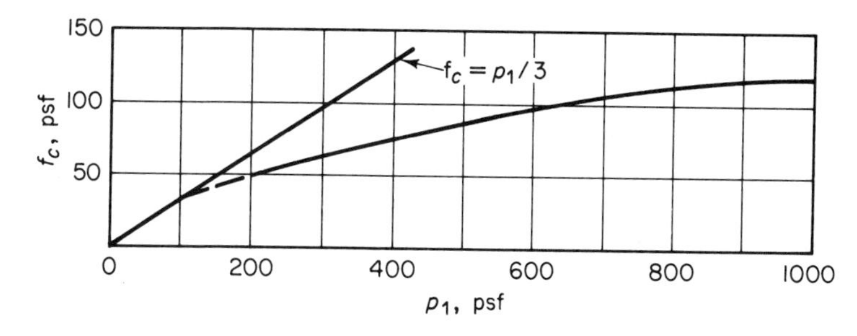

Figure 8-24

TABLE 8-2. PRESSURES FOR BIN OF EXAMPLE 8-1

	Filling			Emptying			
y (ft)	$\frac{\mu' K y}{R}$	$f\left(\frac{\mu' K y}{R}\right)$	p_h psf	$\frac{\mu' K y}{R}$	$f\left(\frac{\mu' K y}{R}\right)$	p_v psf	p_h psf
11.5	0.390	0.326	240	0.601	0.452	432	432
19	0.644	0.474	350	0.994	0.630	602	602
26.5	0.898	0.592	436	1.39	0.750	717	717
34	1.15	0.683	504	1.78	0.831	794	794
41.5	1.41	0.755	557	2.17	0.886	847	847
49	1.66	0.809	597	2.56	0.923	882	882
56.5	1.92	0.854	630	2.95	0.947	905	905
64	2.17	0.886	654	3.35	0.965	923	923

also given. Values of f_{cr} are computed by the ECCS formulas, Eqs. (8-7). Thus, at $y = 64$ ft,

$$\frac{r}{t} = \frac{18 \times 12}{0.312} = 691$$

$$C = \frac{0.315}{\sqrt{0.1 + 0.01 \times 691}} = 0.119$$

$$\rho = \frac{654/144}{30 \times 10^6} \times 691^{3/2} = 0.00275$$

$$C_p = 0.119 + (0.45 - 0.119)\frac{0.00275}{0.00975} = 0.212$$

$$f_{cr} = \frac{0.212 \times 30{,}000}{691} = 9.20 \text{ ksi} < \frac{3}{8} \times 30$$

The filling value of p_h is used to compute ρ for reasons discussed in Sec. 8-2. With the factor of safety of 1.5 the allowable vertical compression is 6.13 ksi.

Hoop tensions, which are largest during emptying, are also given in Table 8-3. However, the computation at $y = 64$ ft does not account for the effect of the meridional tension in the hopper and so is incorrect. The hoop stress at this section is compression; it is computed later in this example.

Thicknesses of the shell at $y = 11.5$ and 64 ft according to Eq. (8-3) and the ECCS Eqs. (8-5), both of which neglect the increase in vertical-buckling strength due to lateral pressure of the bin contents, compare with the corresponding thicknesses from Table (8-3) as follows,

y, ft	Table 8-3	Eq. (8-3)	Eqs. (8-5) ÷ 1.5
11.5	0.125	0.125	0.156
64	0.312	0.375	0.406

TABLE 8-3. STRESSES IN BIN WALL FOR EXAMPLE 8-1

y (ft)	γy psf	F_c[a] (kips/ft)	Roof $DL + LL$ (kips/ft)	Shell (kips/ft)	N_y (kips/ft)	t[b] (in.)	$N_y/12t$ (ksi)	$f_{cr}/1.5$ (ksi)	N_h[c] (kips/ft)	$N_h/12t$ (ksi)
11.5	575	1.29	0.29	0.08	1.66	0.125	1.11	2.45	7.78	5.19
19	950	3.13		0.15	3.57	0.125	2.38	2.84	10.8	7.20
26.5	1325	5.47		0.23	5.99	0.156	3.19	3.20	12.9	6.89
34	1700	8.15		0.31	8.75	0.188	3.88	4.06	14.3	6.34
41.5	2075	11.1		0.38	11.8	0.219	4.49	4.60	15.2	5.78
49	2450	14.1		0.46	14.9	0.25	4.97	5.14	15.9	5.30
56.5	2825	17.3		0.54	18.1	0.281	5.37	5.69	16.3	4.83
64	3200	20.5	0.29	0.65	21.4	0.312	5.71	6.13	16.6	4.43

[a] $F_c = R(\gamma y - p_{ve})$, where the subscript e denotes emptying.
[b] Thicknesses must be increased if allowance for corrosion and/or wear is required.
[c] $N_h = rp_{he}$.

Hopper

At junction with the cylinder,

$$\text{contents} = \frac{\pi\gamma}{24}(D^3 - d^3)\cot\theta = \frac{50\pi}{24}(36^3 - 3.5^3)\cot 42^\circ = 338{,}830$$

$$\text{hopper (assumed } \tfrac{1}{4} \text{ in. thick)} = \frac{36\pi \times 19.94}{2\cos 42^\circ} \times 10.2 = \underline{15{,}480}$$

$$W = 354{,}310 \text{ lb}$$

DIN 1055 requires the pressure p' normal to the hopper wall to be taken at twice the value by Eq. (5–24) using p_v and p_h at filling (Sec. 5–12). From Table 8–2, p_h = 654 psf. Then $p_v = 654/K = 654/0.5 = 1308$ psf. Therefore, $p' = 2(1308 \sin^2 42^\circ + 654 \cos^2 42^\circ) = 1892$ psf, and from Eq. (8–15),

$$N_h = \frac{p'r}{\cos\theta} = \frac{1892 \times 18}{\cos 42^\circ} = 45{,}830 \text{ lb/ft}$$

From Eq. (8–13), with p_v equal to double the filling pressure,

$$N_m = \frac{2 \times 1308 \times 18}{2\cos 42^\circ} + \frac{354{,}310}{2\pi \times 18 \cos 42^\circ} = 35{,}870 \text{ lb/ft}$$

Using the API 650 allowable stress = 0.85 × 21 = 17.9 ksi (Sec. 8–2),

$$t = \frac{45.8}{12 \times 17.9} = 0.213$$

The membrane stress resultant in the hopper is considerably larger than in the cylinder, and both are maximum at the juncture of the hopper and cylinder where there are additional stresses due to bending. Therefore, although a $\frac{7}{32}$-in. plate would give the thickness required by the preceding calculation, use $\frac{1}{4}$ in. instead. Since the hoop stress at this section is actually compressive this computation should be looked upon as a means of establishing a tentative thickness to be verified by an analysis of the localized bending stresses at this section. Thicknesses of the adjacent and lower courses can be based on hoop or meridional tension because of the highly localized nature of the stresses at the hopper-cylinder junction.

Columns

Gravity loads:

$$\text{Roof live load} = \frac{36^2\pi}{4} \times 25 = 25$$

$$\text{contents above cylinder top} = 50 \times \frac{1}{3} \times \frac{36^2\pi}{4} \times 12 = 204$$

$$\text{contents in cylinder} = 50 \times \frac{36^2\pi}{4} \times 60 = 3054$$

contents in hopper = 339

3622 kips

$$\text{weight of roof} = 7.65 \times \frac{\pi}{2} \times 36 \times \frac{18}{\cos 33.7^\circ} = 9.4$$

$$\text{weight of cylinder} = 36\pi \times 60 \times 0.239 \times 40.8 = 66.2$$

weight of hopper = 15.5

91.1 kips

Wind load: An example of design for wind using pressures varying with height as prescribed by ANSI A58.1 (Table 5-19) is given in Sec. 5-16, where it is also pointed out that a uniform 30 psf is often used for moderately high structures. Using this simpler approach and the shape factors of Tables 5-20 and 5-21 gives

		Pressure (kips)	Arm (ft)	Moment (ft-kips)
Roof	30 × 0.6 × 0.5 × 36 × 12 =	3.9	100	390
Cylinder	30 × 0.6 × 36 × 60 =	38.9	66	2567
Hopper	30 × 0.6 × 0.5(36 + 3.5) × 18 =	6.4	29.5	189
Columns	8 × 30 × 2 × 32 × $\frac{14}{12}$ =	17.9	16	286
		67.1		3432

The maximum column load P due to an overturning moment M is given by $P = 2M/Rn$, where R is the radius of the column circle and n is the number of columns. With eight columns on a radius of 18.5 ft,

$$P = \frac{2 \times 3432}{18.5 \times 8} = 46.4 \text{ kips}$$

bin empty plus wind = 91.1/8 - 46.4 = -35 kips (uplift)

bin full, no wind = (3622 + 91.1)/8 = 464 kips

bin full plus wind = 464 + 46.4 = 510 kips

Because of the 33% increase in allowable stress for load combinations involving wind, the critical load is 464 kips.

Assume a connection of the type shown in Fig. 8-16. The cylinder develops very little moment in resisting x-axis rotation of the column in this case, so here the column should be considered hinged with respect to the x axis with most of the moment, say 90%, carried by the column (Sec. 8-10). The column can be assumed fixed with respect to the y axis at this type of attachment.

Try an A572 Grade 50 W14 × 132, for which $d = 14.66$ in., $A = 38.8$ in.^{2}, $S_x = 209$ in.^{3}, $r_x = 6.28$ in., $r_y = 3.76$ in., $r_T = 4.05$ in., and $d/A_f = 0.97$ in.$^{-1}$ Assuming the column is pinned at the base, $K_x = 1$, and because of the bracing strut at midheight (Fig. 8-23), $K_y = 0.5$ (without the strut K_y could be taken as 0.7 because of the rotational restraint noted).

$$\frac{K_x L}{r_x} = \frac{1 \times 36 \times 12}{6.28} = 68.8 \qquad \frac{K_y L}{r_y} = \frac{0.5 \times 36 \times 12}{3.76} = 57.4$$

$$F_a = 21.2 \text{ ksi} \quad \text{and} \quad F_e' = 31.6 \text{ ksi}$$

Because of the strut at midheight, the unsupported length for lateral-torsional buckling is $L/2 = 18$ ft. The W14 $\times$ 132 is compact but does not qualify for the allowable bending stress $F_b = 0.66F_y$ because the unsupported length exceeds $76b_f/\sqrt{F_y} = 76 \times 14.7/\sqrt{50} = 158$ in. = 13 ft. Therefore,

$$\frac{Ld}{A_f} = 18 \times 12 \times 0.97 = 210 \qquad F_b = \frac{12{,}000}{210} = 57 > 0.6 \times 50 \text{ so } F_b = 30 \text{ ksi}$$

$$f_a = \frac{464}{38.8} = 12.0 \text{ ksi} \qquad f_b = \frac{0.9 \times 464 \times 14.66/2}{209} = 14.6 \text{ ksi}$$

The AISC Specification Eqs. (1.6-1), which are the same as Eqs. (6-16) of Sec. 6-4, give

$$\frac{f_a}{F_a} + \frac{f_b}{F_b} \times \frac{C_m}{1 - f_a/F_e'} = \frac{12.0}{21.2} + \frac{14.6}{30} \frac{0.6}{1 - 12.0/31.6} = 1.04$$

$$\frac{f_a}{0.6F_y} + \frac{f_b}{F_b} = \frac{12.0}{30} + \frac{14.6}{30} = 0.89$$

The W14 $\times$ 132 is adequate. (A W14 $\times$ 159 would be required in A36 steel). Also, if the attachment to the bin is made concentric with the cylinder wall, an A572 WF14 $\times$ 74 or an A36 WF14 $\times$ 99 can be used.)

Column attachment

The length of the column attachment must be chosen to keep shear-buckling stresses in the shell within allowable limits. This determines the depth of a ring girder of the type shown in Fig. 8-15a. Using the bottom-course thickness $\frac{5}{32}$ in. (Table 8-3), try $h = 72$ in. with two equally spaced stiffeners between columns. The girder arc-span is $36\pi/8 = 14.1$ ft = 169 in. Then,

$$\frac{h}{t} = \frac{72}{0.3125} = 230 \qquad \frac{h}{\sqrt{rt}} = \frac{72}{\sqrt{216 \times 0.3125}} = 8.76$$

$$a = \frac{169}{3} = 56 \text{ in.} \qquad \frac{a}{h} = \frac{56}{72} = 0.778$$

Yield stress for A283 Grade C steel is 30 ksi. Then from Eq. (8-23b)

$$F_v = \frac{96{,}000}{230^2} \left[\sqrt{1 + 0.0239 \times 8.76^3} + \frac{0.75 + 0.03 \times 8.76}{0.778^2} \right] = 10.5 \text{ ksi} < 0.38 \times 30$$

$$f_v = \frac{464}{2 \times 72 \times 0.3125} = 10.3 \text{ ksi} < 10.5$$

The required stiffener moment of inertia is given by Eq. (8–24),

$$I_s = 2.5ht^3\left(\frac{h}{a} - 0.7\frac{a}{h}\right) = 2.5 \times 72 \times 0.3125^3(1.29 - 0.7 \times 0.778) = 4.09 \text{ in.}^4$$

Use $4 \times \frac{1}{4}$ in. stiffeners on one side of web, $I_s = 0.25 \times 4^3/3 = 5.33$ in.^{4}

Stiffeners are not needed if the bottom-course thickness is increased to $\frac{11}{32}$ in. (0.344 in.) and h is increased to 84 in. Thus:

$$\frac{h}{t} = \frac{84}{0.344} = 244 \qquad \frac{h}{\sqrt{rt}} = \frac{84}{\sqrt{216 \times 0.344}} = 9.74 \qquad \frac{a}{h} = \frac{169}{84} = 2.01$$

$$F_v = \frac{96{,}000}{244^2}\left[\sqrt{1 + 0.0239 \times 9.74^3} + \frac{0.75 + 0.03 \times 9.74}{2.01^2}\right] = 8.16 \text{ ksi}$$

$$f_v = \frac{464}{2 \times 84 \times 0.344} = 8.03 \text{ ksi}$$

Assuming the bottom course to be 8 ft deep, the $\frac{11}{32}$-in. plate weighs $36\pi \times 8 \times 40.8/32 = 1154$ lb more than the $\frac{5}{16}$-in. plate. The 12-in. increase in the depth h results in an additional weight $8 \times 1 \times 132 = 1056$ lb in the columns. The 16 stiffeners needed for the $\frac{5}{16}$-in. plate weigh $4 \times \frac{1}{4} \times 3.4 \times 5.6 \times 16 = 305$ lb. Of course, the choice depends on the extent to which the saving in cost of material is offset by the cost of fabricating and welding the stiffeners. The $\frac{5}{16}$-in. bottom course will be adopted in this example.

Compression ring

In computing the hopper-plate thickness the hopper membrane stress resultant was found to be 35.9 kips/ft. The radial component on the compression ring is $35.9 \sin 42^\circ = 24$ kips/ft, and the ring compression is $24 \times 18 = 432$ kips. Assuming an allowable stress of 16 ksi, the required ring area is $432/16 = 27$ in.^{2}

The effective areas of the $\frac{5}{16}$-in. cylinder plate and $\frac{1}{4}$-in. hopper plate are

$$0.78\sqrt{rt} = 0.78\sqrt{216 \times 0.312} = 6.4 \text{ in.}$$

and

$$0.78\sqrt{\left(\frac{r}{\cos\theta}\right)t} = 0.78\sqrt{\frac{216}{\cos 42^\circ} \times 0.25} = 6.6 \text{ in.}$$

respectively.

The ring section shown in Fig. 8–25 has the same depth as the W14 × 132 column, and the 12 × 1 flange has the same thickness as the column flange. This will enable the ring flange to be welded to the column flange and the ring web to the inside faces of the column flanges. The neutral axis is 5.09 in. from the midline

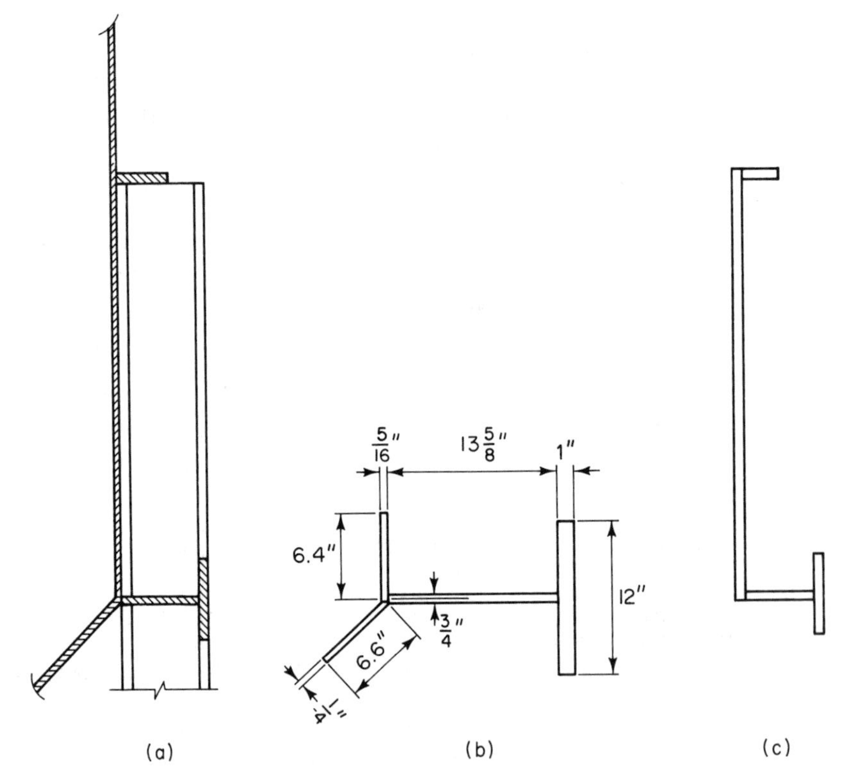

Figure 8-25

of the flange. The area is 26.0 in.^{2} and the moment of inertia 929 in.^{4} Therefore, $r_x = \sqrt{929/26.0} = 5.98$ in., and from Eq. (6-76),

$$\frac{R}{r_x} = \frac{18 \times 12}{5.98} = 36.1 < \frac{367}{\sqrt{36}}$$

$$F_a = 0.67 \times 36\left(1 - \frac{36 \times 36.1^2}{405{,}400}\right) = 21.3 \text{ ksi}$$

$$f = \frac{432}{26.0} = 16.6 \text{ ksi}$$

Ring girder

From Eq. (8-25),

$$M = -464 \times 18\left(\frac{1}{0.785} - \frac{1}{2}\cot 22.5^\circ\right) = -558 \text{ ft-kips}$$

By using a $6 \times \frac{3}{4}$ in. top flange, the girder section is as shown in Fig. 8-25c. The centroid is 22.9 in. above the midline of the $13\frac{5}{8} \times \frac{3}{4}$ plate. The moment of inertia is 36,230 in.^{4} The top-flange stress is the larger:

$$f = \frac{558 \times 12 \times 49.5}{36{,}230} = 9.2 \text{ ksi}$$

The bottom-flange stress is 5.3 ksi compression, which, with the compression-ring stress of 16.6 ksi, gives 21.9 ksi.

Stresses at hopper-cylinder junction

Local bending stresses at the hopper-cylinder transition can be determined by the formulas in Sec. 8-8. Because these formulas neglect the joint-rotation restraint of a ring of the type used here, the results will be approximate. However, they offer some guidance to good proportioning of the cylinder and hopper walls at the transition. It should be noted also that the analysis applies only to the region between columns, and stresses at the column attachments will be somewhat larger (Sec. 8-10).

Values of p_h, p', and M_{mh} for Eq. (8-18), which are already computed, are

$p_h = 654 \text{ psf} = 4.54 \text{ psi}$ $p' = 1892 \text{ psf} = 13.14 \text{ psi}$

$$N_{mh} = 35{,}870 \text{ lb/ft} = 2989 \text{ lb/in.}$$

These are all filling pressures. Then with $t_c = 0.3125$ in., $t_h = 0.25$ in., and $t_s = 0$,

$$T = 0.3125^{2.5} + 0.25^{2.5} = 0.08584$$

$$0.08584a_1 = 0.3125^{1.5}(0.17168 - 0.3125^{2.5}) + 0.3125^2 \times 0.25^2$$

$$a_1 = 0.3093$$

$$0.08584a_2 = 0.25^{1.5}(0.17168 - 0.25^{2.5}) + 0.25^2 \times 0.3125^2$$

$$a_2 = 0.2756$$

The area A_r of the compression ring is $13.75 \times 0.75 + 12 \times 1 = 22.3$ in.^{2} Therefore,

$$f_r\left(2.63 \times \frac{22.3}{\sqrt{216}} + 0.3093 + 0.2756\right) = 216 \times 0.3093 \times \frac{4.54}{0.3125}$$

$$+ 216 \times 0.2756 \times \frac{13.14}{0.25 \cos 42^\circ}$$

$$- 2.63\sqrt{216} \times 2989 \sin 42^\circ$$

$$f_r = -15{,}760 \text{ psi}$$

From Eq. (8-19b),

$$B_c = \frac{-15{,}760}{\sqrt{0.3125}}(0.3093 - 0.3125^{1.5}) - 4.54 \times 216\left(1 - \frac{0.3125^{2.5}}{0.08584}\right)$$

$$- \frac{13.4 \times 216}{\cos 42^\circ} \frac{0.3125^{1.5} \times 0.25}{0.08584} = -6097$$

Since there is no skirt, the bending moment in the hopper wall is equal to the moment in the cylinder wall, so B_h need not be calculated. Therefore,

$$M_c = M_h = 0.289 \times 0.3125(-6097) = -551 \text{ in.-lb}$$

The stress resultant N_y for the cylinder in Table 8-3 is the emptying value. The filling value must be used to be consistent with the values for which f_r and M were computed. This is given by

$$N_y = R(\gamma y - p_{vf}) + \text{roof load} + \text{shell weight}$$

$$= 9(50 \times 64 - 1307) + 0.29 + 0.65 = 18.0 \text{ kips/ft} = 1.50 \text{ kips/in.}$$

where p_{vf} is from Table 8-2 and the roof load and shell weight from Table 8-3.

From Eq. (8-20), for the cylinder,

$$f_m = -\frac{1.50}{0.3125} \pm \frac{6 \times 0.551}{0.3125^2} = -4.80 \pm 33.9 = +29.1, -38.7 \text{ ksi}$$

and for the hopper,

$$f_m = +\frac{2.99}{0.25} \pm \frac{6 \times 0.551}{0.25^2} = +12.0 \pm 53.0 = +65.0, -41.0 \text{ ksi}$$

The hoop stress $f_r = -15.8$ ksi. Then from Eqs. (8-21),

Cylinder f (ksi)		Hopper f (ksi)	
$f_1 = -15.8$	$f_1 = -15.8$	$f_1 = -15.8$	$f_1 = -15.8$
$f_2 = +29.1$	$f_2 = -38.7$	$f_2 = +65.0$	$f_2 = -41.0$
$f_3 = -44.9$	$f_3 = +22.9$	$f_3 = -80.8$	$f_3 = +25.2$

Using the ASME Pressure Vessel Code allowable value $3S_m$ for f, with S_m modified as suggested in Sec. 8-8 for steels not of pressure-vessel quality, the limiting value of f for A283 grade C steel is $3 \times 0.3F_y = 0.9 \times 50 = 45$ ksi. Therefore, the cylinder, for which the largest absolute value of f is 44.9 ksi, is satisfactory, but the hopper is not. The hopper stress can be reduced by increasing t_h or A_r or both. Maximum fs for three values of t_h and two of A_r are as follows:

t_c (in.)	t_h (in.)	A_r (in.^{2})	f_r (ksi)	f_c (ksi)	f_h (ksi)
$\frac{5}{16}$	$\frac{1}{4}$	22.3	-15.8	44.5	80.6
$\frac{5}{16}$	$\frac{5}{16}$	22.3	-15.3	50.2	65.0
$\frac{5}{16}$	$\frac{3}{8}$	22.3	-15.0	53.8	53.3
$\frac{5}{16}$	$\frac{1}{4}$	28.75	-12.6	37.3	70.7
$\frac{5}{16}$	$\frac{5}{16}$	28.75	-13.2	44.7	59.0
$\frac{5}{16}$	$\frac{3}{8}$	28.75	-12.1	44.8	46.3

The combination $t_c = \frac{5}{16}$ and $t_h = \frac{3}{8}$ in. gives a good balance between the two for both ring areas. The ring area 28.75 in.^{2} can be obtained by increasing the web thickness of the smaller ring to 1 in.

Bending stresses at the transition are highly localized. At a distance of about $2.5\sqrt{rt}$ from the transition the moment is zero and is very small (and of opposite site sign) beyond this point.[1] For the bin of this example, these distances are $2.5\sqrt{216 \times 0.3125} \approx 20$ in. for the cylinder and $2.5\sqrt{(216/\cos 42°) \times 0.375} \approx 26$ in. for a $\frac{3}{8}$-in. hopper plate. Thus, there may be cases in which it will be economical to use relatively narrow courses adjacent to the transition, particularly in the hopper, with the adjacent courses sized for the membrane stresses alone.

As noted in Sec. 8-8, local-bending stresses usually are not investigated. This example shows that such an analysis helps avoid situations where local bending may be significant.

Bracing

Use two panels of bracing between adjacent columns (Fig. 8-23). Then with Eq. (8-26) and the wind force $H = 67.1$ kips computed in designing the columns,

$$\phi = \tan^{-1} \frac{18}{14.1} = 51.9°$$

$$T = \frac{2 \times 67.1}{8 \cos 51.9°} = 27.2 \text{ kips}$$

$$A = \frac{27.2}{20 \times 1.33} = 1.02 \text{ in.}^2$$

Use $1\frac{1}{8}$-in. round upset bar, for which $A = 0.994$ in.^{2}

Windward-side buckling (Sec. 8-5)

The weighted average thickness of the 60-ft of bin, from Table 8-3, is $t = 0.207$ in. From Eq. (8-10)

$$\frac{H}{t} = 600 \left(\frac{100 \times 0.207}{36}\right)^{2/3} = 415$$

$$H = 415 \times 0.207 = 86 \text{ ft}$$

Therefore, an intermediate ring girder is not required. According to the wind study discussed in Sec. 8-5, the unstiffened height could be as much as $1.6 \times 86 = 138$ ft.

Example 8-2

The mass-flow skirt-supported bin of Fig. 8-26 is designed for a product with the following properties:

$$\phi = 44°$$

$$\phi' = 32° \text{ on cylinder steel}$$

$$\phi' = 15° \text{ on epoxy-lined hopper}$$

$$\delta = 53^\circ \text{ to } 56^\circ$$

$$\gamma = 48\text{–}51 \text{ pcf}$$

The consolidating pressure curve is shown in Fig. 8–27. The range of both δ and γ is small enough to consider them independent of the consolidating pressure.

The roof live load is 25 psf and the wind load 30 psf. A283 Grade C steel.

Roof

The base angle is $\tan^{-1} 9.5/13 = 36.2^\circ$. By assuming the roof plate to be $\frac{3}{16}$-in. thick, the roof load is 25 + 7.7 = 32.7 psf. From Eq. (7–14),

$$\frac{t}{D} = \frac{(32.7 \tan 36.2^\circ)^{0.4}}{16{,}000 \sin 36.2^\circ} = \frac{1}{2653} \qquad t = \frac{32 \times 12}{2653} = 0.145 \text{ in.}$$

Use $t = \frac{5}{32}$ in.

Rim angle

Using the API formula [Eq. (8–23)],

$$A = \frac{D^2}{3000 \sin \theta} = \frac{32^2}{3000 \sin 36.2^\circ} = 0.58 \text{ in.}^2$$

Use a $3 \times 3 \times \frac{1}{4}$ angle, $A = 1.44$ in.^{2}

Hopper slope

From Fig. 4–18, with $\phi' = 15^\circ$, the hopper apex half angle should be less than about 27° for mass flow. The angle adopted is 25.1° (Fig. 8–26).

Outlet (Sec. 4–9)

Interpolating between Figs. 4–12 and 4–13 with $\phi' = 15^\circ$ and $\delta = 55^\circ$ gives $ff = 1.32$. In Fig. 8–27 the line $f_c = p_1/ff = 0.76p_1$ intersects the curve at about f_c = 35 psf. Then with $\gamma = 48$ pcf Eq. (4–3) gives $d = 2.2 \times 35/48 = 1.60$ ft. Use $d = 2$ ft.

Walker's flow-factor formula, Eq. (4–5b), gives $ff = 1.21$ for which Fig. 8–27 gives $f_c \approx 35$ psf. Equation (4–1) gives $d = 2 \times 35/48 = 1.45$ ft.

The 2-ft opening should be checked for rate of discharge if this is an important consideration (Sec. 4–15).

Cylinder filling pressure p_h

This is smaller than the emptying value, but it is needed to evaluate allowable vertical compression by the ECCS formula. Use the Janssen formula with $K = 0.4$. With $\mu' = \tan 32^\circ = 0.625$, $\mu' K/R = 0.625 \times 0.4/(32/4) = 0.0313$. With $\gamma = 50$ pcf,

$$p_h = \frac{\gamma R}{\mu'} f\left(\frac{\mu' K y}{R}\right) = 640f(0.0313y)$$

Values of f from Table 5–1 give the emptying pressures p_h in Table 8–4.

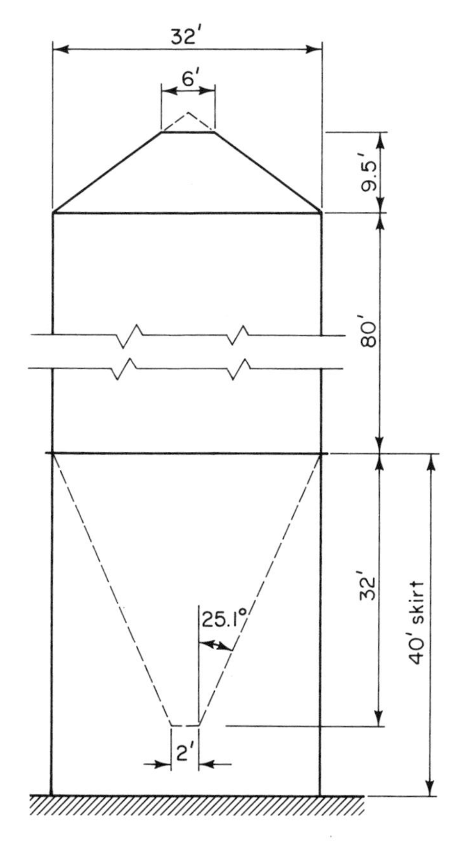

Figure 8-26 Bin for Example 8-2.

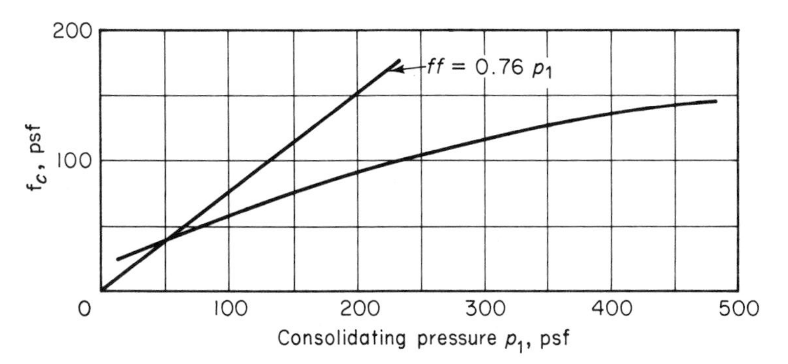

Figure 8-27

TABLE 8-4. PRESSURES FOR BIN OF EXAMPLE 8-2

	Filling			Emptying				
y (ft)	$\frac{\mu' K y}{R}$	$f\left(\frac{\mu' K y}{R}\right)$	p_h psf	$\frac{y}{H}$	$\frac{p_h}{\gamma H}$	p_h psf	$\frac{F_c}{\gamma H R}$	F_c (kips/ft)
12	0.376	0.315	202	0.143	[a]	303	0.082	2.76
20	0.626	0.465	298	0.238	[a]	447	0.147	4.94
28	0.876	0.583	373	0.333	[a]	560	0.217	7.29
36	1.13	0.677	433	0.428	0.244	1025	0.291	9.78
44	1.38	0.748	479	0.524	0.259	1087	0.367	12.3
52	1.63	0.804	514	0.619	0.267	1121	0.441	14.8
60	1.88	0.847	542	0.714	0.268	1126	0.513	17.2
68	2.13	0.881	564	0.810	0.268	1126	0.582	19.6
76	2.38	0.907	581	0.905	0.268	1126	0.643	21.6
84	2.63	0.928	594	1.000	0.268	1126	0.691	23.2

[a]Emptying values of p_h in the upper part of the bin to D below the top taken at 1.5 times the filling values as recommended by Jenike (Sec. 5-14).

Cylinder emptying pressures

Use Jenike-Johanson (Sec. 8-12). With $\mu' KH/R = 0.625 \times 0.4 \times 84/8 = 2.63$, Table 5-10 gives the values of $p_h/\gamma H$ and the resulting values of p_h shown in Table 8-4. Values of F_c/HR from Table 5-12 give F_c, also shown in Table 8-4.

Hopper filling pressures

Use Jenike Eq. (5-39). The distance h from the hopper apex to the transition (Fig. 8-26) is $16/\tan 25.1° = 34.14$ ft. The height h_c of the solid above the transition is 84 ft. Therefore,

$$p' = 50(118.14 - z)\left(1 + \frac{\tan 15°}{\tan 25.1°}\right)^{-1}$$

$$= 31.8(118.14 - z) = 1086\left(3.46 - \frac{z}{h}\right)$$

Values of p' are given in Table 8-5, with the subscript f denoting filling pressures.

Hopper emptying pressure

Use Walker's formulas (Sec. 5-13). The vertical emptying pressure in the cylinder at the transition is needed to compute the hopper emptying pressure. According to Walker's theory, this is the Janssen pressure with $\mu' K$ by Eq. (5-34). Therefore,

$$\epsilon_1 = \frac{\pi}{2} + 32° + \cos^{-1}\frac{\sin 32°}{\sin 55°} = 171.7°$$

$$\mu' K = \frac{\sin 55^\circ \sin 171.7^\circ}{1 - \sin 55^\circ \cos 171.7^\circ} = 0.0653$$

$$p_{vt} = \frac{\gamma R}{\mu' K} f\left(\frac{\mu' K y}{R}\right) = \frac{50 \times 8}{0.0653} f\left(\frac{0.0653 \times 84}{8}\right) = 3040 \text{ psf}$$

where the value of f is from Table 5-1. Then, from Eq. (5-35),

$$\epsilon_2 = \phi' + \sin^{-1} \frac{\sin \phi'}{\sin \delta} = 15^\circ + \sin^{-1} \frac{\sin 15^\circ}{\sin 55^\circ} = 33.4^\circ$$

$$\frac{p'}{p_v} = \frac{1 + \sin \delta \cos \epsilon_2}{1 - \sin \delta \cos(2\theta + \epsilon_2)} = \frac{1 + \sin 55^\circ \cos 33.4^\circ}{1 - \sin 55^\circ \cos 83.4^\circ} = 1.86$$

From Eq. (5-36),

$$K_w = \frac{2}{\tan 25^\circ} \frac{\sin 55^\circ \sin 83.4^\circ}{1 - \sin 55^\circ \cos 83.4^\circ} = 3.85$$

$$p_v = \frac{50 \times 34.14}{2.85} \frac{z}{h} + \left(3040 - \frac{50 \times 34.14}{2.85}\right)\left(\frac{z}{h}\right)^{3.85}$$

$$= 599 \frac{z}{h} + 2441 \left(\frac{z}{h}\right)^{3.85}$$

Values of p_v and p' are given in Table 8-5, with the subscript e to denote emptying pressures, from the transition to the level where they become less than the filling pressures.

Membrane stresses in the bin wall are calculated in Table 8-6. The shell thickness is assumed $\frac{1}{4}$ in. thick throughout in computing the shell weight. Values of f_{cr} are computed as in Example 8-1.

Thicknesses of the shell at y = 12 and 84 ft according to Eq. (8-3) and the ECCS Eqs. (8-5), both of which neglect the increase in vertical buckling strength

TABLE 8-5. HOPPER PRESSURES FOR EXAMPLE 8-2

z(ft)	$\frac{z}{h}$	p'_f psf	p_{ve} psf	p'_e psf
34.14	1	2671	3040	5654
32	0.937	2740	2461	4577
30	0.879	2803	2012	3742
28	0.820	2867	1628	3028
25	0.732	2963	1173	2182
20	0.586	3121		
15	0.439	3281		
10	0.293	3439		
5	0.146	3599		
2.14	0.063	3689		

TABLE 8-6. BIN-WALL STRESSES FOR EXAMPLE 8-2

y (ft)	F_c (kips/ft)	Roof $DL + LL$ (kips/ft)	Shell wt. (kips/ft)	N_m (kips/ft)	t^a (in.)	$N_m/12t$ (ksi)	$f_{cr}/1.5$ (ksi)	$N_h = p_h r$ (kips/ft)	$N_h/12t$ (ksi)
12	2.76	0.26	0.08	3.10	0.125	2.07	2.42	4.8	3.20
20	4.94		0.20	5.40	0.156	2.88	3.20	7.2	3.85
28	7.29		0.29	7.84	0.1875	3.48	3.89	9.0	4.00
36	9.78		0.37	10.4	0.2188	3.96	4.54	16.4	6.25
44	12.3		0.45	13.0	0.25	4.33	5.15	17.4	5.80
52	14.8		0.53	15.6	0.2500	5.20	5.27	17.9	5.97
60	17.2		0.61	18.1	0.2813	5.36	5.86	18.0	5.33
68	19.6		0.69	20.6	0.3125	5.49	6.45	18.0	4.80
76	21.6		0.78	22.6	0.3125	6.03	6.48	18.0	4.80
84	23.2	0.26	0.86	24.3	0.3125	6.48	6.53	18.0	4.80

[a]Thicknesses must be increased if allowance for corrosion and/or wear is required. See Example 8-1 for a comment on hoop tension at the bottom of the cylinder.

due to lateral pressure of the bin contents, compare with the corresponding thicknesses from Table (8-6) as follows:

y, ft	Table 8-6	Eq. (8-3)	Eqs. (8-5) ÷ 1.5
12	0.125	0.156	0.1875
84	0.3125	0.375	0.375

Hopper

At transition,

$$\text{hopper contents} = \frac{\pi\gamma}{24}(D^3 - d^3)\cot\theta = \frac{\pi \times 50}{24}(32^3 - 2^3)\cot 25^\circ = 459{,}810$$

$$\text{hopper weight } (\tfrac{3}{8}\text{ in. thick}) = \frac{\pi \times 32}{2}\,\frac{16}{\sin 25^\circ} \times 15.3 = \underline{29{,}120}$$

$$488{,}930$$

Use Walker's value of p_v at transition (3040 psf) which was calculated in determining the hopper emptying pressure.

From Eq. (8-13),

$$N_m = \frac{3040 \times 16}{2\cos 25^\circ} + \frac{488{,}930}{2\pi \times 16\cos 25^\circ} = 32{,}200\text{ lb/ft}$$

With $p_e' = 5654$ psf from Table 8-5, Eq. (8-15) gives

$$N_h = \frac{p'r}{\cos\theta} = \frac{5654 \times 16}{\cos 25^\circ} = 99{,}820\text{ lb/ft}$$

Use API 650 allowable stress $0.85 \times 21 = 17.9$ ksi:

$$t = \frac{99.8}{17.9 \times 12} = 0.465$$

Use $\frac{1}{2}$ in. (See Example 8-1 for a comment on the hopper hoop tension at the transition). Check the thickness at $z = 28$ ft (Table 8-5):

$$r = 13.12 \text{ ft} \qquad p_e' = 3028 \text{ psf}$$

$$N_h = \frac{3028 \times 13.12}{\cos 25^\circ} = 43{,}800 \text{ lb/ft} \qquad t = \frac{43.8}{17.9 \times 12} = 0.204 \text{ in.}$$

Although p_f' is greater than 3028 psf below this section, r decreases, and N_h does not exceed the value at 28 ft. Therefore, use $t = \frac{3}{16}$ in. below $z \approx 30$ ft.

Skirt

$$\text{hopper plus contents} = \frac{488.9}{32\pi} = 4.86 \text{ kips/ft}$$

$$\text{bin contents} = \frac{\gamma H \pi D^2/4}{\pi D} = \frac{50 \times 84 \times 32}{4} = 33.60$$

$$\text{roof } DL + LL = 0.26$$

$$\text{bin wall} = 0.86$$

$$\text{skirt } (\tfrac{1}{2} \text{ in.} \times 40 \text{ ft high}) = 40 \times 0.0204 = 0.82$$

$$N_m = 40.4 \text{ kips/ft}$$

For $\frac{1}{2}$-in. skirt, $f = 40.4/(12 \times 0.5) = 6.73$ ksi, and $r/t = 16 \times 12/0.5 = 384$:

$$C = \frac{0.315}{\sqrt{0.1 + 0.01 \times 384}} = 0.159$$

$$f_{cr} = \frac{0.159 \times 30{,}000}{384} = 12.4 \text{ ksi}$$

With the ECCS-recommended factor of safety of 1.5, $f_{cr(all)} = 12.4/1.5 = 8.28 > 6.73$ ksi.

Compression ring

$$N_m \text{ for hopper} = 32.2 \text{ kips/ft}$$

$$N_m \sin\theta = 32.2 \sin 25^\circ = 13.6 \text{ kips/ft}$$

$$\text{ring compression } P = rN_1 \sin\theta = 16 \times 13.6 = 218 \text{ kips}$$

Analysis of the compression ring follows the procedure of Example 8-1. For the

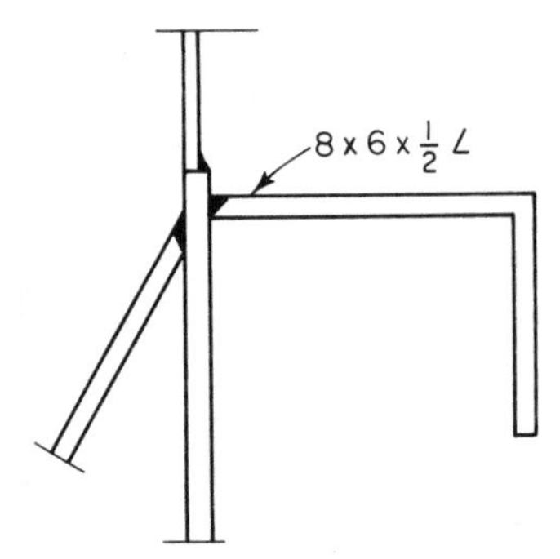

Figure 8-28

detail shown in Fig. 8-28, A = 16.43 in.^{2} and r_x = 3.67 in. Then f = 218/16.43 = 13.3 ksi, while the allowable stress by Eq. (6-76a) computed for F_y = 36 ksi, is 18.3 ksi.

Stresses at transition

Local bending stresses are calculated with Eqs. (8-18)-(8-21) with the following data:

t_c = 0.3125 in. $\quad$ p_h = 1126 psf = 7.82 psi $\quad$ (Table 8-4)

t_h = 0.5 in. $\quad$ p' = 5654 psf = 39.3 psi $\quad$ (Table 8-5)

t_s = 0.5 in. $\quad$ N_{mh} = +32,200 lb/ft = +2683 lb/in.

A_r = 6.75 in.^{2}

From Eqs. (8-18b)-(8-18d),

$$T = 0.4081 \qquad a_1 = 0.4457 \qquad a_2 = a_3 = 0.4606$$

From Eq. (8-18a),

$$f_r = -11.956 \text{ psi}$$

From Eqs. (8-19),

$$B_c = +8880 \qquad M_c = +802 \text{ in.-lb}$$

$$B_h = -6939 \qquad M_h = -1003 \text{ in.-lb}$$

$$B_s = +1392 \qquad M_s = +201 \text{ in.-lb}$$

Stresses in cylinder:

$$N_{mc} = -24.3 \text{ kips/ft} = -2.02 \text{ kips/in.} \qquad \text{(Table 8-6)}$$

$$f_m = \frac{-2.02}{0.3125} \pm \frac{6 \times 0.802}{0.3125^2} = -6.2 \pm 49.3 = -55.5, +43.1 \text{ ksi}$$

Stresses in hopper:

$N_{mn} = +2.68$ kips/in.

$$f_m = \frac{+2.68}{0.5} \pm \frac{6 \times 1.003}{0.5^2} = +5.4 \pm 24.1 = +29.5, -18.7 \text{ ksi}$$

Stresses in skirt:

$N_{ms} = -40.4$ kips/ft $= -3.37$ kips/in.

$$f_m = \frac{-3.37}{0.5} \pm \frac{6 \times 0.201}{0.5^2} = -6.74 \pm 4.82 = -11.6, -1.9 \text{ ksi}$$

The hoop stress $f_r = -12.0$ ksi. Then from Eqs. (8–21),

	Cylinder f (ksi)		Hopper f (ksi)		Skirt f (ksi)	
f_1	-12.0	-12.0	-12.0	-12.0	-12.0	-12.0
f_2	-55.5	+43.1	+29.5	-18.7	-11.6	-1.9
f_3	+43.5	-55.1	-41.5	+6.7	-0.4	-10.1

By using the ASME Pressure Vessel Code allowable value $3S_m$ for f, with S_m modified as suggested in Sec. 8–8 for steels not of pressure-vessel quality, the limiting value of f for A283 grade C steel is $3 \times 0.3F_y = 0.9 \times 50 = 45$ ksi. Therefore, the cylinder, for which the largest absolute value of f_3 is 55.1 ksi, is overstressed. Extending the skirt beyond the transition as a short bottom ring for the cylinder is a good way to reduce the overstress. The extension need not be long since, as already noted in Example 8–1, moments and shears are zero at about $2.5\sqrt{rt}$ from the end of the cylinder, which is about 24 in. in this case. The extension should be about this long to satisfy the conditions which are assumed in deriving the formulas.

With the cylinder, hopper, and skirt each $\frac{1}{2}$ in. thick at their junction, Eqs. (8–16) and (8–17) give the local stresses. From Eq. (8–16), $f_r = -9.1$ ksi. The combined-stress combinations, all of which satisfy the requirement $|f_3| \leqslant 45$ for A283 steel, are as follows:

	Cylinder f (ksi)		Hopper f (ksi)		Skirt f (ksi)	
f_1	-9.1	-9.1	-9.1	-9.1	-9.1	-9.1
f_2	+30.3	-38.1	+37.0	-26.2	-9.3	-4.2
f_3	-39.4	+29.0	-46.1	+17.1	+0.2	-4.9

Wind

	P (kips)	Arm (ft)	M (ft-kips)
On roof cone: $0.6 \times 30 \times 19 \times 9.5 =$	3.2	124	397
On cylinder: $0.6 \times 30 \times 32 \times 120 =$	69.1	60	4146
			4543

$$S = 0.0982 \frac{384^4 - 383^4}{384} = 57{,}680 \text{ in.}^4$$

$$f = 4543 \times \frac{12}{57{,}680} = 0.950 \text{ ksi}$$

Bin empty:

$$\text{roof} + \text{bin wall} + \text{hopper} + \text{skirt} = 0.06 + 0.86 + \frac{29.1}{32\pi} + 0.82 = 2.03 \text{ kips/ft}$$

$$f = \frac{2.03}{12 \times 0.5} = 0.338 \text{ ksi}$$

Therefore, at maximum wind on the empty bin there is uplift on the windward side that varies from a maximum of $12 \times 0.5(0.950 - 0.338) = 3.67$ kips/ft to zero at about 70° on either side of the point of maximum.

Windward-side buckling

The maximum height of bin which can be used without an intermediate wind girder is determined as in Example 8-1. The weighted average thickness in this example is $t = 0.24$ in., which gives $H = 114$ ft. Therefore, an intermediate girder is not needed.

Example 8-3

Design a 1600-ton-capacity coal bunker whose inside dimensions are limited to 32×70 ft. The variation in the unconfined yield strength f_c with consolidating pressure p_1 is shown in Fig. 8-29. Additional properties are $\delta = 50°$, independent of p_1, $\phi = 40°$, $\phi' = 35°$ on gunite, and $\gamma = 48$–50 pcf. Use A36 steel.

The following procedure for determining the dimensions of the outlet is outlined in Sec. 4-13. From Fig. 4-17, $ff = 2.7$ for $\phi = 40°$ and $\delta = 50°$. The line $ff = p_1/f_c = 2.7$ intersects the curve in Fig. 8-29 at $f_c = 160$ psf. From Fig. 4-16, $G(\phi) = 3.4$ for $\phi = 40°$. Then from Eq. (4-6), with $\gamma = 49$ pcf

$$d = \frac{f_c}{\gamma} G(\phi) = \frac{160}{49} \times 3.4 = 11.1 \text{ ft}$$

From Fig. 4-15, with $\phi = 40°$ and $\delta = 50°$, $ff = 1.53$. The line $ff = p_1/f_c = 1.53$ in

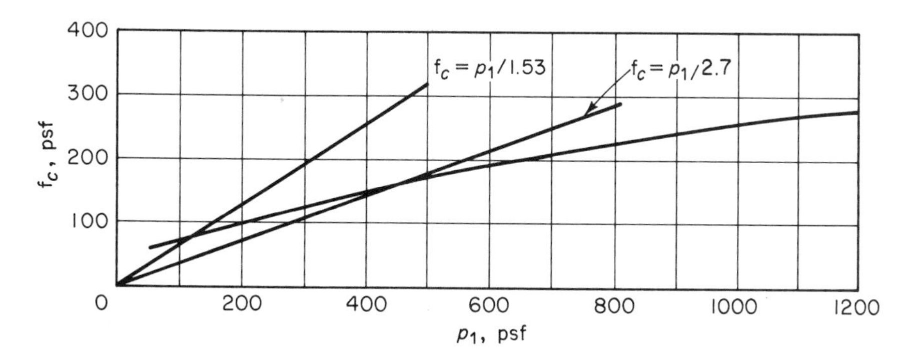

Figure 8-29

Fig. 8-29 intersects the curve at 80 psf. Then from Eq. (4-4),

$$b \geqslant 1.3 \times \frac{80}{49} = 2.12 \text{ ft}$$

Use five outlets 5 × 10 ft each, which gives $d = 11.2$ ft.

The required bunker volume is 1600 × 2000/48 = 66,670 ft^3, which gives 66,670/70 = 952 ft^3/ft. The volume of the hopper shown in Fig. 8-30 is 24(32 + 5)/2 = 444 ft^3/ft. Assuming the angle of repose α to be equal to ϕ, the height of surcharge is 8 tan 40° = 6.7, say 6.5 ft. The volume of surcharge is 2 × 16 × 6.5/2 = 104 ft^3/ft. Therefore, the minimum height of the vertical wall is (952 - 444 - 104)/32 = 12.6 ft. Use 13 ft, and compute pressures for a mean height of 16.25 ft (Fig. 8-30).

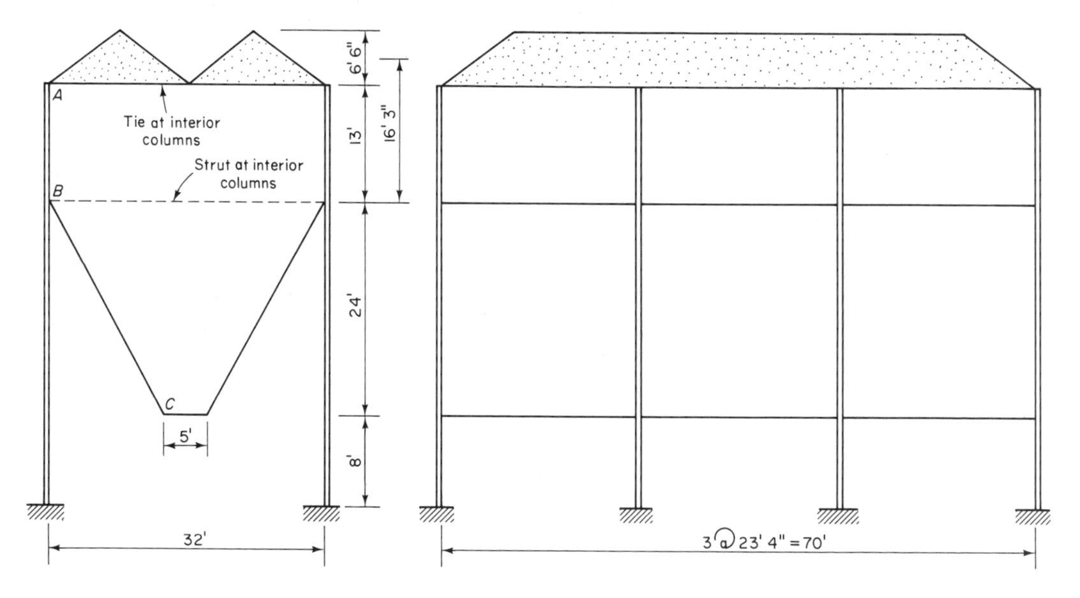

Figure 8-30 Bunker for Example 8-3.

Pressures

Use Rankine's theory for reasons discussed in Sec. 5–9. The active-pressure coefficient is

$$K = \frac{1 - \sin \phi}{1 + \sin \phi} = \frac{1 - \sin 40^\circ}{1 + \sin 40^\circ} = 0.217$$

The pressures $p_v = \gamma y$ and $p_h = Kp_v$ are as follows (Fig. 8–30):

$$\text{At } A: \quad p_v = 50 \times 3.25 = 162 \text{ psf}, p_h = 35 \text{ psf}$$

$$\text{At } B: \quad p_v = 50 \times 16.25 = 812 \text{ psf}, p_h = 176 \text{ psf}$$

$$\text{At } C: \quad p_v = 50 \times 40.25 = 2012 \text{ psf}, p_h = 437 \text{ psf}$$

The hopper side-wall pressures are [Eqs. (5–24) and (5–25)]

$$p'_B = 812 \sin^2 29.4^\circ + 176 \cos^2 29.4^\circ = 328 \text{ psf}$$

$$p'_c = 2012 \sin^2 29.4^\circ + 437 \cos^2 29.4^\circ = 814 \text{ psf}$$

$$q'_B = 0.5(812 - 176) \sin 58.8^\circ = 272 \text{ psf}$$

$$q'_c = 0.5(2012 - 437) \sin 58.8^\circ = 674 \text{ psf}$$

The pressures p_h, p', and q' are shown in Fig. 8–31. Also shown on the surface BC are the weight of $2\frac{1}{2}$ in. of gunnite and a $\frac{1}{4}$-in. plate (40 psf) and the vertical load at C from the weight of coal above the 5-ft bottom of the trough.

Plates

A procedure for determining plate thickness by the large-deflection theory is given in Sec. 6–19. With columns at the third points of the length of the bunker, column spacing is 23 ft 4 in. (Fig. 8–30). Since the plate is continuous at the stiffeners, panel edges can be assumed moment-resistant.

Upper wall (Fig. 8–31a)

Try vertical stiffeners spaced $23.33/5 = 4.67$ ft, which gives an aspect ratio $a/b = 13/4.67 = 2.8$. Therefore, Table 6–6 applies. The following procedure for approximating the trapezoidal load is suggested in Sec. 6–19. By assuming the upper edge of the plate hinged and the other three fixed, the equivalent uniform load by the small-deflection theory is $35 + 0.6(176 - 35) = 120$ psf (Sec. 6–18). This gives an equivalent uniform load for the large-deflection theory of $(120 + 176)/2 = 148$ psf. Then $p = 148/144 = 1.03$ psi, and with the allowable stress $f = 0.75F_y = 27{,}000$ psi suggested in Sec. 8–12,

$$\frac{pE}{f^2} = \frac{1.03 \times 30 \times 10^6}{27{,}000^2} = 0.0424$$

and from Table 6–6,

$$\frac{b}{t}\sqrt{\frac{f}{E}} = 10.16 \qquad \frac{b}{t} = 10.16\sqrt{\frac{30 \times 10^6}{27{,}000}} = 339$$

$$t = \frac{56}{339} = 0.165 \text{ in.}$$

Use $t = \frac{3}{16}$ in. Then $b/t = 56/0.1875 = 299$ for which

$$\sqrt{\frac{p}{E}}\left(\frac{b}{t}\right)^2 = \sqrt{\frac{1.03}{30 \times 10^6}}\, 299^2 = 16.6$$

and from Table 6-6, $\delta/t = 2.04$ and $\delta = 0.38$ in.,

$$\frac{\delta}{b} = \frac{0.38}{56} = \frac{1}{147}$$

Hopper wall (Fig. 8-31b)

Pressures normal to the wall are 328 + 40 sin 29.4° = 348 psf at *B* and 814 + 40 sin 29.4° = 834 psf at *C*. By assuming all four edges fixed, the equivalent uniform load by the small-deflection theory is 348 + 0.62(834 − 348) = 649 psf. There-

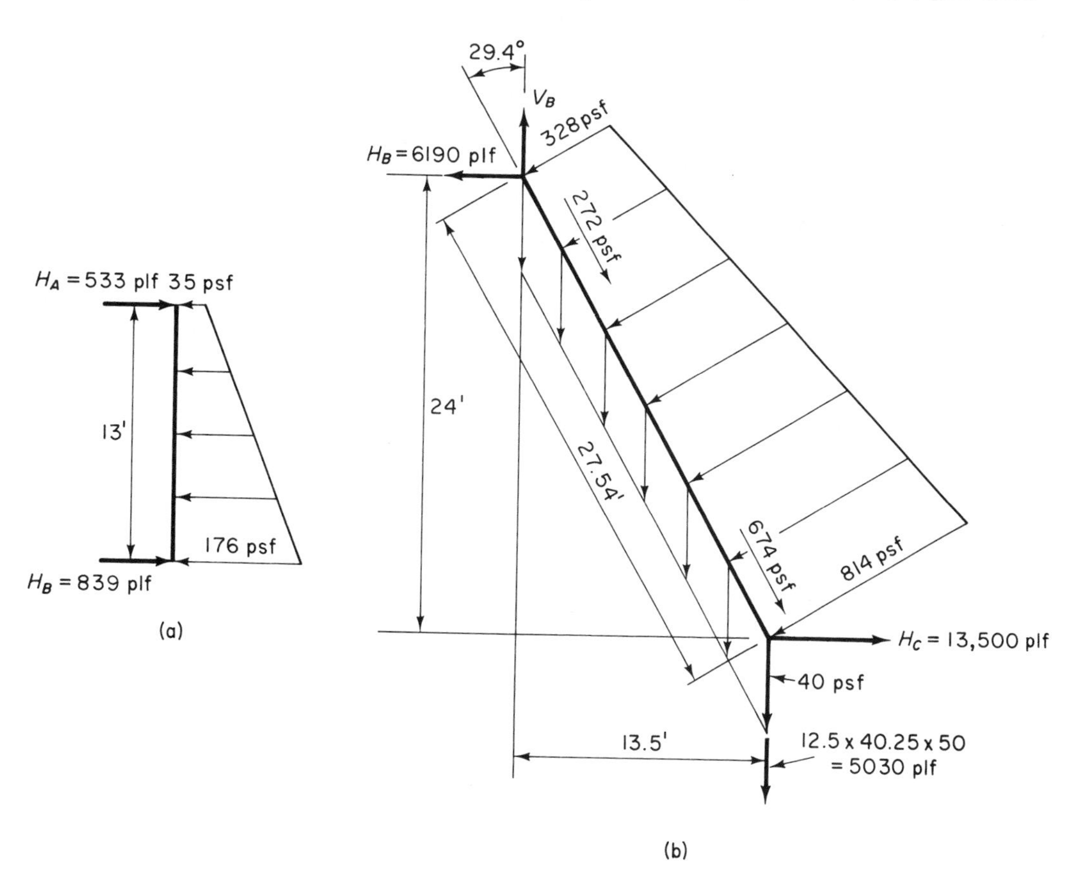

Figure 8-31

fore, the equivalent uniform load for the large-deflection theory is $(649 + 834)/2 =$ 742 psf = 5.15 psi, and

$$\frac{pE}{f^2} = \frac{5.15 \times 30 \times 10^6}{27{,}000^2} = 0.212 \qquad \text{for which } \frac{b}{t}\sqrt{\frac{f}{E}} = 3.66$$

$$\frac{b}{t} = 3.66\sqrt{\frac{30 \times 10^6}{27{,}000}} = 122$$

With stiffeners spaced half that of the upper wall (28 in.), $t = 28/122 = 0.23$ in. Use $\frac{1}{4}$ in.:

$$\frac{b}{t} = \frac{28}{0.25} = 112$$

$$\sqrt{\frac{p}{E}}\left(\frac{b}{t}\right)^2 = \sqrt{\frac{5.15}{30 \times 10^6}}\, 112^2 = 5.20$$

From Table 6–6, $\delta/t = 0.601$ and $\delta = 0.150$ in.,

$$\frac{\delta}{b} = \frac{0.150}{28} = \frac{1}{187}$$

Upper-wall stiffeners (Fig. 8–31a)

Formulas for beams with a trapezoidally distributed load are given in Sec. 7–7. They are derived for a sloping rafter with vertical load (Fig. 7–11), but they are also correct for a beam of length a at any slope with q_1 and q_2 normal to the beam. For the upper-wall stiffener the point of maximum moment is at [Eq. (7–19)]

$$\frac{x_1}{a} = \frac{176 - \sqrt{(176^2 + 176 \times 35 + 35^2)/3}}{176 - 35} = 0.446$$

From Eq. (7–18), with stiffeners spaced 4 ft 8 in.,

$$M = \frac{13^2}{6} \times 0.446 \times 4.67(352 + 35 - 528 \times 0.446 + 141 \times 0.446^2) = 10.5 \text{ ft-kips}$$

$$S = 10.5 \times \frac{12}{22} = 5.73 \text{ in.}^3$$

The W6 × 9 has a section modulus of 5.56 in.^{3} Therefore, it should be adequate with the effective width of wall plate taken into consideration. Assuming a compression-flange stress of 18 ksi, the effective width of plate is given by Eq. (6–30) with $b/t = 56/0.1875 = 299$:

$$\frac{b_e}{t} = \frac{253}{\sqrt{18}}\left[1 - \frac{44.3}{299\sqrt{18}}\right] = 57.6 \qquad b_e = 10.8 \text{ in.,} \qquad \text{say 10 in.}$$

This effective width can be taken half on each side of the flange tips, which gives the cross section shown in Fig. 8–32. The moment of inertia is 28.8 in.^{4}, and

$$f_t = \frac{10.5 \times 12 \times 4.46}{28.8} = 19.5 \text{ ksi}$$

$$f_c = \frac{10.5 \times 12 \times 1.63}{28.8} = 7.1 \text{ ksi}$$

Since f_c is less than the value assumed to determine the effective width, the W6 × 9 is adequate. A WT5 × 7.5 could be used instead of the W6 × 9.

The deflection is given by Eq. (7–20). It is 0.37 in. for which $\delta/a = \frac{1}{420}$.

Hopper-wall stiffeners (Fig. 8–31b)

The point of maximum moment is at

$$\frac{x_1}{a} = \frac{834 - \sqrt{(834^2 + 834 \times 348 + 348^2)/3}}{834 - 348} = 0.465$$

From Eq. (7–18), with stiffeners spaced 28 in.,

$$M = \frac{27.53^2}{6} \times 0.465 \times 2.33(1668 + 348 - 2502 \times 0.465 + 486 \times 0.465^2)$$

$$= 131 \text{ ft-kips}$$

The wall is also subjected to tension from the weight of the wall and the gunite liner and friction on the liner. The horizontal reaction H_c (Fig. 8–31b) is given by

$$24H_c = 5030 \times 13.5 + 40 \times 27.54 \times 6.75 + 328 \times \frac{27.54^2}{6} + 814 \times \frac{27.54^2}{3}$$

$$H_c = 13.5 \text{ kips/ft}$$

The tensile resultant at the point of maximum moment, which is at $x_1 = 0.465 \times 27.54 = 12.8$ ft from C, is

$$T = H_c \sin 29.4^\circ + (5030 + 50 \times 12.8) \cos 29.4^\circ$$
$$+ \tfrac{1}{2} \times 12.8(487 + 674) = 19.0 \text{ kips/ft}$$

The W16 × 45 with a 24-in. effective width of wall (Fig. 8–33) gives

$$f_t = \frac{131 \times 12 \times 10.61}{863} + \frac{19.0 \times 2.33}{19.3} = 19.3 + 2.3 = 21.6 \text{ ksi}$$

$$f_c = \frac{131 \times 12 \times 5.77}{863} - \frac{19.0 \times 2.33}{19.3} = 10.5 - 2.3 = 8.2 \text{ ksi}$$

With this value of f_c and $b/t = 28/0.25 = 112$, Eq. (6–30) gives

$$\frac{b_e}{t} = \frac{253}{\sqrt{8.2}}\left(1 - \frac{44.3}{112\sqrt{8.2}}\right) = 76 \qquad b_e = 19 \text{ in.}$$

14
0.1875
G
W6 x 9
4.46
5.90

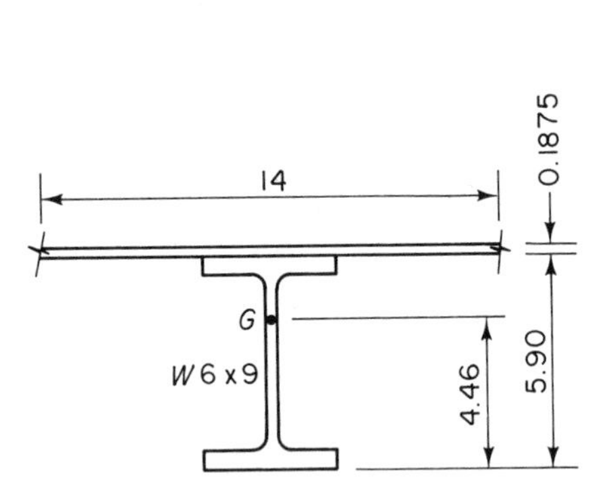

Figure 8-32 Upper-wall stiffener.

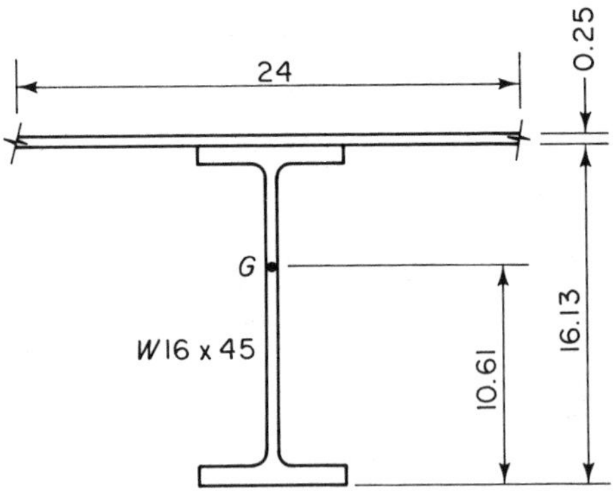

24
0.25
G
W16 x 45
10.61
16.13

Figure 8-33 Hopper-wall stiffener.

for which the effective flange width is 19 + 7 = 26 in. Since this is greater than the value assumed in Fig. 8-33, the section is adequate.

Horizontal stiffener at A

From Fig. 8-31a the horizontal reaction on the wall at A is

$$H_A = 35 \times \frac{13}{3} + 176 \times \frac{13}{6} = 533 \text{ lb/ft}$$

The maximum moment in the stiffener is

$$M = 0.533 \times \frac{23.33^2}{8} = 36.3 \text{ ft-kips}$$

$$S = 36.3 \times \frac{12}{22} = 19.8 \text{ in.}^3$$

Use W12 $\times$ 19, $S = 21.3$ in.^{3}

Horizontal stiffener at B

The horizontal reaction at B of the upper wall is

$$H_B = 35 \times \frac{13}{6} + 176 \times \frac{13}{3} = 839 \text{ lb/ft}$$

The horizontal reaction at B of the hopper wall (Fig. 8-31b) is

$$H_B = H_C + \frac{272 + 674}{2} \times 27.54 \sin 29.4^\circ$$

$$- \frac{328 + 814}{2} \times 27.54 \cos 29.4^\circ = 6190 \text{ lb/ft}$$

The resultant force on the horizontal stiffener is 6190 − 839 = 5351 lb/ft. Therefore,

$$M = 5.35 \times \frac{23.33^2}{8} = 364 \text{ ft-kips}$$

$$S = 364 \times \frac{12}{22} = 199 \text{ in.}^3$$

Use W24 X 84.

In-plane bending of wall

Only the horizontal component of the hopper-wall reaction at B was considered in sizing the horizontal stiffener at B. The vertical component produces beam-type bending in bunkers with low and intermediately high walls. This component can be found from Fig. 8–31b or calculated as half the weight of the stored solid plus the weight of the walls and gunite, which was taken at 40 psf. From Fig. 8–30,

$$V_B = 50(16.25 \times 16 + 9.25 \times 24) + 40(13 + 27.54) = 25{,}700 \text{ plf}$$

Vertical-plane bending will be fixed-ended when the bunker is fully loaded. Therefore,

$$-M_x = 25.7 \times \frac{23.33^2}{12} = 1165 \text{ ft-kips}$$

The maximum moment between any two columns will occur when the bunker is fully loaded between those columns and only partially loaded in the adjacent spans. This moment will be larger than the positive moment $wL^2/24$ but smaller than the negative moment $wL^2/12$ for fully fixed supports. For equal spans it is given with good approximation by $wL^2/14$. Therefore,

$$+M_x = 25.7 \times \frac{23.33^2}{14} = 1000 \text{ ft-kips}$$

The girder must also be checked for bending due to the horizontal reactions H_A and H_B, which give

$$H = 6190 - 839 - 533 = 4820 \text{ plf}$$

With the connection to the columns shown in Fig. 8–37, the moments M_y can be assumed as for M_x:

$$-M_y = 4.82 \times \frac{23.33^2}{12} = 219 \text{ ft-kips} \qquad \text{at column}$$

$$+M_y = 4.82 \times \frac{23.33^2}{14} = 187 \text{ ft-kips} \qquad \text{between columns}$$

Hopper-plate participation in bending will be neglected, and the girder will be assumed to consist of the vertical-wall plate and the horizontal stiffeners at A and B

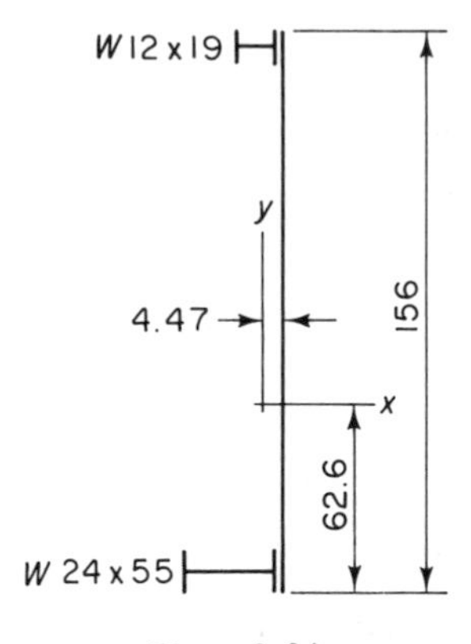

Figure 8-34

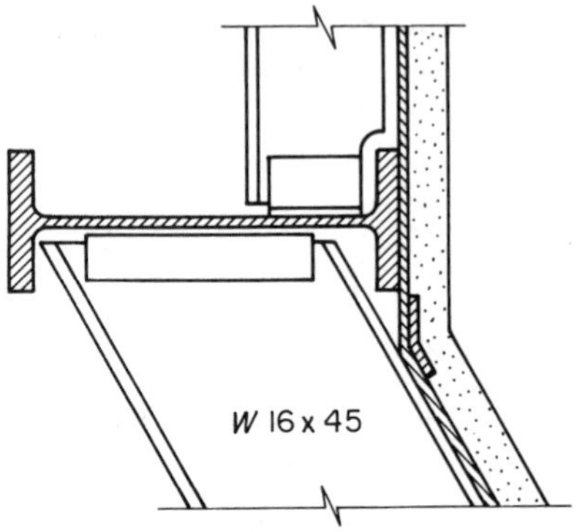

Figure 8-35 Detail of horizontal stiffener at bend line. (Adapted from Ref. 26.)

(Fig. 8-34). The span-depth ratio for vertical-plane bending is 23.33/13 = 1.8, so $f = Mc/I$ can be used. For the axes shown in Fig. 8-33, I_x = 194,700 in.^{4}, and I_y = 4440 in.^{4} These are not principal axes, since neither is an axis of symmetry, so some approximation is involved in using them. The maximum stress is at the column at the lower tip of the outside flange of the W24 × 84:

$$f = -\frac{1165 \times 12 \times 54.6}{194{,}700} - \frac{219 \times 12 \times 18.56}{4440} = -3.4 - 11.0 = -14.9 \text{ ksi}$$

The maximum stress midway between columns, +12.75 ksi, is also at the lower tip of the outside flange of the W24 × 84. Therefore, a smaller section can be used. With a W24 × 55 the maximum stresses are −22.3 ksi at the column and +19.0 ksi midway between columns. Since the reference axes in this analysis are not principal axes, these stresses are approximate. However, any differences would be more than offset with the hopper-wall participation in bending taken into account. Further economy could be achieved by taking this participation into account. However, with the detail shown in Fig. 8-35, the horizontal stiffener cannot be smaller in depth unless the hopper-wall stiffeners are decreased in depth, which would increase their weight. Therefore, the W24 × 55 will be used.

End walls

Design of the end-wall components follows the procedures used in designing the side-wall components except that analysis for vertical bending is not involved.

Hopper bottom

The hopper bottom must be designed to resist the horizontal reaction H_C = 13.5 kips/ft (Fig. 8-31) and to support the weight of the coal between openings. With the five 5 × 10 ft openings spaced 4 ft apart and 2 ft from each end wall to give the 70-ft length of the bunker, a longitudinal stiffener will span 10 ft and 4 ft alternately. With the detail shown in Fig. 8–36, the 4-ft span will be stiff enough to develop fixed-end moments in the 10-ft span. Therefore,

$$M = 13.5 \times \frac{10^2}{12} = 112.5 \text{ ft-kips} \qquad S = 112.5 \times \frac{12}{22} = 61.4 \text{ in.}^3$$

A W18 X 40 would be adequate, but to avoid notching the wall stiffeners, use a W24 X 55 (Fig. 8–36).

The weight of coal on the 4 X 5 ft segment of the bottom between openings is 40.25 X 50 = 2012 psf. This will be carried by a plate welded to the flanges of the longitudinal stiffeners and to 6-in. channels on the 5-ft edges. Assuming the plate to be simply supported, Eq. (6–51) gives $k_{ys} = 0.411$ for $b/a = \frac{4}{5}$. Then with $F = 0.75F_y = 27{,}000$ psi and $p = 2012/144 = 14$ psi, Eq. (6–55) gives

$$\frac{b}{t} = \sqrt{\frac{27{,}000}{0.411 \times 14}} = 68 \qquad t = \frac{48}{68} = 0.7$$

Use $\frac{3}{4}$ in.

Each channel and an effective width of the $\frac{3}{4}$-in plate serves as a support of a 5-ft edge of the 4 X 5 ft segment. Since the plate slenderness $b/t = 48/0.75 = 64$ is

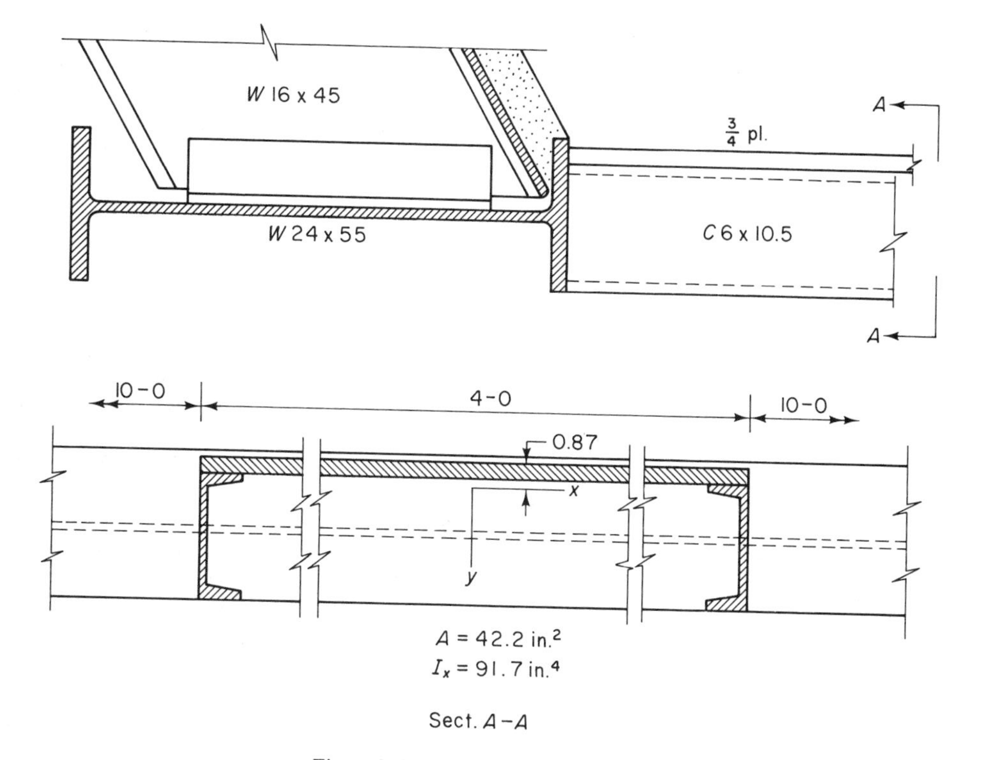

Figure 8–36 Detail of hopper bottom.

not large and the compression bending stress is small, the plate will be fully effective. Therefore, the two channels and the 48-in. plate can be analyzed as a beam spanning 5 ft (Fig. 8–36). This beam also carries a longitudinal tension T in resisting the force $H_C = 13.5$ kips/ft on 14 ft of the longitudinal stringer. Assuming the beam to be simply supported, we obtain

$$M = 4 \times 2.01 \times \frac{5^2}{8} = 25.1 \text{ ft-kips} \qquad T = 12.5 \times 14 = 189 \text{ kips}$$

which, with the section properties shown in Fig. 8–36, gives

$$f = \frac{25.1 \times 12 \times 5.88}{91.7} + \frac{189}{42.2} = 19.3 + 4.5 = 23.8 \text{ ksi}$$

This is slightly in excess of the allowable 22 ksi for A36 steel, but since the moment was calculated for a simple beam and there is considerable end fixity, it can be accepted.

Columns

The load on an interior column, including the weight of the bunker, is 614 kips, all of which, except for the weight of the 13-ft upper wall, is applied at the bend line. The bunker loading system is assumed to be supported independently. The lateral tie at the top and the bend-line strut relieve the columns of the horizontal-stiffener reactions, and the 13-ft connection to the bunker allows the column to be considered fixed for both axes at the upper end of the 32-ft length (Fig. 8–30). Therefore, the effective length is $0.7 \times 32 = 22.4$ ft for which a W14 $\times$ 132 or a W12 $\times$ 152 in A36 steel is adequate. The load on the corner column is 307 kips for which a W14 $\times$ 90 or a W12 $\times$ 79 is adequate. Use the 14-in. sections.

Bend-line strut

The horizontal reaction at the bend line is $6190 - 839 = 5351$ lb/ft (Fig. 8–31). This gives a strut load of $5.35 \times 23.33 = 124.8$ kips. The strut is also loaded transversely by the coal above. An experimental investigation of loads on flow-corrective inserts showed the load on a suspended conical insert (apex up) to be p_v times the base area of the cone.[33] This load is larger during filling, as was confirmed by the tests. Since $p_v < \gamma y$ (except by the Rankine theory, used in this example, where $p_v = \gamma y$), it is conservative to take the strut transverse load at γbh, where b = the width of strut and h = the height of the coal above. With a W14 $\times$ 109, b = 18 in., and the load is $50 \times 1.5 \times 16 = 1200$ lb/ft. The weight of the 18 $\times$ 18 in. gunited member is 400 lb/ft. Therefore, $M = 1.6 \times 32^2/8 = 204.8$ ft-kips. Then $f_a = 124.8/32.0 = 3.90$ ksi, $f_b = 204.8 \times 12/173 = 14.2$ ksi, $F_a = 12.6$ ksi, $F_b = 24$ ksi, $F'_E = 39.2$ ksi, and the AISC interaction formula gives

$$\frac{3.90}{12.6} + \frac{14.2}{24(1 - 3.90/39.2)} = 0.310 + 0.657 = 0.967$$

Of course, this is conservative because it neglects any contribution of the gunite to the strength and stiffness of the member.

Column-top tie

The load on this member is 533 × 23.33 = 12,400 lb in tension. Therefore, only a minimum member, say a W12 × 14, is required.

Connections

Figure 8–37 shows a detail of a connection of the bunker plate, the horizontal stiffener, and the bend-line strut to an interior column. Connection of the top horizontal stiffeners to an interior column is shown in Fig. 8–38a and to the corner column in Fig. 8–38b.

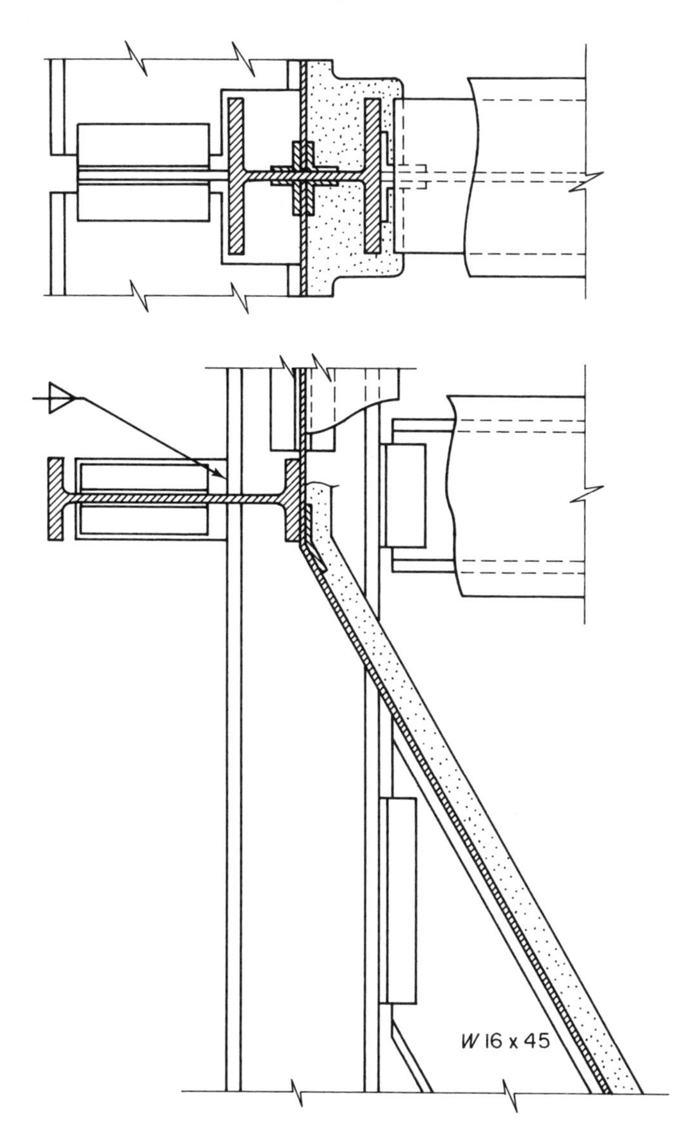

Figure 8–37 Details at column. (Adapted from Ref. 26.)

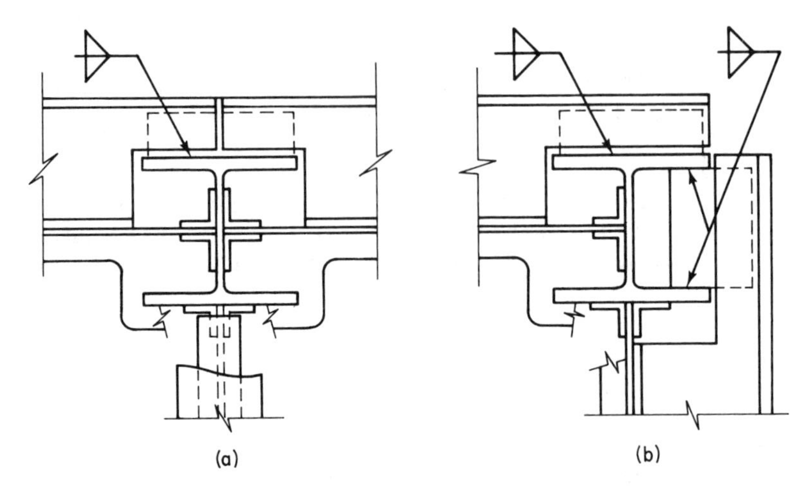

Figure 8–38 Top horizontal stiffeners at columns. (Adapted from Ref. 26.)

Rigid-frame support

As an alternative framing for a bunker of this type, the bend-line strut can be omitted and the columns and a member connecting them at the top designed as a rigid frame. It is not an economical solution in this case.

REFERENCES

1. S. Timoshenko and S. Woinowski-Krieger, *Theory of Plates and Shells,* 2nd ed., McGraw-Hill, New York, 1959.
2. David P. Billington, Thin-Shell Concrete Structures, Sec. 20 in E. H. Gaylord and C. N. Gaylord (eds.), *Structural Engineering Handbook,* 2nd ed., McGraw-Hill, New York, 1979.
3. *API Standard 650,* 5th ed., Welded Steel Tanks for Oil Storage, American Petroleum Institute, Washington, D.C.
4. *Deutsche Normen DIN 4119,* Aboveground Cylindrical Steel Tank Construction, available from American National Standards Institute, New York.
5. J. M. Rotter, N. S. Trahair, and P. Ansourian, Stability of Plate Structures, in *Symposium on Steel Bins for Bulk Solids, Sydney, September 30, 1980, Melbourne, November 12, 1980,* Australian Institute of Steel Construction and Australian Welding Research Association, Milsons Point, 1980.
6. R. S. Wozniak, Steel Tanks, Sec. 23, in E. H. Gaylord and C. N. Gaylord (eds.), *Structural Engineering Handbook,* 2nd ed., McGraw-Hill, New York, 1979.
7. M. Esslinger and K. Pieper, Schnittkräfte und Beullasten von Silos aus überlappt verschraubten Blechplatten, *Der Stahlbau,* Sept. 1973.
8. A. W. Jenike, Denting of Circular Bins with Eccentric Drawpoints, *J. Struct. Div. ASCE,* Febr. 1967.

9. M. H. Mahmoud, Practical Finite Element Modeling of Silo-Material Interaction, in Proc. Technical Program, International Conference on Powder and Bulk Solids Handling and Processing, Rosemont, Ill., May 1977.

10. R. J. Roark and W. C. Young, *Formulas for Stress and Strain,* 5th ed., McGraw-Hill, New York, 1975.

11. J. M. Rotter, Non-axisymmetric Analysis, Lecture 6 of Structural Aspects of Steel Silos and Tanks, Postgraduate Course, School of Civil and Mining Engineering, University of Sydney, Aug. 1981.

12. V. Maderspach, J. T. Gaunt, and J. H. Sword, Buckling of Cylindrical Shells Due to Wind Loading, *Stahlbau,* Sept. 1973 (in English).

13. School of Civil Engineering, An Investigation of the Failure Due to Wind Action of a Group of Six Silos at Boggabri, NSW, *Investigation Report S152,* The University of Sydney, April 1974.

14. P. L. Gould, S. K. Sen, R. S. C. Wang, H. Suryoutomo, and R. D. Lowrey, Column Supported Cylindrical-Conical Tanks, *J. Struct. Div. ASCE,* February 1976.

15. R. S. C. Wang and P. L. Gould, Design of Cylindrical Steel Tanks Subjected to Concentrated Support Loads, *Research Report No. 35,* Department of Civil Engineering, Washington University, St. Louis, 1975.

16. W. Fuchssteiner and O. W. Olsen, Ein Problem der Stahlblechsilos, *Bauingenieur,* Jan. 1979.

17. Alternative Rules for Pressure Vessels, Div. 2, *ASME Boiler and Pressure Vessel Code,* Sect. VIII, the American Society of Mechanical Engineers, New York, 1977.

18. S. Timoshenko and J. M. Gere, *Theory of Elastic Stability,* 2nd ed., McGraw-Hill, New York, 1961.

19. M. Stein and D. J. Yaeger, Critical Shear Stress of a Curved Rectangular Panel With a Central Stiffener, *NASA Tech. Note 1972,* Oct. 1949.

20. N. Mariani, J. D. Mozer, C. L. Dym and C. G. Culver, Transverse Stiffener Requirements for Curved Webs, *J. Struct. Div. ASCE,* Apr. 1973.

21. F. Bleich, *Buckling of Metal Structures,* McGraw-Hill, New York, 1952.

22. P. L. Gould and S. K. Sen, Column Moments in Eccentrically Supported Tanks, *J. Struct. Div. ASCE,* Oct. 1974.

23. R. Cieselski, A. Mitzel, W. Stachurski, J. Suwalski, and Z. Zmudzinski, *Behalter, Bunker, Silos, Schornsteine, Fernsehturme und Freileitungsmaste,* Wilhem Ernst & Sohn KG, Berlin, 1970.

24. W. Fischer, "Silos und Bunker in Stahlbeton," *Veb Verlag Fur Bauwesen,* Berlin, 1966.

25. E. Lightfoot and D. Michael, Prismatic Coal Bunkers in Structural Steelwork, *The Structural Engineer,* Feb. 1966.

26. H. Shah and Lee-Ken Choo, Analysis and Design of Coal Bunkers, *SD & DD Report No. 5,* Sargent and Lundy Engineers, Chicago, 1980.

27. G. Winter, Design of Cold-Formed Steel Structural Members, Sec. 9 in E. H. Gaylord and C. N. Gaylord (Eds.), *Structural Engineering Handbook,* 2nd ed., McGraw-Hill, New York, 1979.

28. A. H. Nilson, Folded Plate Structures of Light Gage Steel, *J. Struct. Div. ASCE,* Oct. 1961.

29. Paul Andersen, Temperature Stresses in Steel Grain-Storage Tanks, *Civil Engineering–ASCE,* Jan. 1966.

30. *Investigation of Radial Pressures of Wheat,* Report by Twin City Testing and Engineering Laboratory, Inc., St. Paul, Minn., 1965.

31. M. N. El-Atrouzy and G. Abdel-Sayed, Shell Roofs and Grain Bins Made of Corrugated Steel Sheets, *Proceedings Second Specialty Conference on Cold-Formed Steel Structures,* St. Louis, Oct. 1973.

32. Don O. Brush and Bo O. Almroth, *Buckling of Bars, Plates, and Shells,* McGraw-Hill, New York, 1975.

33. J. R. Johanson and W. K. Kleysteuber, Flow Corrective Inserts in Bins, *Chem. Eng. Prog.,* Nov. 1966.

9

FOUNDATIONS

9-1. Allowable Bearing Pressures

Flat-bottom bins may be supported on grade, or on spread, mat (also called raft) and pier or pile foundations. Columns of column-supported bins may be supported individually on footings or in groups on ring, strip, grid, raft, and pier or pile foundations.

If subsurface conditions at the site are well known, presumptive allowable bearing pressures (Table 9-1) or allowable pressures according to local building codes may be used to design the foundation. Even where subsurface conditions are not well known, it may be possible to obtain enough information without a program of soil exploration to enable a reasonably conservative bearing pressure to be determined. If soil exploration is required, enough information for many soil deposits can be obtained by split-barrel sampling and standard penetration tests. If settlement is an important consideration, and particularly where columns of column-supported bins may be likely to experience differential settlement, the program of exploration should be planned to give reliable estimates of probable settlements.

Foundations should be located below the depth to which the soil is subject to alternate wetting and drying. This depth varies greatly but is usually confined to the upper 5-10 ft. Volume changes seldom occur below the groundwater level. Foundations should also be located below the level to which heaving results from freezing of the soil. A map of extreme frost penetration is given in Fig. 9-1.

TABLE 9-1. SUGGESTED PRESUMPTIVE ALLOWABLE BEARING PRESSURES

Class of material	Allowable pressure (tons/ft^2)
Soft or broken bed rock, soft limestone	20
Compacted, partially cemented gravel, sand and hardpan overlying rock	20
Gravel and sand-gravel mixtures	6
Loose gravel, compact coarse sand, hard dry clay	4
Loose coarse sand, compact fine sand (confined), sand-gravel mixtures	3-4
Loose medium sand (confined), stiff clay	1.5-2
Soft broken shale, soft clay	1-1.5

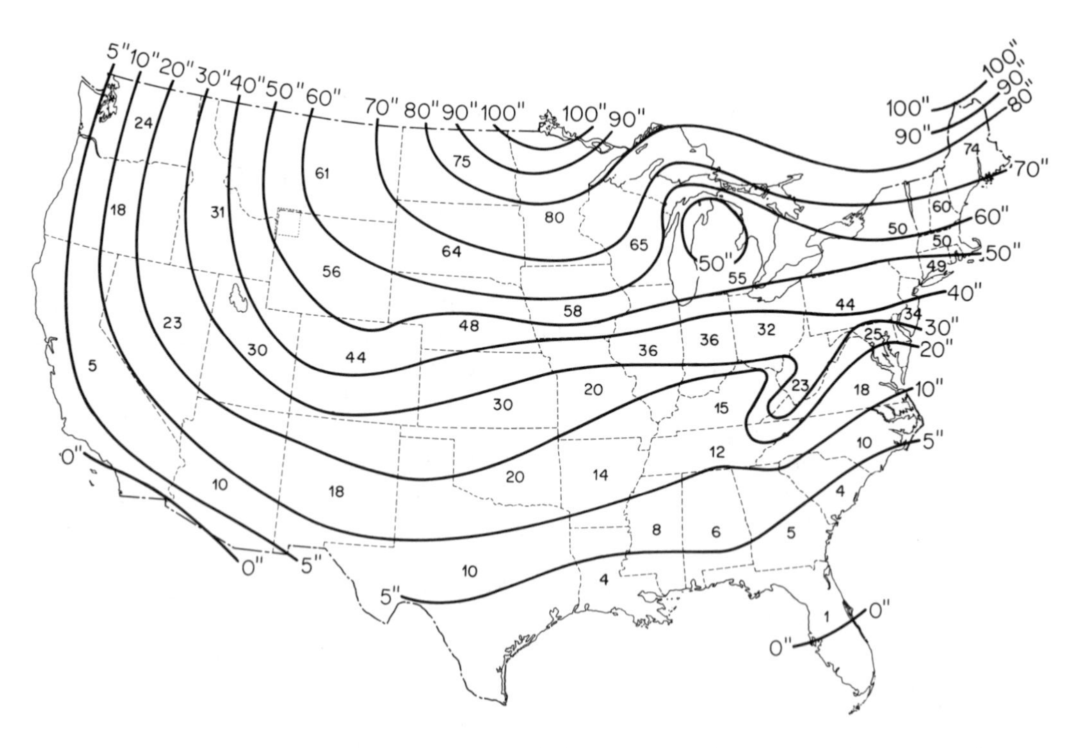

Figure 9-1 Extreme frost penetration, inches, based on state averages. (National Weather Service, NOAA, U.S. Dept. of Commerce.)

9-2. Flat-Bottom Bins

Flat-bottom bins may be supported on grade if bearing conditions are satisfactory. Loam and inorganic matter should be replaced with granular materials compacted in layers and topped with sand, graded gravel, or crushed stone. If a sand cushion is placed on a crushed-stone fill, the stone should be graded from coarse at the bottom to fine at the top to prevent percolation of the sand through the stone. Clay or lumps of earth should not be in contact with a steel bottom. Drainage is important for stability of the soil and prevention of corrosion of the bottom plate and should be provided under the bottom as well as in the surrounding area. The recommendations of API 650 are shown in Fig. 9-2. Flat-bottom bins may also be grade-supported on reinforced-concrete slabs, with or without a steel bottom.

If the soil at the surface has inadequate bearing capacity, a concrete ringwall may be used to confine the weaker material to enable it to transmit the load to a stronger soil below. If the ringwall is proportioned so that the soil pressure on its base equals the soil pressure at the same elevation under the bin bottom (Fig. 9-3), the required width b of the ringwall is

$$b = \frac{w_s}{\gamma H + 50d} \tag{9-1}$$

where w_s = weight of shell plus roof load, lb/ft of circumference
γ = bulk density of solid in bin, pcf
H = height of bin, ft
d = height of ringwall, ft

The coefficient 50 of the term $50d$ in the denominator of this equation is the difference between the unit weight of concrete and the unit weight of the soil, assumed to be 150 and 100 pcf, respectively. This term is explained in Ref. 1.

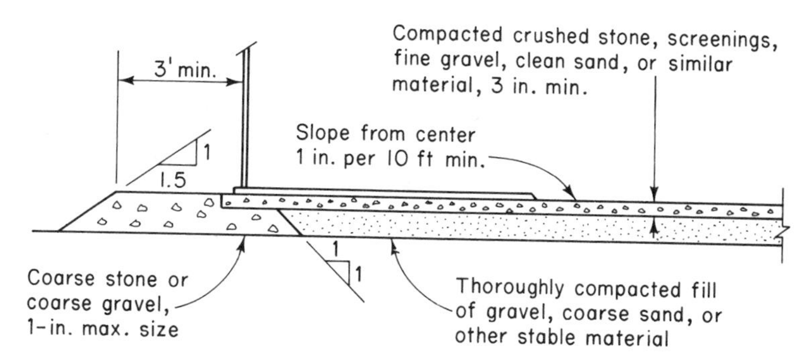

Figure 9-2 Foundation for ground-supported flat-bottom bin.

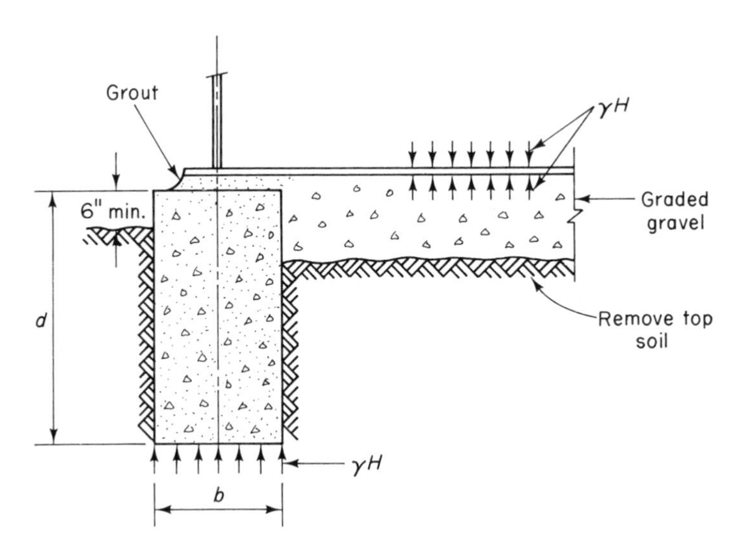

Figure 9-3 Foundation ringwall.

The ringwall should extend at least 6 in. above grade and 6 in. below frost line. Circumferential reinforcement must be provided to resist the hoop tension from the lateral pressure of the confined soil. The lateral pressure is calculated as for a retaining wall, which, by Eq. (5-1), gives $p_h = K_a p_r$. With $K_a = 0.3$ and $p_r = \gamma H$, the hoop tension T per foot height of ringwall is

$$T = 0.15\gamma HD \tag{9-2}$$

where D is the diameter of the bin.

If the bearing capacity of the soil is inadequate for ground support, a flat-bottom bin may be supported on a spread footing (Fig. 9-4a) or, if satisfactory soil is at too large a depth for economical use of a spread footing, on a mat with ringwall (Fig. 9-4b) or a pier or pile foundation.

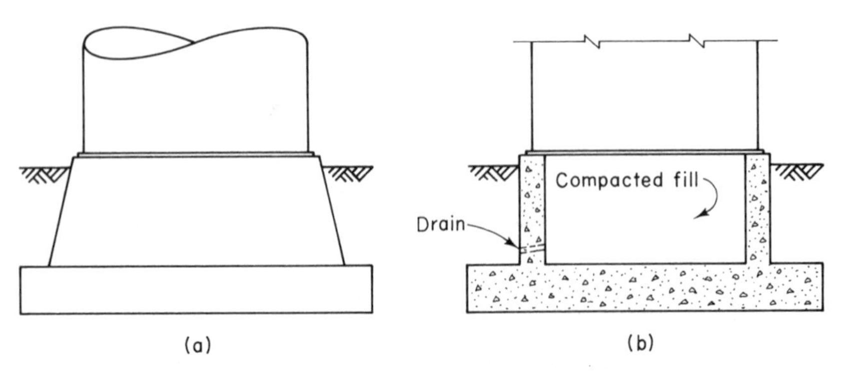

Figure 9-4 (a) Pier foundation, (b) slab-and-ringwall foundation.

9-3. Footings on Sand

The allowable soil pressure for a footing on sand or other granular material depends on the density of the material, the least width B of the footing, the depth D_f of the footing, and the depth of the groundwater table. The relative density is most readily determined by a penetration test. Figure 9-5 gives allowable soil pressures for three ratios of D_f/B as functions of the number of blows N in a standard penetration test.[1] The lines radiating from the origin give soil pressures that provide a factor of safety of 2, while the horizontal lines give pressures that correspond to a 1-in. settlement. These pressures apply for groundwater levels at or below a depth $D_f + B$ below the ground. If the water is expected to rise above this level, allowable pressures from Fig. 9-5 should be multiplied by the factor C_w given by

$$C_w = 0.5 + \frac{0.5D_w}{D_f + B} \tag{9-3}$$

where D_w = the depth to groundwater level measured from the surface of the surcharge surrounding the footing.

9-4. Footings on Clay

The allowable bearing pressure q_a for a footing on clay may be taken at

$$q_a = 0.83q_u \left(1 + \frac{0.2B}{L}\right)\left(1 + \frac{0.2D_f}{B}\right) \tag{9-4}$$

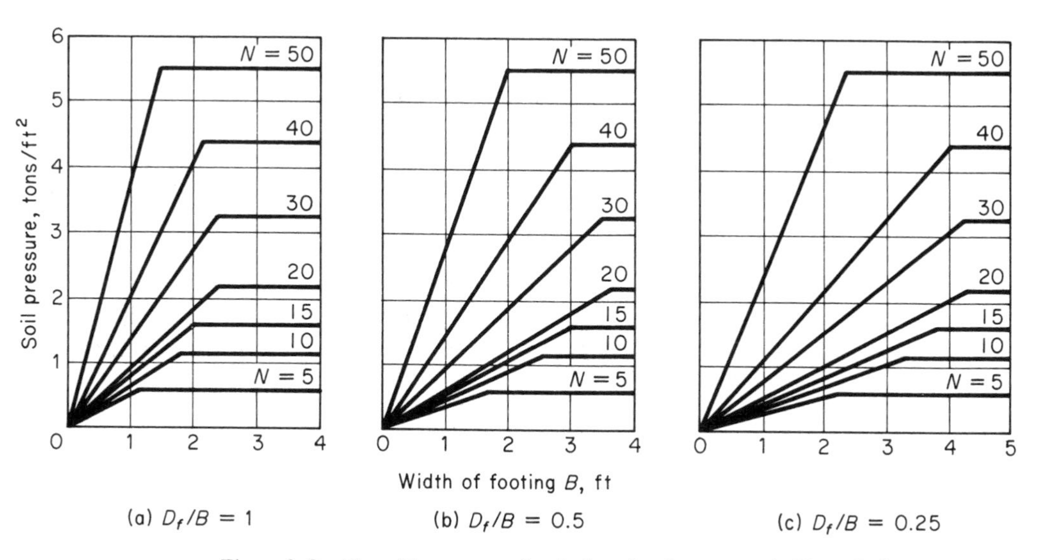

Figure 9-5 Allowable pressures for shallow footings on sand. (From Ref. 1.)

where L = the length of footing and q_u = the average unconfined compressive strength of the clay within a depth equal to the width of the largest footing of the group. This formula gives a factor of safety of 3 on the ultimate bearing capacity. Differential settlements of adjacent footings designed for this allowable pressure are not likely to exceed $\frac{3}{4}$ in. if q_u is greater than 2 tons/ft^2 and the distance between footings is enough to eliminate overlap of the underlying stress patterns.

9-5. Raft Foundations

The allowable pressure of a raft foundation on sand is governed by settlement because allowable pressures based on bearing capacity are always very large. If a settlement of 2 in. can be tolerated, the allowable soil pressure may be taken as twice the values given by the horizontal lines in Fig. 9-5. In terms of N this gives

$$q_a = 0.22N \qquad 5 \leqslant N \leqslant 50 \tag{9-5}$$

This value should be corrected for the water table as for footings on sand [Eq. (9-3)]. Equation (9-5) gives the allowable net pressure $q_n = Q/A - \gamma D_f$, where Q = the total load including the raft, A = the base area of the raft, and γ = the unit weight of the soil.

The area A of a raft foundation on clay may be determined from Eq. (9-4), where q_a is taken as the allowable net pressure already defined.

9-6. Pier Foundations

A pier foundation is often used where a soil stratum capable of supporting the load is at a considerable depth below the surface. Piers may be constructed by excavating to a suitable stratum and building the pier inside or by filling a machine-drilled hole with concrete. Drilled-hole diameters range from 12 in. to 12 ft. A belling attachment can form a bell with a diameter as much as three times the diameter of the shaft.

Since the settlement of a pier on sand is less than that of a shallow footing on a comparable sand, the allowable pressure may be determined from the horizontal lines in Fig. 9-5. This pressure should be corrected for the water table by Eq. (9-3).

The allowable pressure for a pier founded on clay may be taken as

$$q_a = 1.25q_u \left(1 + \frac{0.2B}{L}\right) \tag{9-6}$$

This formula gives a factor of safety of 3 against a bearing-capacity failure.

9-7. Pile Foundations

A number of formulas relating the driving resistance of piles to their static-load supporting capacity have been developed.

Piles driven into sand generally act in friction, point bearing, or a combination of the two. The ultimate capacity is often estimated by the *Engineering News* formula

$$Q_u = \frac{WH}{s + c} \tag{9-7}$$

where Q_u = ultimate capacity, lb
W = weight of hammer, lb
H = drop of hammer, in.
s = final pile penetration, in.
c = 1 for gravity hammer, 0.1 for other hammers

A factor of safety of 6 is suggested. However, this formula has shown poor correlation and wide scatter in comparisons with load tests, and its use is not recommended.[1,2] A better formula is the following by Janbu[2]:

$$Q_u = \frac{WH}{k_u s} \tag{9-8}$$

where $k_u = C_d(1 + \sqrt{1 + \lambda/C_d})$
$C_d = 0.75 + 0.15\,(W_p/W)$
$\lambda = (WHL/AEs^2)$
W_p = weight of pile, lb
L = length of pile, in.
A = net cross-sectional area of driven pile, in.^{2}
E = modulus of elasticity of pile, psi

A factor of safety of 3 is recommended.

The ultimate capacity of piles driven into clay cannot be reliably predicted from pile-driving formulas, because the largest part of the pile capacity is not related to stress transmission in the pile during driving. The capacity of a group of friction piles in clay can be computed from

$$Q_{ug} = (B + L)lq_{u1} + 2.5q_{u2}\left(1 + \frac{0.3B}{L}\right)BL \tag{9-9}$$

where Q_{ug} = ultimate capacity of group
B, L = width, length of pile group out to out
l = length of pile
q_{u1} = average unconfined compressive strength of clay within length l
q_{u2} = average unconfined compressive strength of clay within a distance B below pile tips

A factor of safety of 3 is recommended.

Piles driven through weak or compressible soil to a layer of stiff or hard clay support load by point bearing. If the underlying layer is sand, support is partly by point bearing and partly by skin friction from the sand. Although point-bearing capacity in these cases can be estimated by a pile-driving formula, it is much better to determine it by load tests. However, the load-test capacity of a single pile is partly from skin friction in the overlying material, and since the friction is likely to be lost with consolidation of the soft material, the load test should be corrected by the amount carried by friction. The portion of load carried by friction can be determined by a load test on a second pile driven several feet short of the firm material.

If a pile foundation is designed without the benefit of load tests and a pile-driving formula is used, a requirement in the specifications that piles be driven to the required resistance according to the formula used gives the designer some assurance that the result will be adequate.

The center-to-center spacing of piles in a group is usually 30 in., although they are customarily spaced about three times the butt diameter. Piles usually project 3 or 4 in. into the footing or mat. The distance from the center of outside piles to the edge of a footing should be at least 18 in. Reinforcement should clear the pile tops by about 3 in. It is important that piles be cut off at the correct elevation so that concrete above the pile will not be unreinforced because of reinforcement having been bent around the pile.

Settlement. A pile foundation with a compressible stratum below the pile tips may settle. If the stratum is extensive enough to suggest that settlement may be excessive, it should be evaluated.[1,2]

9-8. Continuous Footings

Column-supported bins which must be supported on soils with a low bearing capacity may require strip, grid, mat, or ring foundations. A strip footing (also called combined footing) supports two or more columns in a row. A grid foundation consists of intersecting strip footings with columns at the intersections. A mat or raft foundation is a continuous footing supporting several rows of columns in each direction. A ring foundation is a continuous footing in the form of (usually) a circle.

The pressure at any point of a continuous footing supported by a uniform soil is approximately proportional to the settlement at that point, and since the settlement is equal to the deflection of the footing, the pressure is not uniformly distributed. However, it is usually assumed to be uniform, and the shears and moments in a strip footing are computed as for a continuous beam carrying a uniformly distributed load consisting of the soil pressure with the column loads as the reactions. Similarly, a mat foundation is designed as a uniformly loaded flat-slab floor.

Nonuniform distributions of soil pressure can be taken into account by assuming the soil to be an elastic medium, with the contact pressure at any point given by ky, where k is the modulus of the subgrade reaction and y is the deflection.[3] Analyses based on the assumption that k is constant over the supporting area have been developed for strip, grid, mat, and ring foundations.[4-6]

The subgrade modulus k is difficult to determine. It depends on the stress-deformation characteristics of the soil, the size and shape of the loaded area, and the positions of other loaded areas. Furthermore, it is seldom, if ever, uniform over the foundation area. Therefore, empirical allowances for variability of pressure, unequal settlements of columns, and other uncertainties are often used. Thus, strip footings may be designed for the larger of the internal forces at any section computed for a uniform soil pressure and one consisting of triangular distributions for each column (Fig. 9-6). Another procedure is to use more than the theoretical amount of reinforcement based on a uniform distribution and to use the same amount at the top of the footing as at the bottom.

A segment between columns of a ring foundation with equally spaced columns loaded equally can be analyzed as a circular-segment beam supported at the columns and fixed against rotation about the radii at the columns. Such a beam is subjected to torsional moments as well as bending moments. Formulas for a number of cases are given in Ref. 7. If the columns are assumed not to prevent twist of the footing at the columns, the bending moments M_0 at the column and M_1 midway between columns are given by

$$M_0 = PR\left(\frac{1}{\theta} - \frac{1}{2}\cot\frac{\theta}{2}\right) \tag{9-10a}$$

$$M_1 = PR\left(\frac{1}{\theta} - \frac{1}{2\sin\theta/2}\right) \tag{9-10b}$$

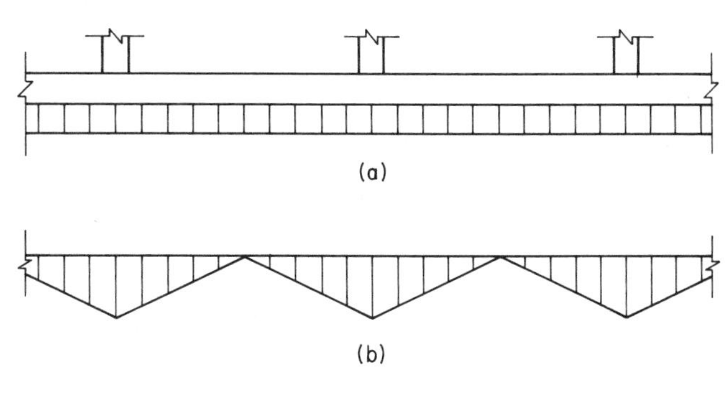

Figure 9-6

where P = load on one column
R = radius of column circle
θ = angle between radii at columns

Positive moment produces tension at the bottom of the ring. The torsional moments are zero at both these points. The maximum torsional moments are at approximately the quarter points and range from about 9% of the maximum bending moment for a ring supporting four columns to only 6% for one supporting eight columns. They are small enough to be neglected.

Analysis of a ring foundation with soil pressures ky is given in Ref. 6. The boundary conditions are the same as for Eqs. (9–10). Tables of coefficients for determining the bending moments, torsional moments, deflection, and angle of twist are given for rings with 4, 6, 8, and 12 columns for two values each of the parameters kR^4/EI_x and GJ/EI_x. The moments M_0 for rings with 6, 8, and 12 columns differ from those by Eq. (9–10a) by a maximum of 2%, while values of M_1 for 6 columns range from 85 to 89% and for 12 columns from 97 to 98% of those by Eq. (9–10b). For a ring with 4 columns, values of M_0 are from 85 to 93% and of M_1 51 to 62% of the corresponding values by Eqs. (9–10). Therefore, Eqs. (9–10) give acceptable results, with the possible exception of a 4-column bin for which a ring foundation is unlikely in any case.

The moments by Eqs. (9–10) differ only slightly from the corresponding moments for a fixed-end straight beam $R\theta$ long. Thus, the ratio of M_0 to the moment $PR\theta/12$ varies from 1.04 for a ring with four columns to 1.01 for one with eight columns, while the corresponding ratios of M_1 to the moment $PR\theta/24$ are 1.08 and 1.02. Therefore, each segment of a ring foundation supporting the number of columns likely to be used for a column-supported bin can be designed as a uniformly loaded, fixed-end beam of length equal to the arc length between columns.

REFERENCES

1. R. B. Peck, W. E. Hanson, and T. H. Thornburn, *Foundation Engineering,* 2nd ed., Wiley, New York, 1974.
2. H. G. Larew, T. H. Thornburn, and H. O. Ireland, Soil Mechanics and Foundations, Sec. 5 in E. H. Gaylord and C. N. Gaylord (Eds.), *Structural Engineering Handbook,* 2nd ed., McGraw-Hill, New York, 1979.
3. M. Hetenyi, *Beams on Elastic Foundations,* University of Michigan Press, Ann Arbor, 1946.
4. F. Kramrisch and P. Rogers, Simplified Design of Combined Footings, *J. Soil Mech. Div. ASCE,* Oct. 1961.

5. Suggested Design Procedures for Combined Footings and Mats, Report of ACI Committee 436, *J. ACI,* Oct. 1966.
6. E. Volterra and R. Chung, Constrained Circular Beams on Elastic Foundations, *Trans. ASCE,* Vol. 120, 1955.
7. R. J. Roark and W. C. Young, *Formulas for Stress and Strain,* 5th ed., McGraw-Hill, New York, 1975.

SUBJECT INDEX

A

Angle of friction:
  effective, 41, 43
  external, 39
  internal, 40
Angle of repose, 38
Arching, 64, 66
  under initial pressure, 70–71
Axial compression with bending:
  allowable stress:
    AISC, 153–54
    AISI, 154
  curved members, 156–57

B

Beam-column (*see* Axial compression with bending)
Beams, 142–48
  compact section (definition), 143
  deflection, 147–48
  design for postbuckling strength, 161–63
  lateral buckling, 143–47, 220–21
    AISC allowable stress, 146

Beams (*cont.*)
  lateral buckling (*cont.*)
    AISI allowable stress, 145
    equivalent radius of gyration, 143–44
  local buckling, 143–44
  shape factor, 142–43
  shear:
    AISC allowable stress, 147
    AISI allowable stress, 147
Bending with axial compression (*see* Axial compression with bending)
Blending, 60–61
Bolts, structural:
  high-strength, 29–30
  tensile properties, 35
  unfinished, 29
Bulk density, 37, 44
Bulk solids:
  angle of repose, 38–39
  compressibility, 37–38, 45, 48
  corpuscular, 46
  definition, 37
  density, 37, 44
  fibrillar, 47
  granular materials, 38
    powder fraction, 38

Bulk solids (*cont.*)
- laminar, 47
- powders, 44–46, 59
- principal pressures, 40–41
- typical properties, 52–53

Bunkers:
- example, 322–34
- longitudinal beams, 2
- rectangular, 2
- stiffeners, 2
- ties, 2

C

Caking, 59
Columns (*see* Compression members)
Column-supported bins, 1–2, 279–84
- bracing, 284
- column eccentricity, 284
- example, 301–13
- insert plates, 279, 281–84
- knuckle at transition, 274–75
- ring girder, 279–81
  - example, 308–11

Compression junction:
- stresses, 275–77
  - examples, 311–13, 320–21
- *see also* Compression ring

Compression members, 148–52
- AISC allowable stress, 150
- AISI allowable stress, 150
- curved, 154–56
- design for postbuckling strength, 163–64
  - examples, 164–66
- effective length, 150
- local buckling, 151
- overall buckling, 148–52
  - angles, 149
  - channels, 149
  - hat section, 149
- twist-bend buckling, 151–52
  - example, 152–53

Compression ring, 273–74
- allowable stress, 202–3, 273
- examples, 237–38, 244, 256, 309–10, 319–20
- stresses at cylinder-cone junction, 275–77
  - examples, 311–13, 320–21
- *see also* Rings

Conical-roof rafters:
- effective width of roof plate, 221
- fixed at compression ring:
  - example, 241–44
  - formulas, 240–41
- intermediate rafters cantilevered, 225
- lateral-torsioned buckling, 220–21
  - Fischer's tests, 220–21
- open-web joist, 300
- simply supported at compression ring:
  - example, 234–37
  - formulas, 223–25
- spacing, 221–22
- with supported center ring:
  - example, 226–29
  - formulas, 222

Conical roofs:
- self-supported, 208
  - example, 219
  - membrane forces, 214–15
  - with radial stiffeners, 208
  - tests under hydrostatic pressure, 216
  - thickness, 217–19
- supported:
  - bracing, 257
  - examples, 226–31, 231–39, 241–45
  - minimum thickness, 221–22
  - rafters (*see* Conical-roof rafters)
  - on rafters and trusses, 208–10
  - ultimate strength, 219–20

Conveyors:
- apron, 7
- belt, 8
- factors affecting selection, 4
- screw, 5
  - limiting depth of solid, 6
  - helicoid, 5
  - sectional flight, 5
  - types, 6
- vibratory, 7

Corrosion allowance, 298
Corrugated-sheet bins, 295
Coulomb's formula, 99
Curved members, axial compression along chord, 154–56
- with bending, 156–57

Cylindrical shells:
axial compression, 172–79
allowable imperfections, 176–77
allowable stresses, 261–64, 297–98
Boardman formula, 175
ECCS formulas, 175, 262–64
factor of safety, 178–79
inelastic buckling, 176
length effect, 177
Miller formula, 176
Pflüger formula, 175
Steinhardt-Schulz formula, 173–74
test results, 174, 178
axial compression with internal pressure:
ECCS formula, 183–84
factor of safety, 184
lap joints, 264–66
bending, 179
concentrated loads at top, 295
eccentric drawoff:
DIN provisions, 95–96, 267
Jenike formula, 267
economical width of rings, 299
fabrication and erection, 299
hoop tension (allowable), 260–61
minimum thickness, 264
stiffened (*see* Longitudinally stiffened cylindrical shells)
stress resultants, 259–60
temperature stresses, 292–93
vertical compression, 100–1, 119 (*see also* Cylindrical shells, axial compression)
wind buckling, 270–71

D

Denting due to eccentric drawoff, 266–67
Dome-roof rafters:
deflection, 249
fixed at compression ring:
example, 254–56
formulas, 246–48
net load, 248
vertical buckling, 248–49
lateral-torsional buckling, 248
simply supported at compression ring:
example, 250–52
formulas, 245
vertical buckling, 248–49
Dome-roof rafters (*cont.*)
with supported center ring, 245
see also Dome roofs
Dome roofs:
erection by winching or blowing, 301
local-buckling pressure, 248
rafters (*see* Dome-roof rafters)
self-supported, 209–13
buckling pressure, 211–12
example, 214
membrane forces, 211
tests, 212
thickness, 213
stiffened, 213
supported 245–49
bracing, 256–57
examples, 249–56
ultimate strength, 219–20
unsymmetrical load, 249
Doming, 62

E

Earthquake loads, 136–39
ACI 313–77, 139
allowable stresses, 294
reduced dynamic effect, 139
resonance coefficients, 137
seismic-risk zones, 138
Uniform Building Code, 137
Eccentric outlets, 95–96, 266–69
Effective length of columns, 150
Effective transition, 4, 91
Effective width:
flat plates in tension, 172–73
roof plate on rafters, 221
stiffened elements, 159–61
unstiffened elements, 159–61
Effective yield locus, 41
Elevators:
bucket, 24–25
screw, 24
Emptying pressures:
eccentric outlets, 95–96, 266–70
DIN formula, 95
finite-element analysis, 267–69
Jenike formula, 267
funnel flow:
ACI 313–77, 94

Emptying pressures (*cont.*)
funnel flow (*cont.*)
DIN 1055, 94
pneumatic, 98
Soviet Silo Code, 94
tests, 92, 105-6
mass-flow in cylinder:
Jenike-Johanson, 115-17, 125-26
Walker, 110-11, 114, 125-26
mass-flow in hopper:
Jenike, 117-18, 121, 125-26
Walker, 111-12, 114, 123, 125-26
mass-flow tests, 122-28
nonuniform in flat-bottom bins, 268-69
rate of discharge, 82-84
see also Pressures in bins
Erection, 298-301
rim stiffening, 300
Expanded-flow bin, 4, 58

F

Fabrication, 298-301
Feeders:
apron, 12
bar-flight, 14
belt, 12, 63
gate, 11-12
reciprocating-plate, 13
rotary drum, 16-17
rotary plow, 16
rotary table, 16, 63
rotary vane, 16-17
screw, 14-15, 63
selection of, 4, 17-18
vibrating, 13-14
Filling pressures, 93-94
DIN 1055, 94
dustlike materials, 97-98
eccentric filling, 269-70
impact, 73-74
mass-flow cylinder:
Jenike-Johanson, 115, 125-26
Walker, 109, 113, 125-26
mass-flow hopper, Walker, 109-10, 123, 125-26
Reimbert, 97
tests, 96, 104, 125-26
see also Pressures in bins
Flat-bottom bins:
advantages, 1, 62
flow boundary, 80-82
flow factor, 70
foundations, 294, 339-40
live-storage, 80-82
Flow factor (*see* Hopper outlets)
Flow function, 42-44, 65-67
Flow inducer:
aeration, 19
air jet, 19
inflatable panel, 19-20
rod, 19
rotating paddle, 19
sledgehammer, 19
stirrers, 19
vibrators, 20-22
Flow pressures (*see* Emptying pressures)
Flow-promoting device (*see* Flow inducer)
Flow properties, 44-51
archability, 46-47
blending, 60-61
caking, 59
conditions affecting, 45-51
abrasion, 51
corrosiveness, 51
degradation, 50
gradation, 50, 59
moisture content, 48-49
segregation, 50, 60
temperature, 49-50
flaky materials, 44
floodability, 46-47
floodable materials, 44
flowability, 45-48
free-flowing materials, 44
hanging up, 44
mine-run ores, 60
powders, 44-46
Foundations:
allowable bearing pressure, 327-38
flat-bottom bins, 339-40
footings on clay, 341-42
footings on sand, 341
frost penetration, 338
grid, 344
mat, 344
pier, 342

Foundations (*cont.*)
pile, 343–44
raft, 342
ring, 344–46
ringwall, 339–40
strip, 344
Frost penetration in soil, 338
Funnel flow:
definition, 3–4
types, 60, 91
Funnel-flow bins, 106
advantages, 51, 55–56
design recommendations, 296
flow factor, 70
live storage, 80
see also Emptying pressures, Filling pressures, Hopper outlets, Hoppers

G

Grain, 53
Granular materials, 38
typical properties, 52–53

H

Hanging up, 44
Homogenizing bin, 98
Hopper gates, 11–12
Hopper outlets, 64–73
arching, 64–66
under initial pressure, 70–71
doming, 62, 64–65
eccentric, 95–96
effective area, 62–63
examples, 77–79, 302, 314
flow factor:
definition, 65
flat-bottom bin, 70
funnel-flow bin, 70
mass-flow bin, 67–70
piping, 71–72
flow function, 42, 65
flow-no-flow criteria, 67
flow rate, 82–84
impact pressure during filling, 73–74
piping, 62, 71–72
under initial pressure, 73
size determination, 64–65, 71
Hoppers:
chisel, 57, 61, 76
conical, 57, 61, 76
membrane forces, 271–72
stresses at junction with cylinder, 275–77, 311–13, 320–21
eccentric, 61
flow factor (*see* Hopper outlets)
funnel-flow, 79
mass-flow:
chisel, 76
conical, 75–76
design recommendations, 297
examples, 77–79
pyramidal, 76
surface finish, 77
transition, 77
wedge, 76
outlets, (*see* Hopper outlets)
pyramidal, 61, 76
membrane forces, 285–88
wall bending stress, 288
rate of discharge, 82–84
transition, 57, 71
vibrating, 22–23
wall pressures, 106–12, 117
wedge, 57, 61, 76
see also Emptying pressures, Filling pressures
Hydraulic radius, 93
noncircular cross sections, 96–97

I

Impact breaker, 59
Impact pressure during filling, 73–74

J

Jannsen's formula, 92–93
Jenike flow factors, 67–70
Jenike mass-flow formulas, 114–21

K

Knuckle, 274–75

L

Laterally loaded plates (*see* Plates loaded normal to surface)

Level indicators, 23–24
Liners, 4
Local buckling:
stiffened elements, 143–44, 159–61
unstiffened elements, 143–44, 159–61
see also Beams, Compression Members, Cylindrical shells
Longitudinally stiffened cylindrical shells:
axial compression, 179–83
broad panels, 181
buckling stress, 179
elastic support of stiffeners, 181
load-strain relations, 183
narrow panels, 181
postbuckling strength, 180
test results, 181
uptimate strength, 181
economy, 300

M

Mass flow, definition, 3
Mass-flow bins:
characteristics, 56–58
design recommendations, 297
flow factor:
Jenike, 67–68
Walker, 69
flow-factor charts, 67–70
hopper proportions, 75–77
Jenike formulas, 114–21
peak pressure, 116
switch, 116
tests, 122–29
Walker formulas, 109–14
see also Emptying pressures, Filling pressures, Hopper outlets, Hoppers
Membrane forces:
in conical hopper, 272
in cylinder, 259–60
in pyramidal hopper, 285–88
in shell of revolution, 271–72
Mine-run ores, 60
Mohr circle, 87
Multicell bins, 292

O

Openings in bin walls, 294–95

P

Performance specification, 298–99
Perishable solids, 59
Piping, 62, 71–72
under initial pressure, 73
Plates in compression:
buckling coefficients, 157–58
effective width, 159–61
postbuckling strength, 158–61
stiffened longitudinally, 166–71
eccentric load, 169–70
with transverse stiffeners, 171–72
Plates in tension:
effective width, 172
example, 227–28
Plates loaded normal to surface:
allowable stresses, 185
nonuniform load, 198
examples, 199
plastic-moment strength, 185
trapezoidal load, 192–93
allowable stress, 192
example, 193
triangular load, 189–92
allowable stress, 192
deflection, 192
example, 192–93
uniform load (large-deflection theory), 193–200
allowable stresses, 195
deflection, 194–96
examples, 197–98
uniform load (small-deflection theory), 186–89
allowable stress, 186
deflection, 187
examples, 188–89
Pneumatic emptying, 98
Powders, 44–48, 59
Pressure coefficient (*see* Pressures in bins, Wind)
Pressures in bins:
homogenizing bins, 98
peaks, 90–91
plastic equilibrium, 85, 89
pressure coefficient:
cohesive material, 86
cohesionless material, 86

Pressures in bins (*cont.*)
pressure coefficient (*cont.*)
Nanninga, 89
Rankine, 86, 93
switch, 90–91
see also Emptying pressures, Filling pressures
Pressure test cell, 38
Principal pressures, 40–41

R

Rafters (*see* Conical-roof rafters, Dome-roof rafters)
Rate of discharge, 82–84
Rectangular bin:
allowable stresses, 290–91
deep-walled, 290
example, 322–24
folded-plate analysis, 290
shallow-walled, 99, 289–90, 296
wall stiffeners, 289
see also Bunkers, Hoppers
Reimbert formulas, 97
Rim angle, 300, 301
Ring girder (*see* Column-supported bins)
Rings:
loaded normal to plane, 200–2
example, 230–31
loaded radially, 202
buckling, 202–3
example, 237–38
moment-resistant connections, 203–5, 244
Roofs:
conical (*see* Conical roofs)
dome (*see* Dome roofs)
erection, 300–1
slope, 208
umbrella, 208

S

Segregation, 60
Seismic loads (*see* Earthquake loads)
Shallow bins, 99, 289–90
definition, 100
design recommendations, 296
Shear cell, 39, 41
Shells:
edge effects, 259
membrane theory, 217
minimum thickness, 264
see also Cylindrical shells, Hoppers
Skirt-supported bins, 1–2
example, 313–22
Snow loads, 135–36
Specifications, 298–99
Spherical dome (*see* Dome roofs)
Square bin, advantages, 1
Steel:
brittle fracture, 27
protection against, 28
ductility, 27
factors affecting mechanical properties, 27
Steel for bins:
clad, 29, 33–34
non-ASTM, 29
pipe and tubing, 29–30
plates and shapes, 28, 30–31
sheet and strip, 28, 31
tensile properties, 32–35
Stiffened flat plates (*see* Plates in compression)
Stiffened shells (*see* Longitudinally stiffened cylindrical shells)
Switch, 116

T

Temperature stresses, 292–93
Tension rings, 277–79
example, 238–39
see also Rings
Time yield locus, 42
Transition stresses (*see* Compression ring)

U

Umbrella roof, 208 (*see also* Conical roofs)
Unconfined yield strength, 51

V

Vertical compression in cylinder wall, 100–1, 104–5
ACI 313-77, 105

Vertical compression in cylinder wall (*cont.*)
DIN, 105
Jenike formula (mass-flow), 119
Vibrating hopper, 22–23
Vibrators (*see* Flow inducer)

W

Walker flow factor, 69–70
Walker mass-flow formulas, 109–14
Warping constant, 144
Welding electrodes, 31
Wind, 129–35
buckling of cylindrical shells, 270–71, 313, 322
closely spaced bins, 133–34
on cylindrical bin (examples), 134–35, 307, 322
dynamic pressure, 129
exposure categories, 131
Wind (*cont.*)
fastest-mile speed, 130
mean recurrence interval, 129
open-top bins, 132
pressure coefficient, 129
for bins, 131
for members, 131
return period, 129
shape factor, 129
stagnation pressure, 129
tests on cylinders, 132–34
velocity pressure, 129
table, 131

Y

Yield locus, 40
effective, 41
time, 42

AUTHOR INDEX

A

Abdel-Sayed, G., 295, 336
Ahmed, S. R., 132, 133, 134, 141
Airy, W., 85, 139
Almroth, B. O., 295, 336
Andersen, Paul, 292, 336
Andrews, C. K., 5, 25
Ansourian, P., 261, 263, 334

B

Bauer, W. G., 11, 25
Billington, D. P., 259, 334
Blanchard, M. H., 109, 122, 123, 124, 141
Bleich, F., 144, 149, 150, 152, 160, 166, 205, 281, 335
Brush, Don O., 295, 336

C

Carr, Ralph L., Jr., 37, 38, 44, 45, 46, 47, 48, 50, 54
Carroll, P. J., 13, 26
Carson, J. W., 92, 94, 106, 114, 115, 117, 126, 140
Chajes, A., 149, 205
Chandrasekaran, A. R., 139, 141
Choo, Lee-Ken, 290, 330, 333, 334, 335
Chung, R., 345, 347
Cieselski, R., 286, 335
Clague, K., 124, 125, 141
Colijn, H., 7, 25, 67, 68, 69, 71, 72, 73, 84
Culver, C. G., 280, 281, 335

D

Davenport, A. G., 130, 141
Dumbaugh, G. D., 22, 26
Dym, C. L., 280, 281, 335

E

Eisenhart Roth, M. V., 11, 16, 17, 18, 19, 22, 26
El-Atrouzy, M. N., 295, 336
Ellis, S. P. J., 19, 26
Engh, T. A., 19, 26
Esslinger, M., 132, 133, 134, 141, 264, 265, 334

F

Fischer, M., 220, 258
Fischer, W., 189, 190, 191, 192, 206, 288, 335
Fuchsteiner, W., 275, 335

G

Gaunt, J. T., 271, 335
Gaylord, C. N., 143, 145, 148, 152, 154, 156, 159, 205
Gaylord, E. H., 143, 145, 148, 152, 154, 156, 159, 205
Gere, J. M., 166, 172, 202, 205, 211, 257, 280, 335
Giunta, J. S., 80, 81, 84
Gould, P. L., 273, 281, 282, 283, 284, 335

H

Hanson, W. E., 339, 341, 343, 344, 346
Herber, K., 219, 241, 249, 256, 257, 258
Hetenyi, M., 345, 346
Hickerson, W. L., 7, 25
Hildebrand, F. B., 173, 205
Homes, A. G., 94, 96, 101, 104, 106, 140
Horne, M. R., 169, 205

I

Ireland, H. O., 343, 344, 346

J

Jain, P. C., 139, 141
Janssen, H. A., 85, 92, 139
Jenike, A. W., 39, 41, 42, 44, 54, 64, 65, 67, 72, 76, 80, 84, 85, 91, 92, 94, 106, 114, 115, 116, 117, 119, 126, 139, 140, 267, 334
Johanson, J. R., 85, 91, 92, 94, 106, 114, 115, 117, 119, 126, 139, 140, 332, 336
Johnston, B. G., 150, 166, 171, 176, 205, 245, 248, 258
Jungbluth, O., 212, 257

K

Kahmer, H., 184–206
Kalyanaraman, V., 159, 160, 164, 205
Ketchum, M. S., 85, 139
Kleysteuber, W. K., 332, 336
Klöppel, K., 155, 205, 208, 212, 213, 257
Koenen, M., 93, 140
Kovtum, A. P., 103, 140
Kramrisch, F., 345, 347
Kulak, G. J., 179, 206

L

Larew, H. G., 343, 344, 346
Lee, C. A., 11, 16, 19, 20, 21, 25
Lightfoot, E., 289, 335
Lowry, R. D., 273, 281, 283, 335

M

Maderspach, V., 271, 335
Mahmoud, M. H., 267, 335
Mariani, N., 280, 281, 335
Martens, Peter, 96, 106, 140
Marulic, W. J., 179, 206
Michael, D., 289, 335
Miller, C. D., 174, 176, 178, 206
Mitzel, A., 286, 335
Montgomery, C. J., 179, 206
Morgan, E. J., 183, 184, 206, 214, 216, 257
Mozer, J. D., 280, 281, 335
Munse, W. H., 28, 36
Myers, J. L., 22, 26

N

Nanninga, N., 89, 90, 139
Narayanan, R., 169, 205
Newmark, N. M., 174, 175, 178, 206
Nilson, A. H., 290, 335

O

Olsen, O. W., 275, 335

P

Peck, R. B., 339, 341, 343, 344, 346
Pekoz, T., 159, 160, 164, 205
Pflüger, A., 175, 206
Pieper, K., 94, 98, 107, 108, 140, 264, 265, 334
Platonov, P. N., 103, 140
Protte, W., 155, 205

R

Rathe, J., 175, 176, 206
Reif, A., 184, 206
Reimbert, A., 92, 96, 97, 99, 140
Reimbert, M., 92, 96, 97, 99, 140
Reisner, W., 11, 16, 17, 18, 19, 22, 26
Reissner, E., 173, 205
Roark, R. J., 186, 202, 206, 269, 281, 335, 345, 347
Roberts, I., 85, 139
Rogers, P., 345, 347
Roos, E., 208, 212, 213, 257
Rotter, J. M., 261, 263, 270, 334, 335

S

Saal, H., 175, 184, 206
Safarian, S. S., 94, 140
Schroeder, H. H., 132, 133, 134, 141
Schulz, V., 173, 174, 177, 206
Seide, P., 166, 167, 183, 184, 205, 214, 216, 257
Sen, S. K., 273, 281, 283, 284, 335
Shah, H., 290, 330, 333, 334, 335
Shang, J. C., 179, 206
Sharp, M. L., 167, 205
Shipp, R. T., 155, 205
Sinden, A. D., 21, 26
Sridharan, S., 181, 182, 183, 206
Stachurski, W., 286, 335
Stein, M., 166, 167, 205, 280, 335
Steinhardt, O., 173, 174, 177, 206
Stephens, M. J., 179, 206
Sturm, R. G., 179, 206
Suryoutoma, H., 273, 281, 283, 335
Suwalski, J., 286, 335
Sword, J. H., 271, 335

T

Thom, H. C. S., 130, 141
Thornburn, T. H., 339, 341, 343, 344, 346
Timoshenko, 166, 172, 190, 192, 193, 202, 205, 206, 211, 257, 259, 280, 313, 334, 335
Trahair, N. S., 261, 263, 334
Turitzin, A. M., 85, 91, 139

V

Vandepitte, D., 175, 176, 206
Volterra, E., 345, 347

W

Walker, A. C., 181, 182, 183, 206
Walker, D. M., 69, 70, 84, 109, 122, 123, 124, 127, 140, 141
Walters, J. K., 91, 139
Wang, R. S. C., 273, 281, 282, 283, 335
Weingarten, V. I., 183, 184, 206, 214, 216, 257
Wenzel, F., 94, 140
Wilson, W. M., 174, 175, 178, 206
Winter, G., 149, 159, 160, 161, 164, 205, 290, 335
Woinowsky-Krieger, S., 190, 192, 193, 206, 259, 313, 334
Wood, R. H., 186, 206
Wozniak, R. S., 261, 267, 273, 275, 295, 334
Wright, H., 70, 84, 124, 125, 141

Y

Yaeger, D. J., 280, 335
Young, W. C., 186, 202, 206, 269, 281, 335, 345, 347

Z

Zmudzenski, Z., 286, 335